CHRIS ATTON

pp. 15-16 , 19

163

164

218-9, 220

173-4 236-7

252-4

Democratic Communications

Critical Media Studies

Series Editor
Andrew Calabrese, University of Colorado

This series covers a broad range of critical research and theory about media in the modern world. It includes work about the changing structures of the media, focusing particularly on work about the political and economic forces and social relations which shape and are shaped by media institutions, structural changes in policy formation and enforcement, technological transformations in the means of communication, and the relationships of all these to public and private cultures worldwide. Historical research about the media and intellectual histories pertaining to media research and theory are particularly welcome. Emphasizing the role of social and political theory for informing and shaping research about communications media, Critical Media Studies addresses the politics of media institutions at national, subnational, and transnational levels. The series is also interested in short, synthetic texts on key thinkers and concepts in critical media studies.

Titles in the series

Governing European Communications: From Unification to Coordination by Maria Michalis

Knowledge Workers in the Information Society
edited by Catherine McKercher and Vincent Mosco

To Chris
A fine colleague
and friend. Here's
to meeting f2f
again soon —!
20 March 2008

Democratic Communications

Formations, Projects, Possibilities

JAMES F. HAMILTON

LEXINGTON BOOKS

A division of
ROWMAN & LITTLEFIELD PUBLISHERS, INC.
Lanham • Boulder • New York • Toronto • Plymouth, UK

LEXINGTON BOOKS

A division of Rowman & Littlefield Publishers, Inc.
A wholly owned subsidiary of The Rowman & Littlefield Publishing Group, Inc.
4501 Forbes Boulevard, Suite 200
Lanham, MD 20706

Estover Road
Plymouth PL6 7PY
United Kingdom

British Library Cataloguing in Publication Information Available

Library of Congress Cataloging-in-Publication Data

Hamilton, James F., 1959–
Democratic communications : formations, projects, possibilities / James Hamilton.
p. cm.
Includes bibliographical references and index.
ISBN-13: 978-0-7391-1866-5 (cloth : alk. paper)
ISBN-10: 0-7391-1866-8 (cloth : alk. paper)
1. Alternative mass media—History. 2. Mass media—Political aspects. I. Title.
P96.A44H36 2007
302.2309—dc22

2007040225

Printed in the United States of America

The paper used in this publication meets the minimum requirements of American National Standard for Information Sciences—Permanence of Paper for Printed Library Materials, ANSI/NISO Z39.48–1992.

Contents

Preface vii

Acknowledgements xi

Introduction: The Problem of the Mainstream and the Alternative 1

Part One – Market Formations

Introduction to Part One 29

1 Providentialism and Rationalist Empiricism in Early Modern England 33

2 The Emergence of Broadcasting and the Rationalization of Participation 61

Part Two – Struggling Against the Market

Introduction to Part Two 89

3 Philanthropy, Professionalization, and Social-Reform Communications 93

4 Community Media Projects and Their Containment Through the Mass-Culture Critique 121

5 Modernism and the Aestheticization of Dissent 161

Part Three – Toward New Formations

Introduction to Part Three 197

6 Market Radicalism and the Struggle of Participation 199

7 Democratic Communications
as Critical, Collective Education 233

Afterword: Utopia and Inspiration 265

Bibliography 271

Index 321

About the Author 337

Preface

In 1996, a group of media activists met in a San Francisco Bay-area apartment to discuss the political potential of microradio. As a meeting attendee noted, the group's guest was engineer and microradio activist Tetsuo Kogawa, who talked "about the social and political uses of microradio technology" while "simultaneously soldering and scotch-taping a series of Radio Shack parts, worth about fifteen dollars, to a paper plate. Within fifteen minutes, he had completed his soldering and announced that he had finished 'building' his transmitter." After three more minutes of fine-tuning, "he announced that we were 'on the air'." More startling than this transformation of readily available pieces into a radio transmitter was the group's loss at what to do with it. As the attendee recalled, "We adults were trying to figure out how to sound like some left-wing equivalent to [veteran U.S. television news broadcaster] Walter Cronkite, and were, of course, silenced by the prospect." By contrast, the teenage daughter of one of the attendees "was the only person in the room who was so unintimidated by the 'new' medium as to be able to talk spontaneously over the air." The adults' difficulty was that they were so used to listening "passively to what was broadcast over the airwaves, and only when 'permitted' or 'invited' to speak over an approved station were we able to prepare ourselves for such an event."[1]

This episode vividly illustrates the primary premises of this book. First, it suggests that means of engaging in forms of democratic communications do not necessarily require immense commitments of time and money, contrary to many claims of the necessity for large-scale projects. Second, the episode suggests a much more substantial barrier to the practice of democratic communication, which is the difficulty conceiving of new possibilities. As Fredric Jameson recently pointed out about the political perversity of the present situation generally, it is easier to imagine the end of the world than it is to imagine a response to the market economy which creates vast inequalities around the world.[2] In a similar way, it is easier to imagine the rejection of media practice than it is to imagine different bases for it altogether. The intention of this book is to help extend an ability to see greater possibilities in the present as well as to more broadly imagine what might yet be possible.

This book is a critical exploration of Anglo-American historical traditions and practices of democratic communications. The gall and perhaps naiveté of addressing such an immense topic is matched only by the necessity of attempting to do so. Its point of departure is the observation that, despite being grounded in often very distinct social formations, communications practices too-easily polarized as "mainstream" and "alternative" are often much more similar than different, not only at the level of form and style, but also at the level of organization, technology and social intention. The argument of similarity should not be taken to mean that there are no differences. Instead, the point is that the key distinctions between mainstream and alternative are social, cultural, and historical rather than formal, essential, and universal.

To the degree that this similarity exists (I am well-aware that this is a premise to be substantiated rather than a fact to be accepted), basing a politically progressive communications project on the foundational separation of the mainstream from the alternative overlooks ways in which the now-compromised progressive practice limits if not undercuts the progressive intention. At the same time that this commonality may limit progressive potential, it can also work the other way, suggesting the degree to which ostensibly mainstream practices may enable progressive intentions in ways yet to be fully recognized. Posing this situation in seemingly ambiguous terms underscores the many ways in which democratic communication is best seen as a problem to engage instead of an answer to be found, and thus open to history and human practice, and thus dependent on our ability to collectively imagine what we cannot yet see.

The focus of the book deserves some explanation. While focusing predominantly on the United States, it also grounds this case in its early modern Anglo precedents. Although one would be hard-pressed to discount the exceptional determinacy of the Anglo-American traditions described here (coinciding as they do until very recently with dominant political and economic power globally for the past three-hundred years), the focus of this book on them is not intended to assert that these traditions are the only ones or even the most important ones, nor that there is some unitary, essential Anglo or American tradition out there simply to be recovered. Rather, as the book emphasizes at many points, this tradition is constructed, selective, and exclusionary in many ways. Limiting the discussion to such traditions also does not imply the lack of important scholarship regarding others. Nor does the use of a nationalist framework negate the value of "un-national," international, transnational, comparative, and/or non-Western approaches to communication research, as a number of recent works attest.[3] Indeed, and for example, the study of alternative media and democratic communications continues to broaden productively in just this way.[4] The value of focusing on these traditions despite the insufficiency of their claim to be "The Tradition" is how doing so brings to the fore the insufficiently recognized depth at which they work, to the point where even communications projects that see themselves as having diametrically opposed intentions take these to be foundational, obvious, and thus beyond question. By simply helping reveal the deep determinations of these traditions, selective as they are, and, by doing so, help enhance an ability to raise questions about what has largely been taken for granted, the book will have accomplished what it sets out to do.

The conclusion of the book also deserves some explanation. Because it does not end with a laundry list of specific steps that all progressive organizations should take, it may also be underwhelming if not capitulating for readers eager for more dramatic and "practical" summation and recommendations. Yet, such an urge displays naiveté in its call for a universal formula, while it discounts the historical rather than essential nature of democratic communications. Such an urge also suggests its own too-often nostalgic basis in Enlightenment if not millennial conceptions of social change as the single, convulsive replacement of one system by another, which itself remains open for debate.[5] The biggest problem, however, with such an urge is how it equates anything short of a convulsive, total revolution with reformism and thus by definition a waste of time. If so, such a position discounts the tremendous and increasing amount of creative, radical work taking place throughout the world today. Much of this work remains resolute despite setbacks and immense challenges, thus drawing strength by working in the timeframe of what Raymond Williams called the long revolution of the steady expansion of popular capabilities of self-determination.[6] To the degree that this book assists this process in whatever small way, it will also have fulfilled its intention.

The work and contributions of countless people were drawn upon for this project, but a few deserve special mention. Chris Atton has been a long-standing and valuable intellectual colleague as well as a steady source of encouragement. Patricia Gibbs initiated and led an effort to produce a special issue on alternative media that appeared in *Media History* in 2001, and our collaboration in that early project has played a key part in all my subsequent work. Many thanks as well to *Media History* editors (Tom O'Malley, in particular) for seeing its value. Paul Hillier plowed through large portions of often rough drafts of this book, offering insightful readings and suggestions. Noah Arceneaux and Tom Hove commented valuably on portions of it. Kim Eun-Gyoo, with whom I co-authored research on *OhmyNews*, demonstrates a level of commitment to the practice of democratic communications in ways I hope someday to embody. Resources and projects at the Grady College of Journalism and Mass Communication—in particular, the support from my department and intellectual comradeship from the Critical Media Studies Interest Group—made possible the conditions in which this research could be done. Michael Castengera's online newsletter kept me abreast of key industry developments and trends. Kent Middleton offered insights regarding the legal protections of news-service materials, with suggestions about dependable sources for industry profitability offered by Hugh Martin. Kyoo-Hoon Han, Aaron Hite, Murendehle Juwayeyi, Sarah McLean, Melissa Medori, Woongjae Ryoo, Danielle Wormley, and Hye-Jin Yoon contributed diligent and invaluable research assistance.

Above all, Greg, Laura, and my wife Cynthia deserve great thanks for sticking with me and picking up the slack due to all the nights and weekends with my nose buried in my work. Now I can finally answer their question "When are you ever going to finish that book?"

Notes

1. Louis Hiken, "Forward," in Lawrence Soley, *Free Radio: Electronic Civil Disobedience* (Boulder, Colo.: Westview Press, 1999), x.

2. Paraphrased by Henry Giroux, "Forward," in Sut Jhally, *The Spectacle of Accumulation; Essays in Culture, Media & Politics* (New York: Peter Lang, 2006), xii.

3. John Downing, *Internationalising Media Theory: Transition, Power, Culture: Reflections on Media in Russia, Poland and Hungary, 1980–95* (London and Thousand Oaks: Sage, 1996); Marc Raboy, ed., *Public Broadcasting for the 21st Century* (Luton: University of Luton Press, 1996); Jeremy Tunstall and David Machin, *The Anglo-American Media Connection* (Oxford: University of Oxford Press, 1999); James Curran and Myung-Jin Park, eds., *De-Westernizing Media Studies* (London: Routledge, 2000).

4. Dirk Hoerder and Christiane Harzig, *The Press of Labor Migrants in Europe and North America, 1880s to 1930s* (Bremen: Labor Newspaper Preservation Project, 1985); John D.H. Downing, Tamara Villarreal Ford, Genève Gil, and Laura Stein, *Radical Media; Rebellious Communication and Social Movements* (New York: Sage, 2001).

5. John Foran, ed., *The Future of Revolutions: Rethinking Radical Change in the Age of Globalization* (London: Zed Books, 2003); John Holloway, *Change the World Without Taking Power* (London: Pluto Press, 2002).

6. Raymond Williams, *The Long Revolution* (New York: Columbia University Press, 1961).

Acknowledgements

I gratefully acknowledge the following permissions to use portions of work that have appeared previously.

Portions of the Introduction copyright 2001 from "Theory Through History: Exploring Scholarly Conceptions of U.S. Alternative Media," *The Communication Review* 4 (2001): 305–326, by James Hamilton. Reproduced by permission of Taylor & Francis Group, LLC., http://www.taylorandfrancis.com. Also drawn upon for the Introduction is James Hamilton, "Rethinking Communication, Media, and Activism," *International Journal of Media and Cultural Politics* 2, no. 2 (July 2006): 220–225. Reprinted by permission of Intellect, Ltd., http://www.intellectbooks.co.uk/.

Portions of Chapter 1 appear by permission of Sage Publications Ltd. from James Hamilton, "Remaking Media Participation in Early Modern England," *Journalism: Theory, Practice, Criticism* 4, no. 3 (August 2003): 293–313. Copyright (© Sage Publications, 2003).

Portions of Chapter 2 appeared as James Hamilton, "Unearthing Broadcasting in the Anglophone World," chapter in *Residual Media*, edited by Charles Acland (Minneapolis, Minn.: University of Minnesota Press, 2007), 283–300. Used by permission of the University of Minnesota Press.

Portions of Chapter 3 draw upon data compiled by The Foundation Center and available at http://foundationcenter.org/findfunders/statistics/: "Distribution of Foundation Grants by Subject Categories, circa 2005," "Distribution of Foundation Grants by Subject Categories, circa 2004," and "Top 50 Recipients of Foundation Grants for Media and Communications, circa 2005." Used with permission of The Foundation Center, New York. Copyright © 2004 and 2005 by the Foundation Center.

Portions of Chapter 4 appeared as James Hamilton, "Rationalizing Dissent?: Challenging Conditions Of Low-Power FM Radio," *Critical Studies in Media Communication* Volume 21:1 (March 2004): 44–63. Reprinted with permission of National Communication Association/Taylor & Francis. Copyright National Communication Association (www.natcom.org) of Washington, DC. Also drawn upon for this chapter was "Low-Power FM Radio Licensing and the Spatial Production of Community," a paper presented to "Back to the Future: Community Radio in the Age of Clear Channel," a research panel in the Broadcast and Internet Radio Division, Broadcast Education Association, Las Vegas, Nevada, 28 April 2006.

Portions of Chapter 5 appeared as "Visualising Critique: Montage as a Practice of Alternative Media," *Media History* 7, no. 2 (December 2001): 159–170. Used by permission of Taylor & Francis Group, LLC., http://www.tandf.co.uk/journals.

Portions of Chapter 6 are reprinted by permission of Sage Publications Ltd. from Eun-Gyoo Kim and James Hamilton, "Capitulation to Capital? OhmyNews as Alternative Media," *Media, Culture & Society* 28, no. 4 (July 2006): 541–560. Copyright (© Sage Publications, 2006).

Portions of Chapter 7 appeared as "Common Forms for Uncommon Actions: The Search for Political Organization in the California Dust Bowl," *American Journalism* 16, no. 1 (Winter 1999): 79–103. Reprinted with permission of the American Journalism Historians Association. Also drawn upon for this chapter is James Hamilton, "The Internot?," *M/C Reviews* (April 2000), online at http://reviews.media-culture.org.au/modules.php?name=News&file=article&sid=1757. Reprinted with permission of M/C – Media and Culture.

Introduction

The Problem of the Mainstream and the Alternative

Faith in communications to overcome persistent problems and inequalities in society has been a long-standing feature of modern life.[1] Increasingly, such faith is expressed in terms of a perceived opposition between, on the one hand, mainstream communications institutions (whether commercial or statist) and their enforcement of a narrow, self-serving agenda of protecting power and privilege and, on the other hand, alternative media and their ability to bring previously hidden information to public awareness and to give a voice to the voiceless.[2]

With the advent of the Internet and its capabilities of self-publishing and transnational distribution, such views are even more numerous and insistent. "Thanks to digital technologies," argues one proponent, "consumers are becoming creators. The marketing and entertainment industry no longer has exclusive control over creating content."[3] A second proponent asserts "we're about empowering this generation of young people in their 20s to engage in a dialogue of democracy and to tell the stories about what's going on in their lives using the dominant medium of our time."[4] A third argues that people today "want to be able to use the information in a larger community—to talk about, to debate, to question, and even to meet the people who think about the world in similar or different ways."[5]

While the views expressed in these seemingly revolutionary sentiments fit very easily within the alternative side of the mainstream/alternative opposition, the sources of these views decidedly do not. The first example is from the Word of Mouth Marketing Association, which seeks more effective ways to sell products and services. The second example is from former U.S. Vice President Al Gore, who is helping oversee the establishment of Current TV, a commercial cable-television and Internet company that accepts video programming from viewers in exchange for payment. The third example is from Rupert Murdoch, chairman and chief executive officer of News Corporation, who, in a 2005 speech to the American Society of Newspaper Editors, advocated interactivity as a path to greater profitability.

The jarring combination of seemingly alternative capabilities advocated by mainstream interests is mirrored by seemingly mainstream intentions (such as making audience-friendly content to maximize audience size, or having a powerful impact on its audience) advocated by alternative-media organizations. The Direct Action Media Network seeks to make complex accounts more understandable by putting "coverage of social justice actions into both historical and contemporary context so that any audience will find the events and issues covered accessible."[6] Guerrilla NewsVideo seeks to "combine high-impact imagery, tracks from top recording artists and interviews with leading experts about important issues underexposed by the big media outlets. They aim to rock as hard as they inform, shock and inspire."[7]

Various hybrid examples and projects blur the distinctions between the mainstream and alternative even further. For example, content in one side is often used in the other, such as the prominent and politically progressive website TruthOut.org, "visited more than four million times per month by readers seeking the straight story and wanting to be involved in building a better tomorrow."[8] Yet, while it posts some original contributions and commentary, the bulk of its content consists of items taken from mainstream communications organizations such as *The New York Times*, the *Washington Post*, and the Associated Press.[9] Indeed, news stories developed in one side often become central in the other through reference, paraphrase and summary, republication, and commentary. Prominent examples include *The Progressive* and its publishing in 1979 of plans for an atomic bomb as an intervention in the campaign for nuclear disarmament; the August 1996 three-part "Dark Alliance" series in the *San Jose Mercury News* that explored and documented structural relationships and interests between the CIA's support of contra rebels and the U.S. crack-cocaine epidemic of the 1980s; the late-2005 publication by *The New York Times* of a story about the U.S. National Security Administration domestic spying program done without judicial oversight; and the early 2007 revelation over the politically motivated firings of U.S. attorneys.[10]

And how might one categorize the aforementioned Current TV? It promotes itself as an "outpost in a locked-down television landscape" where "the channel goes two ways. Here, you are viewer and producer." Yet, the form of this empowerment is "short pods" of video ranging "from international dispatches to profiles of cool people to intelligence on new trends"—a content mix quite at home in consumerist magazines for young, affluent audiences.[11] Indeed, all too conceivable in the current situation is a Protest Channel comprised of on-the-street video footage shot by global-justice activists with running commentary, interviews and analysis, sandwiched as it would likely be between a channel devoted to home improvement and one to golf.

Relationships between the mainstream and the alternative take an even wider variety of forms. People work in one area and later work in the other. Writers and filmmakers have appeared sequentially in one then the other, or in both simultaneously. More general commonalities include forms of organization; forms of presentation such as documentary, polemic, personal reflection, and critical review; and forms of relationship as announcer, guide, discussant, interviewer, and analyst or as reader, viewer, spectator, financial contributor, occasional correspondent, and occasional subject.[12]

It would be easy to oversimplify an understanding of these and many other such examples, which are neither isolated nor exceptional. Claims that commercial and traditional public-service media have completely absorbed alternative uses—or that the alternative has eclipsed and rendered the mainstream impotent—simplify the wide range of variations. To go beyond simplistic claims, other arguments employ more nuanced explanations of these commonalities by emphasizing pragmatic, social-Darwinian survival of the fittest in the consumer marketplace; hobbyists' amateurish mimicry of professional practices; a restoration by the alternative of original intentions that had been perverted by the mainstream; an extension of communications capabilities by the alternative that had been limited and controlled by the mainstream; or the dialectical reversal by the alternative of mainstream dominance (such as in claims of the master's weapons as turned against the master). Explanations also note the importance of inventor-driven research and development or avant-garde experimentation, which are then combined variably with any of the explanations already noted.

Despite these and other ways in which such commonalities between mainstream and alternative communications practices have been understood, even the nuanced explanations are based on the supposition of an original state of separation of the mainstream/dominant from the alternative/oppositional, and of one side acting on a separate, already existing opposite side to challenge, resist, or remake it. However, it has become increasingly clear that, despite all the evidence to the contrary, the continued uncritical acceptance of this supposition leads to a debilitating lack of ability to build a concrete, specific historical understanding, in all its complexity, not only of compromising modes of democratic communications but also of insufficiently recognized resources.

Even though Raymond Williams's early 1960s assessment that "we have not yet seen in the world a democratic system of communications" is still accurate, the general outlines of such a system that he alludes to remain relevant. Such a system "is something that belongs to the whole society"; it dispenses with "the idea that communication is the business of a minority talking to, instructing, leading on, the majority"; and it also refuses "the ideology of people who are interested in communications only as a way of controlling people, or of making money out of them."[13] Contained within the last compound phrase of this formulation, however, is the crux of the current theoretical and practical problem. The progressive objection to using communications as a means of capitalist exploitation is easily understood. But the objection to the intention of controlling people is less clear-cut, particularly given the developments noted above. Is it possible for progressive movements dedicated to extending democratic principles and practice to engage in a media practice that adopts what amounts to undemocratic measures of success—how many people did we get to believe what we wanted them to?

As flawed as they are, however, such assertions and the examples that precede them press us insistently to reconsider a key supposition of democratic communications—that a clear opposition indeed exists between mainstream communications practices and those labeled the alternative. If one takes the innumerable examples seriously, however, it may be more useful to regard the two not as polarized, distinct opposites, but as variations within a single, more general, yet highly contradictory social formation. Such a rethinking suggests in turn a premise more complex as well as more profound than that expressed by the mainstream/alternative opposition: that, to a significant degree, both "sides" rely on quite similar, deeply held ideas about means of communications and their roles in society.

If an a priori, foundational separation is not automatically accepted, what emerges as most striking—as well as least understood—about alternative media and democratic communications as conceived and practiced today, in all their variety and with hopes that I share, is not so much their differences from the posited mainstream as their deep commonalities.[14] And such commonalities must be squarely confronted if we are to convincingly address the need for democratic communications in the deep, multiple crises that peoples of the world currently face.

Materialism and the Production of Conditions

While a commonality between mainstream and alternative communications practices can be extensively demonstrated by citing countless examples, exploring why such a commonality exists, how it came about, and what its implications are for democratic communications requires a wider-ranging approach. The first step in such an approach is to avoid projecting present terms into the past and thus mistakenly portray a new development as a long-standing tradition. Such a tendency is typical of common forms of ahistorical, essentialist, functionalist analysis. As such thinking appears in the study of alternative media, the predominant error (particularly pronounced in the United States) is to produce an historical lineage of alternative-media practitioners, ranging from colonial American printers and agitators to such instances as abolitionist groups, Populist and labor publications, suffragette activism, labor radio, and anticorporate activists. By ostensibly recovering a long-standing tradition that originated in the revolutionary founding of the United States (notwithstanding the debate whether the founding itself can be legitimately described as "revolutionary"), the hope is to establish the legitimacy if not "mainstreamness" of present-day alternative media. However, not only does such an approach essentialize human activity, it ignores the constitutive role of specific contexts in producing the very phenomenon of interest.

Contrasting with an ahistorical, functionalist projection of present terms (such as alternative media) into the past is a materialist approach, which recognizes naturalized certainties ("there have always been alternative media . . .") as human, creative projects. Harvey captures the essence of a materialist perspective in his early-1970s redefinition of space as a topic of geographical inquiry.

> There are no philosophical answers to philosophical questions that arise over the nature of space—the answers lie in human practice. The question "what is space?" is therefore replaced by the question "how is it that different human practices create and make use of distinctive conceptualizations of space?"[15]

At the core of such a perspective is an awareness of theory as itself a product of the human practices it seeks to explain. "What is distinctive about this kind of criticism," writes Anderson, "[. . .] is that it includes, indivisibly and unremittingly, *self*-criticism [emphasis original]," thus providing not only a theoretical account of society and history, but also and at the same time a social and historical account of the production of itself and its topics of inquiry.[16]

Given the proposition of a deep commonality between the popularly posited mainstream media and the alternative media, how might this proposition be best understood and investigated? Further, characterizing this deep commonality as part of an age of capitalism suggests a long-standing, perhaps constituent practice. How might such a fundamental practice be best conceived?

While an historical approach is needed to address these questions in the required scale, common varieties of historical explanation as applied to communications systems and practices are of little use for such a purpose. The difficulty lies not so much in the claims made by these various explanations, but in their very structure of exclusive epochs and in some cases their teleological assertions, which prevent these explanations from recognizing persistence, hybridity, and commonality of communications within specific historical contexts. Whether the trend over time is seen as the struggle of democracy against absolutism, of women against patriarchy, of the world's peoples against racism, of popular culture against elite culture, of liberals against traditionalists, of a unified nation against fragmented sectional identities, of working classes against the market and the state, or epochal changes from oral to written and electronic culture, from authentic to mass culture, or from modernity to postmodernity, the difficulty is not only that such frameworks operate at an epochal level (thus preventing a more detailed understanding of specific, concrete, and often contradictory actions and developments). In order to be convincing, they must be presented as a development in which one group or movement vanquishes another or one epoch overtakes another as a process of mutually exclusive change.[17]

Additional difficulties with such explanations arise as well. While explanations that foreground groups or classes in society focus valuably on issues of human practice, and thus counter claims that developments through time were natural, automatic, and inevitable, such groups or classes are often portrayed as essential or pre-formed prior to the struggle as portrayed, thus sidestepping questions of groups' formation. Explanations built around changing modes of representation are in some ways more valuable for an argument about commonality and hybridity, in that such modes are seen as being more generally available, thus opening the possibility that distinct if not opposing social groups made use of the same practices. However, these kinds of explanations tend not only to portray modes of representation as extra-human inevitabilities (such as in various forms of technological determinism), they sidestep the question of why such modes of representation emerged in the first place and why one supplanted another, thus also falling back on versions of functionalism or essentialism.[18] For these reasons, oft-used historical explanations of communications systems—while informing this book in important ways—lack the ability to provide a useful framework for a study with the present intentions.

By contrast, a more illuminating entry point to the study of alternative media is the conditions of the appearance of the explicit term "alternative media." Far from being functionally equivalent to underground media, radical media, and labor media, and thus able to be used as an umbrella term to label any oppositional cultural practice anywhere at any time, cultural theorists such as Williams remind us how changes in language and cultural form mark the concrete work in specific contexts of reorganizing and reforming social relations.[19] In a materialist approach of the kind characterized above, this specific context must be recovered for a full understanding of the emergence and specificity of the term "alternative media."

Scholarly Conditions of Emergence

The scholarly conditions of the emergence of alternative media are best understood in relation to the field of communications study. Throughout the twentieth century and despite earlier attempts at a more expansive understanding of communication in society, studies of communications and social movements were conducted within what one could call a mass-society thesis, which can be generally regarded as a class-based assumption of the necessity of essential distinctions between elites and general populations. It was expressed most clearly in research programs on communications and social movements within studies of propaganda and its effects.[20]

Despite the waxing and waning of many challenges throughout the history of the United States, beliefs in American pluralism (as well as that in many other countries) came under general critique during the social and political upheavals of the 1960s.[21] Such upheavals were foreshadowed in social criticism of the 1950s that exposed the anomie of middle-class life in the "American Dream," as well as in the newly emerged popular youth culture, the resurgence of feminism, and the coalescence of long-standing organized resistance to racism that later emerged as the civil-rights movement.[22]

These upheavals affected every sphere of American life, including views about the nature and role of communications in social movements. As the reigning assumptions of the mass-society thesis (powerful media and passive people) were unmercifully critiqued, hopes of a powerful vanguard spearheading social change or, conversely, fears about unruly masses manipulated by charismatic demagogues gave way in the academy to an interest in popular as opposed to elite focuses of study. This ferment made possible sympathetic and sustained attention to subaltern peoples and social movements. As respectful of popular abilities as the mass-society perspective had been suspicious of them, what could be called a subcultural perspective on the study of communications and social movements soon became a way for scholars to recognize and legitimize heretofore unrecognized and unacknowledged popular practices, including cultural production in such forms as the ethnic press, immigrant press, labor press, underground press of the 1960s, and other kinds of social-movement media.

Such scholarly work was spearheaded by social historians and sociologists who drew upon many influences. The critical perspective on U.S. historiography anchored by Progressive-era historians such as Charles Beard emerged with renewed vigor along with a general acknowledgement of the new social history, inspired by E.P. Thompson and British Marxist historiography, by Annales historians, and by politically engaged scholars in the United States who drew upon Beard and other indigenous radical traditions.[23] Indeed, Hobsbawm's early work on resistance movements in early modern England prefigured not only the protests of the 1960s but the present-day interest in social movements.[24] The influence of an indigenous, radical sociology formed by such people as C. Wright Mills inspired the New Left not only with its radical critique of American consumer society and imperial foreign policy, but in its explicit independence from Marxism-Leninism and communist party politics.[25]

As a result, fewer such studies took for granted elite programs of social control and the inevitable absorption into the dominant culture. Instead of seen as manipulated by elite outsiders for self-serving purposes, social movements were increasingly and sympathetically recognized for their variably autonomous and broadly social accomplishments. Sociological as well as historical studies of ethnicity and other differences were unhinged from fears about cultural dilution, and instead embraced ethnicity as one of a number of necessary defenses against dominant, consumerist values. Studies of black, feminist, and labor media began to enjoy a new legitimacy in academia.[26] Media-focused studies in the new social history argued for a conception of working-class audiences as actively determining their own relation to mainstream media as well as being producers of their own media products for progressive, political ends.[27] As recast within this framework, few more dramatic reinterpretations from the application of the mass-society thesis and its interest in propaganda effects exist than those dealing with American colonial agitators. In the context of world-war propaganda and fear of foreign infiltration, Sam Adams was portrayed as a powerful, manipulative demagogue. But, by the early 1970s, pamphleteer Tom Paine was portrayed as a conduit through which a popular social movement was expressed, with artisans in both the colonies and in Britain simply finding their own voice through Paine's writings.[28]

Such a valorization of the popular was and remains a needed and valuable move toward recognition and solidarity. However, it nevertheless introduced the necessity of making a corresponding, all-too-often essentialized division between all facets of the movement and those of its opposition, including communications.

Activist Conditions of Emergence

The conceptual division between the movement and the mainstream in order to validate social movements' legitimacy remains very influential today. Yet, it differed substantially from activist practice. Compared with scholars' redefinition and reevaluation of social movements in the immediate aftermath of the resistance movements of the 1960s, the explicit appearance of alternative media in popular practice outside of the academy was based in opposite if not audacious claims that the dominant had been incorporated into the oppositional and thus made irrelevant.

The self-reflexive appearance of alternative media in the United States can be traced to the early 1970s and the remaking of the 1960s underground counterculture into a post-1960s alternative culture. One of the earliest efforts to institutionalize this remaking was by members of the U.S. Underground Press Syndicate, which put forth at a 1973 meeting a key proposal to change the organization's name to the Alternative Press Syndicate. As the name-change motion states, the status of "ideas and lifestyles" that had been espoused in the 1960s by the underground press and "shared only by small pockets of people" had, by the early 1970s, changed profoundly.

> The underground ideas and lifestyle have emerged into the mainstream of thought everywhere, in books, plays, films, advertising. Even the established press has adopted our new journalistic style and our role as questioners of authority. [. . .] Meanwhile, increased acceptance, public visibility, and a general

> shift in editorial content away from shocking the public towards "serving the people" has made "underground" a misnomer.[29]

Despite naïvely overstating the degree to which the underground press originated these innovations and the degree to which these points actually constituted key differences between the underground and the establishment prior to the 1970s, the rationale for the proposed name change stems from an exaggerated sense that the rebellions of the 1960s had indeed won the battle with the establishment.

The claim also carried clear implications for practice, including the adoption of commercial goals and imperatives and, as a result, their limits and pressures. As one underground-press editor recalled later about the name-change proposal, "what's left in 1973 [at the U.P.S. conference] is, more than anything else, alternative journalists looking for an audience." As he continued in a paraphrase of what he took as the reigning sentiment at the time, "'there is hope for a journalism in America that continues to be antagonistic towards the status quo, but more in the way it approaches stories than either in jargon or style. And what that needs is a firm advertising base'."[30]

Over and above the argument about which side had won, the renaming of this group is best seen as a recognition of an explicit commonality between the mainstream and the alternative culturally and communicatively. Such a move was joined by others. Additional, early uses of "alternative media" include the titles of membership directories and indexes compiled by the newly named Alternative Press Syndicate and a similar effort to index such publications in England, and the appearance of various directories and how-to-do-it pamphlets and manuals on alternative media.[31] Others include the emergence of alternative urban weekly newsmagazines as a niche, commercial publication targeted to young, hip readers and, by 1978, of a trade group that represented such publications in the United States called the Association of Alternative Newsweeklies (A.A.N.).[32]

At the root of the early 1970s emergence of alternative media in popular practice in the Anglo-American context was the sense that divisions between mainstream and alternative were at best superfluous if not irrelevant. Although claims that such episodes were simply evidence of the co-optation of the alternative by the mainstream (or, conversely, the mainstream by the alternative), such an assessment discounts if not ignores the rationales given by participants themselves, as well as the corresponding changes clearly evident. In this way, the realms of communications and popular (consumerist) culture served as a site of deep, variable commonality.

Vanguardism, Essentialism, and Technological Determinism

Since the 1970s, while popular practice sought to merge if not supercede the establishment/underground distinction, scholarly work continued to be structured around a focus on the uniqueness of alternative social movements and media from the mainstream. Scholarly acknowledgement of explicitly named alternative media began in the 1980s, and was initiated in a specific set of conditions which set key limits and pressures on its study and practice. These conditions can be characterized in terms of two contradictory developments. First, despite the waning of general, large-scale social movements, issue-based movements experienced an upsurge, presaging what some see today as coalitions of such movements bound together by opposition to neoliberal programs of globalization.[33] Second, and at the same time, political dissent continues to be variably incorporated into the market as a lifestyle via the rhetoric of capitalism, individualism, and style. For example, much advertising paradoxically urges consumers to buy mass-produced products in order to liberate themselves from mass society.[34] Even underground cultures overtly hostile to capitalism inspire cutting-edge marketing campaigns for commercial products.[35] More recent U.S. examples include street art from the barrios of East Los Angeles increasingly appearing in mass-marketing campaigns that enshrine "that badass of rebel cool, the *cholo*, or Chicano gang member." As one commentator put it, "skinning the image of badness and selling it as off-the-rack rebellion is the stuff of marketers' dreams."[36] And urban street protest marches form the backdrop for a 2005 magazine fashion photo spread documenting the "bohemian peacenik aesthetic [that] inspired designers this season."[37]

These and related conditions encourage scholarly work and practice of alternative media and democratic communications while also circumscribing it. On one hand, studies of alternative media are a collective recognition of the long history of non-market-driven media activity, as well as a collective response to the accelerating consolidation of commercial media industries.[38] They also imply if not advocate a goal of becoming a mirror image of commercial media in strength and kind, in the hopes of consolidating a similarly powerful oppositional force.[39]

However, the conception of alternative media that such literature relies upon remains limited by certain residual, core assumptions that continue from its initial formation. First is an admiration of the ability of alternative media (and the vanguards who create it) to lead social change. This assumption appears in many studies of alternative media which consist of biographical descriptions of their founders, their publications, and increasingly of the technologies they use.[40] Not only does the initial account of alternative media in the United States (a popular rather than scholarly collection of biographical accounts) define all kinds of oppositional and social-movement media practice as alternative, it begins with newspapers' role in American colonial resistance, in keeping with the aim to incorporate alternative media within mainstream American society.[41]

In addition to sharing vanguard assumptions, such studies also tend to essentialize their respective social groups, portraying them as fully formed prior to their cultural work. Not only does this assumption downgrade the importance of the very topic they otherwise argue is important (with alternative media as simply a passive reflection, neutral carrier, or simple expression of the group's identity and claims), such a position is increasingly difficult to support in view of claims of the cultural rather than natural basis of ethnicities and identities.[42] This interpretive framework manifests itself for example in historical studies of feminist media that share a vanguard perspective while they essentialize the movement, although not without contradiction.[43] These and similar formulations assume a pre-formed and often uniform feminist sphere, thereby essentializing and making superficial not only the process by which such a sphere is articulated, but seeing it in isolation from other kinds of oppression.[44]

In addition to the shortcomings of vanguardism and essentialism, studies that work at a societal level often also rely on a technological determinism that locates the enabling condition for alternative media in technologies of reproduction and distribution. With the emergence of satellite transmission, handheld camcorders, and the Internet, such technological explanations have become more numerous.[45] While techniques of reproduction are indeed important, studies that privilege them as a cause take a vanguard perspective by fetishizing media products and decontextualizing processes of their making.

A comparative historical investigation of media and revolutions in the United States and Western Europe makes clear how difficult the vanguard assumptions of the mass-society thesis are to transcend and how technological determinism hampers efforts to focus on cultural and historical rather than essential and technical processes.[46] The introductory essay dismisses behaviorist notions of media causing revolutions as well as idealist notions that media and social processes of communication are of little importance, but it nevertheless has trouble formulating a position that doesn't rely on either one. On one hand, it works hard to portray the important role of media and communication in organizing social relationships. Toward this end, it cites work by such scholars as Habermas and his insight about "the importance of the nature of the dialogue in which members of the public engaged, rather than the sociological identity of the participants or the specific content of their debate," Chartier's similar assertion that one must study the form of communication as well as the content, as well as related arguments made by others.[47] Yet throughout the essay the author uses "form" as a synonym for technology (such as books, newspapers, and pamphlets) instead of as a way of talking about kinds of social relationships. The result of this confusion is an argument constructed about the power of technologies. "Whatever their differences," concludes the author, "revolutionary upheavals are defined in part by the accelerated circulation of words and images that characterizes them"—a comment about technologies rather than about the contextual practice of social relationships in which technologies are developed and used, or about the "words and images" that can be seen to constitute such relationships and practices.[48]

Alternative Media Today

The most recent work on alternative media is distinctive in its significant efforts to supercede vanguardism, essentialism, and technological determinism, largely through a structuralist critique of such premises. One particularly influential body of work has emerged from scholars working within a broadly defined sociological approach which draws variably upon the work of people such as Castells, Melucci, Bourdieu, and Laclau and Mouffe.[49] Work in the related topic area of social movement studies that explicitly addresses social movements' communications practices applies such varied approaches as collective behaviorism, resource mobilization, a focus on political process, social constructionism, historical materialism, and cultural studies.[50] Such studies have made admirable and invaluable contributions by deepening theoretically the study of alternative media and democratic communications as social processes embedded in larger contexts, and they have shed great light on present examples and recent developments—particularly in relation to new social movements and their use of the Internet.[51]

However, a primary focus on structural analysis—social networks and/or of textual relationships—defers an exploration of the broad historical processes that produced the necessary conditions for these networks and relationships. The strengths as well as the limitations of such an approach can be seen in one such proposed definition of alternative media as "practices of symbolic production which contest [. . .] the concentration of symbolic power in media institutions."[52] It and similar efforts focus productively on culture as a practice rather than as a text, and on the positioning of the practice in relation to the dominant rather than simply on its isolated meaning. Arguing that such practices need to be interpreted as responses to a particular state of affairs points the way toward other similarly valuable kinds of contextualization. One must also account for the production of the very conditions that contextualize the act and thus that not only make it meaningful, but possible. Indeed, the most fundamental level of determination can be seen as not the act nor its contextualization in its moment, but the conditions that make possible both the act and its contextualization—and, even more fundamentally, the production of those very conditions.

Ostensibly to provide the historical understanding largely absent from such sociological studies, a second important literature is characterized by an overtly historical approach. Such studies have recovered and contributed many significant insights into specific episodes and their contextual contingencies and intricacies.[53] However, when assessed together, narrowly focused biographies of individual publishers, writers, journalists, publications, or organizations create a patchwork collection of accounts, making more difficult a broader understanding of the general media practices and necessary conditions upon which these individual cases rely. Like many sociological studies, they contextualize the act and its conditions, but rarely if ever account for the production of those conditions and their determinations.

By contrast, a variety of scholars have begun to address the broader, historical formations of media practice and the production of their conditions. Whether concerned with news, activist art and the Internet, self-published 'zines, novels, workers' theater, progressive radio, documentary, or self-reflexive attention to the theoretical contexts of alternative-media practice, such works have begun to reconnect particular forms and cases with broader traditions, resources, movements, and developments.[54] Despite many differences, they all bring to the study of their topic a sense of culture as practice, context, and history rather than as—or supplementing an understanding of—culture as text, form, and structure. Such an awareness has only just begun to emerge in research in alternative media. For example, work by Downing and by Atton has initiated a reconstruction of the historical, intellectual milieu of alternative-media practice, locating a crucial source in anarchism.[55] Such a direction points to a particularly valuable and necessary kind of investigation, which has yet to more fully emerge.

From Technologies to Formations

Asserting that mainstream and alternative communications share a significantly common formation does not necessarily mean that no differences exist. It would be a gross oversight to pretend that the presence of these commonalities means that there are no political differences at stake. To take just one example (and despite areas of commonality), Greens are not the same as social democrats, libertarians, or neoliberal conservatives. Yet, while ideological differences stake out areas of difference, all sides define, publicize and promote themselves to wider publics through their newspapers, newsletters, and 'zines; their journalists, commentators, and columnists; their analyses, documentaries, exposés, and parodies; their advertising and public relations; their protest marches; and, more recently, their listservs, online petitions, websites, blogs, vlogs, and podcasts.[56] Such a commonality of practice demands that questions about their implications be squarely addressed, but done so contextually so as not to gloss over their differences.

A way to address commonalities of communications practices is via the issue of technology, but in a way much different than has typically been the case. What must first be resisted is the use of a form of technological determinism to make the case that, for example, progressive movements should not publish newspapers (assumed to be intrinsically a "mainstream" medium) because the use of that medium will inevitably lead to the movement's compromise if not co-optation by the mainstream. An opposite claim is equally disabling: namely, that technologies are entirely blank slates, with their effects determined entirely by their users and uses.[57] The difficulty is that, in both cases, technologies are seen as separate from contexts and often, in the case of communications, synonymous with "media," thus adding to an often severe conceptual confusion.[58]

Williams's wide-ranging work on communications offers a way to unpack and specify such variations. To unravel these complexities, he distinguishes between techniques ("a particular skill or application of a skill"), a technical invention ("a development of such a skill or the development or invention of one of its devices"), and a technology ("the body of knowledge appropriate to the development of such skills and applications and, second, a body of knowledge and conditions for the practical use and application of a range of devices").[59] This finer-tooled understanding highlights the way in which what is habitually referred to as a technology, when fully constituted, is actually the combination of a technical apparatus, the conditions which brought it into being, and the social knowledge and organization that directs its use—all of which emerge from and enabled by what Williams calls "the whole complex of social productive forces and relationships within which they are developed and used."[60]

Such a position does not deny that technical apparatuses have their specific capabilities that place limits and pressures on uses. For example, a printed page is not the same as a televisual display. At the very least, each requires very different productive capacities in their respective societies (papermaking and ink-printing versus electronic imaging, reproduction and display), which also presumes different social decisions regarding which capacities to develop at the expense of others. Each also requires different kinds of learned skills to fully use, with the printed page requiring a higher, more abstract level of training in terms of literacy, compared with that required by the seemingly unmediated presentation of direct, bodily human speech and movement that distinguishes televisual works. Sense can be more easily made of the televisual absent such specialized training.[61] But these limits and pressures derive from and are exerted and felt within a specific time, place, and set of conditions. Recognizing the depth at which such limits and pressures operate is to escape essentialist claims about the ostensibly independent determination by technologies, how such capabilities are developed and deployed, and for what kinds of purposes—individual as well as social—and with what kinds of implications. There is no technical reason why a novel cannot be published on a billboard or a shopping list produced as a television program. Rather, the reason why we do not regularly see such inanities (except, perhaps, in activist or artistic expression which seeks to call attention to the conventional foundation of cultural forms and practices) is due to the limits and pressures deriving from—again, to use Williams's words—the whole complex of social productive forces and relationships.[62]

An example of the value of recognizing and specifying key differences and complex relationships between techniques, technical inventions, and technologies can be seen when assessing one commonly proposed strategy of making communications more democratic: to broaden ownership. The argument in general is that, only if more of "us" owned television stations, radio stations, and newspapers, then this broader range of owners would automatically contribute to progressive change (the nature of which is amenable to "us"). Such a position ignores the taking-on of ownership, operation, and production within an already existing set of conditions and relationships at once regulatory, economic, political, social, and technological (in the sense explained above). Current efforts to remake these conditions—at least at the policy level—recognize the incompleteness of such a solution.[63] The abstraction of ownership from the more general social organization substitutes essences for relationships and contexts. Whether portrayed as a cause or an effect, such a "solution" assumes that abstracted elements are addressable on their own. In these ways, this approach "is essentially a cancellation of the contemporary world," seeking universals in lieu of an actual historical understanding "in the sense of real times and real places" in which people "act and react, struggle and concede, co-operate, conflict and compete."[64]

Cultures of Truth

This book proposes a perspective and approach to a more productive way of addressing alternative media as a form of democratic communications. Despite being enabled, inspired, and deeply indebted to earlier work on the topic, it is a response to the drawbacks and shortcomings of a lack of a sense of the historicity of alternative media as a scholarly subject and as a practice; a shying away from devoting serious attention to clear areas of commonality in communications practices between different formations; and a too-often uncritical use of vanguardism, essentialism, and technological determinism. In contrast, this book complements sociological textual/structuralist analyses and biographical, historical case studies by offering a cultural and historical investigation of the formative conditions of practice relied upon by the dominant and the oppositional. In doing so, it also detects, explores, and assesses the implications of these resources as they continue to be remade, and suggests ongoing possibilities for alternative media and democratic communications.

Instead of focusing on texts, structures, technologies narrowly defined, and/or individual cases, the argument here focuses on forms, modes, contexts, and practices of authoritative representation. In this view, social formations rely ultimately on claims of truth as the essential ground for proposal and critique. Truth as an authoritative account of the world is not simply found, but is actively produced within and by virtue of a specific set of conditions that consist of a complex of what are habitually separated into the political, economic, organizational, cultural or textual. By contrast to their essential separation, the complex of conditions of the production of truth includes popular epistemologies (accepted ways that truth can be legitimately found and established), popular cultural forms (accepted ways that truth can be legitimately embodied and expressed), and popular modes of organization and activity (accepted ways that truth can be legitimately spread and shared), all made possible by and operating within the highly variable, expanding capitalist social order. Regarding such separate spheres as a mutually produced, historical complex of practices provides a flexible way of understanding the mutually productive connections between forms of expression, traditions of interpretation, and institutions. Each facet that comprises such complexes is not a separate sphere or static and self-contained, but part of an extremely varied cultural and historical formation, the interrelations and mutual constitutions of which is the focus of this book.

In order to adequately reconstitute such relationships and complexes, this book takes a transdisciplinary approach, working with concepts and perspectives drawn from a number of fields. By using a combination of contextualized summary, documentary analysis of specific projects, and synthesis of relationships between these levels, the book attempts, again in Williams's words, "to discover the nature of the organization" that composes what are being called here cultures of truth.[65] An approach that seeks to focus on the historical constitution of practices and the making and operation of relationships that make them possible and meaningful is fraught with difficulty, not the least of which is that no ready-made, preset procedure that one might mechanically follow.[66] However, such an approach presents the only possible means of addressing adequately the historical depth and the implications in so many areas of cultures of truth as they are conceived of here.

The exploration here of projects and formations of democratic communications suggests the scope of insight that might be gained through such an approach. The first part of the book, "Market Formations," sketches in broad relief the emergence and implications of key communications formations in modern market economies and societies. Such formations constitute important bases for communications practiced by social movements from radical to reactionary. Chapter 1, "Providentialism and Rationalist Empiricism in Early Modern England," discusses the early modern transition from truth as personal and transcendental to natural and able to be generally apprehended, and its fit within the emergent institutionalization of capitalism. Small, private publishing and state-supported publishing together with an nascent commercial practice comprised the institutional basis of print capitalism that, with polemic, comprised a rationalist formation of truth production in emergent market economies and newly formed nation states.

Chapter 2, "The Emergence of Broadcasting and the Rationalization of Participation," examines the rise of and connections between emergent industrial market democracies and broadcasting as a cultural form of embodying newly emerged publics. In the nineteenth century amid the rapid, uneven expansion of capitalism and the gradual, partial extension of the franchise in the United States, the problem of communications became how to engage a diverse, far-flung "public." Broadcasting within the dominant culture of naturalism emerged as the prototypical mode of what was later to be called "mass communication"—pioneered first in print by evangelical Christian social activists, and later realized electronically as individualized entertainment. Despite its innovations of organization, distribution, and uniformity, and the exploration of broadcasting by a wide range of politically progressive formations, the class-bound, vanguard, and consumerist traditions of its use complicated then and continue to complicate today the popular practice of democracy.

The second part of the book, "Struggling Against the Market," speaks not only to a growing awareness of the complicity of received forms of communications with the inequalities of capitalist societies, but an emergent effort to find ways of organizing communications outside of market imperatives. Chapter 3, "Philanthropy, Professionalization, and Social-Reform Communications," discusses patronage and its more recent cousin philanthropy as critical responses to capitalism, in that they seek to assert control in terms other than those provided by the economic market and consumer society. While constituting an important means of support outside at least direct market control, such pressures have tended to rationalize alternative-media projects as naturalist forms of documentary in support of at best a liberal critique of the dominant.

Chapter 4, "Community Media Projects and the Containment of the Mass-Culture Critique," discusses the anchoring of the practice of communications in communities, which have their own historical form as a response to centralizing pressures of capitalist consolidation. In the United States, these ranged from the radical republican cooperative commonwealth, to utopian colonies, national unions, and individual localities that, by the early twentieth century, emerged as bulwarks against the perceived onslaught of national mass culture. An example addressed in some detail is the recently introduced low-power FM radio service, which consists of a non-corporate effort constituted by rationalist conceptions and preoccupations. However, through industry and legislative dilution, its practice is increasingly rationalized by statist regulation through the enforcement of a standardized, corporatized format and operation.

Chapter 5, "Modernism and the Aestheticization of Dissent," discusses a wholesale rethinking of rational-empiricist cultures of truth within capitalist societies, which emerged in earnest within the context of world war and scientific discovery. Dada and successive waves of modernist avant-garde artistic practice highlighted not only the naiveté of rationalist empiricism, but also its role in legitimating repressive authority, bureaucracy, and instrumental rationalization. What can be called a critical-modernist culture of truth was also enabled by the recognition of communication as productive instead of inertly transparent or mechanically mediative, a view that began to consolidate in nineteenth-century philosophical speculation tied to linguistics and to critical social theory.[67] However, the popular understanding of much of this effort remained and remains compromised by being seen as expressions of aestheticist or essentialist subcultures, as biased distortions, or as manipulative propaganda.

Part Three, "Toward New Formations," seeks to historicize and critique premises and developments documented in the prior discussions as a way of suggesting important, yet still uncertain tendencies and directions. Chapter 6, "Market Radicalism and the Struggle of Participation," examines the current practice of self-publishing variously called "citizen journalism," "stand-alone journalism," and "consumer-generated media." While seeking to counter bureaucratic, professional elitism, such journalistic and marketing efforts have pioneered a rationalized form of popular participation yoked to commercial imperatives in such ways as members of the public functioning as a low-cost research-and-development department or a flexible workforce. At the same time, the continued market pressure to enlarge the scope of popular participation constitutes the possibility of a reformation of market communications institutions in ways still only dimly apparent.

Chapter 7, "Democratic Communications as Critical, Collective Education," rethinks the organizational emphasis in much theorizing and practice of alternative media on production and the resulting strategic goals of how to finance, build, organize, and maintain as large an organization as possible, as well as the equally foundational definition and separation of moments of production from consumption so common in communications study. While others have correctly noted how research in alternative media has largely neglected a focus on the "consumption" stage in favor of a focus on production, the argument here recasts the sense of consumption into a matter of composition.[68] Critical pedagogies seek to escape rationalist pressures toward promoting and defending isolated, individual views, naturalist tendencies toward uncritical empiricism and institutionalization, and modernist aestheticism by seeking to build a collective, critical awareness of ways to meld projects of alternative media and media democratization with forms of popular culture and practices of popular education. The example of migrant labor-camp newspapers in 1930s California help develop an historical conception of direct autonomous composition as the practice of democratic communications while also extending the sense of the "product" thus produced into modes of organization and collective work. While the current institutionalization of such a project in the form of media literacy tends to isolate such a practice in the form of textual analysis in classrooms, it signals an important, recent development with clear progressive possibilities.

The final part, "Afterward: Utopia and Inspiration," suggests in tentative terms not only the value of utopian thinking about democratic communications, but also a way of conceiving of it as the social production of inspiration.

Reconceiving Alternative Media

Adapting a point Williams made about culture and expression to the argument of this book, to fail to see the debilitating drawbacks of received thinking about alternative media and democratic communications "is to begin the whole difficult process of discovering and describing relations between all these practices [. . .] head down and standing on one foot." The best work on alternative media and democratic communications demonstrates that "such feats of agility are not impossible, and have indeed been performed." But, as Williams notes further, "it would be more reasonable to get back on both feet again and to look at our actual productive activities without assuming in advance" that alternative media are essentially separate, distinct, or different from the mainstream.[69]

To do so, one must work toward a qualitative change in the understanding and practice of democratic communications, as distinct from the received modes as is radical nonviolence from assembling and equipping an opposing army. The hope of this book and the primary contribution of critical scholarship generally is to retheorize accepted assumptions and explore how new ways of seeing can open up new understandings of conditions and new routes for practice—as well as the capacity to continually renew both.

Notes

1. Daniel Czitrom, *Media and the American Mind; From Morse to McLuhan* (Chapel Hill, N.C.: University of North Carolina Press, 1982).

2. The phrase "our media, not theirs" is a clear, current example of this foundational separation. In addition to its being used by an international group of scholars and activists (http://www.ourmedianet.org/), it also forms part of the title of Robert W. McChesney and John Nichols, *Our Media, Not Theirs; The Democratic Struggle Against Corporate Media* (New York: Seven Stories Press, 2002).

3. Kerry Stranman, "Consumer Created Content," *Word of Mouth Marketing Association*, <http://www.womma.org/content/consumercreatedcontent.pdf> (19 Sept. 2005).

4. Jesse Hamlin, "Gore's New Media Venture Seeks to Blend TV, Internet; Former VP Says His Current Network Gives Viewers a Voice," *San Francisco Chronicle*, 5 April 2005, A1.

5. Rupert Murdoch, speech to the American Society of Newspaper Editors on 13 April 2005, *News Corporation*, <http://www.newscorp.com/news/news_247.html> (23 June 2005).

6. "What is DAMN?," *Direct Action Media Network*, <http://www.tao.ca/earth/damn/index2.html> (13 May 2002).

7. "About GNN," *Guerrilla News Network*, <http://www.guerrillanews.com/profile/about.html> (13 May 2002).

8. "TruthOut—Donations," *TruthOut*, <https://secure.entango.com/donate/> (24 Sept. 2005).

9. Similar organizations include Common Dreams News Center (http://www.commondreams.org/).

10. "The H-Bomb Secret: How We Got It—Why We're Telling It," *The Progressive*, November 1979; "Dark Alliance," *The Lycaeum*, <http://www.lycaeum.org/drugwar/DARKALLIANCE/> (10 Jan. 2006); Gary Webb, *Dark Alliance: The CIA, the Contras, and the Crack Cocaine Explosion* (New York: Seven Stories Press, 1998); James Risen and Eric Lichtblau, "Bush Lets U.S. Spy on Callers Without Courts," *New York Times*, 15 December 2005, A1; Paul McLeary, "How TalkingPointsMemo Beat the Big Boys on the U.S. Attorney Story," *CJR Daily*, <http://www.cjrdaily.org/behind_the_news/how_talkingpointsmemo_beat_the.php> (23 March 2007); Terry McDermott, "Blogs Can Top the Presses; Talking Points Memo Drove the U.S. Attorneys Story, Proof That Web Writers With Input from Devoted Readers can Reshape Journalism," *Los Angeles Times*, 17 March 2007, A1.

11. "Current TV // About Current," *Current TV*, <http://current.tv/about/> (24 Sept. 2005).

12. Among the few studies that address such complexities is Nina Eliasoph, "Routines and the Making of Oppositional News," *Critical Studies in Mass Communication* 5, no. 4 (December 1988): 313–34.

13. Raymond Williams, "Communications and Community," in *Resources of Hope*, ed. Robin Gable (London: Verso, 1989), 29. The essay was originally published in 1961.

14. This insight builds on the recognition of the hybridity of alternative media. See John D.H. Downing, Tamara Villarreal Ford, Genève Gil, and Laura Stein, *Radical Media; Rebellious Communication and Social Movements* (Thousand Oaks, Calif.: Sage, 2001), ix; and Chris Atton, *An Alternative Internet; Radical Media, Politics and Creativity* (Edinburgh: Edinburgh University Press, 2004), 9–11. Recognition and analysis of connections and commonalities include Raymond Williams, "The Press and Popular Culture: An Historical Perspective," in *Newspaper History from the Seventeenth Century to the Present Day*, ed. George Boyce, James Curran and Pauline Wingate (London: Constable, 1978), 41–50; and Michael Denning, *The Cultural Front: the Laboring of American Culture in the Twentieth-Century* (New York: Verso, 1996).

15. David Harvey, *Social Justice and the City* (Baltimore, Md.: Johns Hopkins University Press, 1973), 13–4.

16. Perry Anderson, *In the Tracks of Historical Materialism* (London: Verso, 1983), 11.

17. These varieties of historical explanation of communications systems are summarized in James Curran, "Media and the Making of British Society, c. 1700–2000," *Media History* 8, no. 2 (December 2002): 135–54; and Lawrence Grossberg, Ellen Wartella, D. Charles Whitney, and J. Macgregor Wise, *Mediamaking; Mass Media in a Popular Culture*, 2nd. ed. (Thousand Oaks, Calif.: Sage, 2006), 34–64. For discussions of history writing and explanation as an ongoing problem forever to be confronted rather than as a closed matter of achieving neutrality and accuracy, see Hans Kellner, *Language and Historical Representation; Getting the Story Crooked* (Madison, Wisc.: University of Wisconsin Press, 1989) and John Nerone, "Theory and History," *Communication Theory* 3, no. 2 (May 1993): 148–57.

18. Raymond Williams, *Television; Technology and Cultural Form* (New York: Schocken, 1975).

19. Raymond Williams, *Marxism and Literature* (Oxford: Oxford University Press, 1977), 138; Paul Jones, *Raymond Williams's Sociology of Culture; A Critical Reconstruction* (New York: Palgrave Macmillan, 2004), 92–126.

20. Hanno Hardt, *Critical Communication Studies; Communication, History & Theory in America* (London: Routledge, 1992), 77–122; James Hamilton, "Theory Through History: Exploring Scholarly Conceptions of U.S. Alternative Media," *The Communication Review* 4, no. 3 (2001): 305–26; Richard Flacks, "The Question of Relevance in Social Movement Studies," in *Rhyming Hope and History; Activists; Academics, and Social Movement Scholarship*, ed. David Croteau, Williams Hoynes, and Charlotte Ryan (Minneapolis, Minn.: University of Minnesota Press, 2005), 3–19.

21. Stuart Hall, "The Rediscovery of 'Ideology': Return of the Repressed in Media Studies," in *Culture, Society and the Media*, ed. Michael Gurevitch, Tony Bennett, James Curran, and Janet Woollacott (New York: Routledge, 1988), 56–90; Richard Lee, *Life and Times of Cultural Studies; The Politics and Transformation of the Structures of Knowledge* (Durham, N.C.: Duke University Press, 2003), 173–91.

22. Staughton Lynd, "Towards a History of the New Left," in *The New Left; A Collection of Essays*, ed. Priscilla Long (Boston, Mass.: Porter Sargent, 1969), 1–13; Todd Gitlin, *The Sixties; Years of Hope, Days of Rage* (New York: Bantam, 1987).

23. Charles A. Beard, *An Economic Interpretation of the Constitution of the United States* (New York: Macmillan, 1913); Charles A. Beard, *Economic Origins of Jeffersonian Democracy* (New York: Macmillan, 1915); James A. Henretta, "Social History as Lived and Written," *American Historical Review* 84, no. 5 (December 1979): 1293–323; Henry Abelove, Betsy Blackmar, Peter Dimock, and Jonathan Schneer, eds., *Visions of History* (New York: Pantheon, 1984); Lynn Hunt, ed., *The New Cultural History* (Berkeley, Calif.: University of California Press, 1989); Paul Buhle, ed., *History and the New Left: Madison, Wisconsin, 1950–1970* (Philadelphia, Pa.: Temple University Press, 1990); Jon Wiener, "Radical Historians and the Crisis in American History, 1959–1980," in *Professors, Politics and Pop* (New York: Verso, 1991), 175–216.

24. Eric Hobsbawm, *Social Bandits and Primitive Rebels; Studies in Archaic Forms of Social Movement in the 19th and 20th Centuries* (Glencoe, Ill.: Free Press, 1959); Eric Hobsbawm, *The Age of Revolution, 1789–1848* (Cleveland, Ohio: World Publishing Co., 1962).

25. C. Wright Mills, *The Power Elite* (New York: Oxford University Press, 1956); C. Wright Mills, *The Sociological Imagination* (New York: Oxford University Press, 1959); C. Wright Mills, *The Marxists* (New York: Dell, 1962); Gitlin, *The Sixties*, 11–30; Tom Hayden and Dick Flacks, "The Port Huron Statement at 40," *The Nation*, 5 August 2002, 18–21.

26. Oliver Zunz, "The Synthesis of Social Change; Reflections on American Social History," in *Reliving the Past; the Worlds of Social History*, ed. Oliver Zunz (Chapel Hill, N.C.: University of North Carolina, 1985), 81–92; John Bodnar, *The Transplanted; A History of Immigrants in Urban America* (Bloomington, Ind.: Indiana University Press, 1985); David Brody, "Labor History in the 1970s: Toward a History of the American Worker," in *The Past Before Us; Contemporary Historical Writing in the United States*, ed. Michael Kammen (Ithaca, N.Y.: Cornell University Press, 1980), 252–69.

Examples of studies of movement media include Carter R. Bryan, "Negro Journalism in America Before Emancipation," *Journalism Monographs* no. 12 (Sept. 1969); Roland E. Wolseley, *The Black Press, U.S.A.* (Ames, Iowa: Iowa State University Press, 1971); Martin Dann, ed., *The Black Press 1827–1890; The Quest for National Identity* (New York: G.P. Putnam, 1971); Lee Finkle, *Forum for Protest: The Black Press During World War II* (Rutherford, N.J.: Farleigh Dickinson University Press, 1975); Penelope Bullock, *The Afro-American Periodical Press, 1838–1909* (Baton Rouge, La.: Louisiana State University Press, 1981); Henry Lewis Suggs, *The Black Press in the South, 1865–1979* (Westport, Conn.: Greenwood Press, 1983); Julius E. Thompson, *The Black Press in Mississippi, 1865–1985* (Gainesville, Fla.: University Press of Florida, 1993); Frankie Hutton, *The Early Black Press in America, 1827–1860* (New York: Greenwood Press, 1993); Richard Digby-Junger, "*The Guardian*, *Crisis*, *Messenger*, and *Negro World*: The Early-20th-Century Black Radical Press," *Howard Journal of Communications* 9, no. 3 (July–Sept. 1998): 263–82; Henry Lewis Suggs, *The Black Press in the Middle West, 1865–1985* (Westport, Conn.: Greenwood Press, 1996); and Jane Rhodes, *Mary Ann Shadd Cary: The Black Press and Protest in the Nineteenth Century* (Bloomington, Ind.: Indiana University Press, 1998). Labor-press studies entered academic legitimacy in the wake of Joseph Conlin, ed., *The American Radical Press 1880–1960*, 2 vols. (Westport, Conn.: Greenwood Press, 1974). Studies of the feminist press were spurred by such initial efforts as Anne Mather, "A History of Feminist Periodicals, Pt. 1," *Journalism History* 1, no. 3 (Autumn 1974): 82–5; Anne Mather, "A History of Feminist Periodicals, Pt. 2," *Journalism History* 1, no. 4 (Winter 1974–75): 108–11; and Anne Mather, "A History of Feminist Periodicals, Pt. 3," *Journalism History* 2, no. 1 (Spring 1975): 19–23, 31.

27. Examples include Roy Rosenzweig, *Eight Hours for What We Will: Workers and Leisure in an Industrial City, 1870–1920* (Cambridge: Cambridge University Press, 1983) and Lizabeth Cohen, "Encountering Mass Culture at the Grassroots: the Experience of Chicago Workers in the 1920s," *American Quarterly* 41, no. 1 (March 1989): 6–33.

28. John C. Miller, *Sam Adams; Pioneer in Propaganda* (New York: Little, Brown & Co., 1936); Philip Davidson, *Propaganda and the American Revolution, 1763–1783* (Chapel Hill, N.C.: University of North Carolina Press, 1941); Eric Foner, *Tom Paine and Revolutionary America* (New York: Oxford University Press, 1976), xvii.

29. Reproduced in Abe Peck, *Uncovering the Sixties; The Life and Times of the Underground Press* (New York: Pantheon, 1985), 286.

30. Peck, *Uncovering the Sixties*, 289.

31. A serial publication from the Alternative Press Syndicate variously titled *Alternative Media Magazine*, *Alternative Press Review*, and the *Alternative Journalism Review* began to appear in 1974, a continuation of the *Underground Press Review* by the same organization. The first edition of the *Directory of Alternative Media Periodicals* (Brighton: Smoothie Publications) was published in 1970.

32. Susan Krieger, *Hip Capitalism* (Beverly Hills, Calif.: Sage, 1979); Phyllis Kaniss, *Making Local News* (Chicago, Ill.: University of Chicago Press, 1991), 133–59; Thomas Frank, *Conquest of Cool; Business Culture, Counterculture, and the Rise of Hip Consumerism* (Chicago, Ill.: University of Chicago Press, 1997); Thomas Frank, "Liberation Marketing and the Culture Trust," in Patricia Aufderheide, Erik Barnouw, Richard M. Cohen, Thomas Frank, Todd Gitlin, David Lieberman, Mark Crispin Miller, Gene Roberts, and Thomas Schatz, *Conglomerates and the Media* (New York: New Press, 1997), 173–90; Jay Walljasper, "Age of the Mega-Alternatives," *Utne Reader*, July–August 1997, 91–4; "About AAN," *Association of Alternative Newsweeklies*, <http://aan.org/gyrobase/Aan/ViewPage?oid=2086> (2 Dec. 2005).

33. Barbara Epstein, *Political Protest and Cultural Revolution* (Berkeley, Calif.: University of California Press, 1991); Stanley Aronowitz, *The Death and Rebirth of American Radicalism* (New York: Routledge, 1996); Howard Zinn, *A People's History of the United States* (New York: New Press, 1997), 443–64; John Holloway and Eloína Peláez, eds., *Zapatista!: Reinventing Revolution in Mexico* (London: Pluto Press, 1998); Roger Burbach, *Globalization and Postmodern Politics: from Zapatistas to High-Tech Robber Barons* (London: Pluto Press, 2001); Catherine Eschle, *Global Democracy, Social Movements, and Feminism* (Boulder, Colo.: Westview, 2001); and T.V. Reed, *Art of*

Protest; Culture and Activism from the Civil Rights Movement to the Streets of Seattle (Minneapolis, Minn.: University of Minnesota Press, 2005).

34. Frank, *Conquest of Cool*.

35. Stephen Duncombe, *Notes from Underground; Zines and the Politics of Alternative Culture* (London: Verso, 1997), 141–73.

36. Mark Dery, "Brown Power," *Print*, September/October 2005, 96.

37. "Give Peace a Chance," *Harper's Bazaar*, March 2005, 328–39.

38. John Nichols and Robert W. McChesney, *Tragedy and Farce: How the American Media Sell Wars, Spin Elections, and Destroy Democracy* (New York: New Press, 2005); Robert W. McChesney and Ben Scott, eds., *Our Unfree Press: 100 Years of Radical Media Criticism* (New York: New Press, 2004); Robert W. McChesney, *The Problem of the Media: U.S. Communication Politics in the Twenty-First Century* (New York: Monthly Review Press, 2004).

39. Robert Cirino, "An Alternative American Communications System," in *American Media and Mass Culture; Left Perspectives*, ed. Donald Lazere (Berkeley, Calif.: University of California Press, 1987), 568–76; Michael Albert, "Mass Media: Theirs and Ours," *Z Magazine*, <http://zena.secureforum.com/Znet/ZMag/articles/oldalbert6.htm> (19 Jan. 2006).

40. Lauren Kessler, *The Dissident Press; Alternative Journalism in American History* (Newbury Park, Calif.: Sage, 1984); Peck, *Uncovering the Sixties*; Jeff Land, *Active Radio; Pacifica's Brash Experiment* (Minneapolis, Minn.: University of Minnesota Press, 1999); Deirdre Boyle, *Subject to Change; Guerrilla Television Revisited* (New York: Oxford University Press, 1997); Parry D. Teasdale, *Videofreex; America's First Pirate TV Station & the Catskills Collective that Turned It On* (Hensonville, N.Y.: Black Dome Press, 1999); Penelope Bullock, *The Afro-American Periodical Press, 1838–1909* (Baton Rouge, La.: Louisiana State University Press, 1981); Kathleen Endres and Therese L. Lueck, *Women's Periodicals in the United States; Social and Political Issues* (Westport, Conn.: Greenwood, 1996); Digby-Junger, "The *Guardian*, *Crisis*, *Messenger*, and *Negro World*"; Allen Ruff, *"We Called Each Other Comrade": Charles H. Kerr & Company, Radical Publishers* (Urbana, Ill.: University of Illinois Press, 1997); Rodger Streitmatter, *Voices of Revolution: The Dissident Press in America* (New York: Columbia University Press, 2001).

41. David Armstrong, *A Trumpet to Arms: Alternative Media in America* (Boston, Mass.: South End Press, 1981).

42. Peter Kivisto, "Overview: Thinking About Ethnicity," in *The Ethnic Enigma; The Salience of Ethnicity for European-Origin Groups*, ed. Peter Kivisto (Philadelphia, Pa.: Balch Institute Press, 1989), 11–23; Stuart Hall and Paul du Gay, eds., *Questions of Cultural Identity* (London: Sage, 1996); Richard Jenkins, *Rethinking Ethnicity: Arguments and Explorations* (Thousand Oaks, Calif.: Sage, 1997); Matthew Frye Jacobson, *Whiteness of a Different Color: European Immigrants and the Alchemy of Race* (Cambridge, Mass.: Harvard University Press, 1998); Gary Taylor, *Buying Whiteness: Race, Culture, and Identity from Columbus to Hip Hop* (New York: Palgrave Macmillan, 2005).

43. Amy C. Branner, "Culture Work: the Production of Radical Feminism and Underground Print Media at *Off Our Backs*" (M.A. thesis, University of Missouri, 1996); Marilyn Crafton Smith, "Women's Movement Media and Cultural Politics," in *Women and Mass Communication*, ed. Pamela J. Creedon (Newbury Park, Calif.: Sage, 1989).

44. Lisa McLaughlin, "Beyond 'Separate Spheres': Feminism and the Cultural Studies/Political Economy Debate," *Journal of Communication Inquiry* 23, no. 4 (October

1999): 327–54; Ranu Samantrai, *AlterNatives; Black Feminism in the Postimperial Nation* (Stanford, Calif.: Stanford University Press, 2002).

45. DeeDee Halleck, "Deep Dish TV: Community Video from Geostationary Orbit," *Leonardo* 26, no. 5 (October 1993): 415–20; Wendy Harcourt, ed., *Women@Internet: Creating New Cultures in Cyberspace* (London: Zed Books, 1999); Andrea Langlois and Frédéric Dubois, eds., *Autonomous Media; Activating Resistance & Dissent* (Montréal, Québec: Cumulus Press, 2005).

46. Jeremy Popkin, "Media and Revolutionary Crisis," in *Media and Revolution; Comparative Perspectives*, ed. Jeremy Popkin (Lexington, Ky.: University of Kentucky Press, 1995), 12–30.

47. Jürgen Habermas, *The Structural Transformation of the Public Sphere: an Inquiry into a Category of Bourgeois Society*, transl. Thomas Burger and Frederick Lawrence (Cambridge, Mass.: MIT Press, 1989); Roger Chartier, *The Cultural Origins of the French Revolution*, transl. Lydia Cochrane (Durham, N.C.: Duke University Press, 1991); Benedict Anderson, *Imagined Communities: Reflections on the Origin and Spread of Nationalism* (London: Verso, 1983); James Carey, *Communication as Culture; Essays on Media and Society* (Boston, Mass.: Unwin Hyman, 1989).

48. Popkin, "Media and Revolutionary Crisis," 27.

49. Manuel Castells, *The Rise of the Network Society* (Oxford: Blackwell, 2000); Alberto Melucci, *Challenging Codes: Collective Action in the Information Age* (Cambridge: Cambridge University Press, 1996); Pierre Bourdieu, *Language and Symbolic Power* (Cambridge: Polity, 1991); Ernesto Laclau and Chantal Mouffe, *Hegemony and Socialist Strategy: Towards a Radical Democratic Politics*, transl. Winston Moore and Paul Cammack (London: Verso, 1985); and Ernesto Laclau, *New Reflections on the Revolution of Our Time* (London: Verso, 1990).

50. Croteau et al., eds., *Rhyming Hope and History*; Reed, *Art of Protest*, 286–87.

51. Chris Atton, *Alternative Media* (Thousand Oaks, Calif.: Sage, 2002); Atton, *An Alternative Internet*; Wilma De Jong, Martin Shaw, and Neil Stammers, eds., *Global Activism, Global Media* (London: Pluto Press, 2005); Nick Couldry and James Curran, eds., *Contesting Media Power: Alternative Media in a Networked World* (Lanham, Md.: Rowman and Littlefield, 2005); Steve John and Stuart Thompson, eds., *New Activism and the Corporate Response* (New York: Palgrave Macmillan, 2003); Martha McCaughey and Michael D. Ayers, eds., *Cyberactivism: Online Activism in Theory and Practice* (New York: Routledge, 2003); Graham Meikle, *Future Active: Media Activism and the Internet* (New York: Routledge, 2002); Andy Opel and Donnalyn Pompper, eds., *Representing Resistance: Media, Civil Disobedience, and the Global Justice Movement* (Westport, Conn.: Praeger, 2003); Clemencia Rodriguez, *Fissures in the Mediascape; an International Study of Citizens' Media* (Cresskill, N.J.: Hampton, 2001); Richard Rogers, *Information Politics on the Web* (Cambridge, Mass.: MIT Press, 2004); Daya Kishan Thussu, ed., *Electronic Empires: Global Media and Local Resistance* (London: Arnold, 1998); Wim Van de Donk, Brian D. Loader, and Paul G. Nixon, and Dieter Rucht, eds., *Cyberprotest; New Media, Citizens and Social Movements* (New York: Routledge, 2004).

52. Nick Couldry, "Mediation and Alternative Media, or Relocating the Centre of Media and Communication Studies," *Media International Australia*, no. 103 (May 2002): 25.

53. In addition to Boyle, *Subject to Change*, Land, *Active Radio*, Ruff, *"We Called Each Other Comrade,"* and Streitmatter, *Voices of Revolution*, see Vincent J. Roscigno and William F. Danaher, *The Voice of Southern Labor; Radio, Music, and Textile Strikes, 1929–1934* (Minneapolis, Minn.: University of Minnesota Press, 2004).

54. In addition to Duncombe, *Underground*, Reed, *Art of Protest*, and Downing et al., *Radical Media*, examples include Todd Gitlin, *The Whole World Is Watching: Mass Media in the Making & Unmaking of the New Left* (Berkeley, Calif.: University of California Press, 1980); Kevin Barnhurst and John Nerone, *The Form of News; A History* (New York: Guilford, 2001); Annmarie Chandler and Norie Newmark, eds., *At a Distance; Precursors to Art and Activism on the Internet* (Cambridge, Mass.: MIT Press, 2005); Barbara Foley, *Radical Representations; Politics and Form in U.S. Proletarian Fiction, 1929–1941* (Durham, N.C.: Duke University Press, 1993); Colette Hyman, *Staging Strikes: Workers' Theatre and the American Labor Movement* (Philadelphia, Pa.: Temple University Press, 1997); Matthew Lasar, *Pacifica Radio: The Rise of an Alternative Network* (Philadelphia, Pa.: Temple University Press, 1999); and Paula Rabinowitz, *They Must Be Represented; The Politics of Documentary* (London: Verso, 1994).

55. John D.H. Downing, "The Independent Media Center Movement and the Anarchist Socialist Tradition," in *Contesting Media Power: Alternative Media in a Networked World*, ed. Nick Couldry and James Curran (Lanham, Md.: Rowman and Littlefield, 2003), 243–57; and Atton, *Alternative Internet*, 101–2.

56. For a study of the early appropriation of advertising methods in a social movement, see Margaret Finnegan, *Selling Suffrage; Consumer Culture & Votes for Women* (New York: Columbia University Press, 1999).

57. Williams, *Television*, 126–34.

58. Raymond Williams, "Means of Communication as Means of Production," in *Problems in Materialism and Culture* (London: Verso, 1980), 50–3.

59. Raymond Williams, "Communications, Technologies and Social Institutions," in *What I Came To Say* (London: Hutchinson Radius, 1989), 173.

60. Williams, "Means of Communication," 52.

61. Williams, "Means of Communication."

62. Williams, "Means of Communication," 52.

63. An example in the United States is Free Press, which describes itself as "a nonpartisan organization working to involve the public in media policymaking and to craft policies for more democratic media." "Free Press: Media Reform Through Outreach, Activism, Lobbying and Networking," *Free Press*, <http://www.freepress.net/> (12 Jan. 2006).

64. Williams, *Television*, 129.

65. Raymond Williams, *The Long Revolution* (New York: Columbia University Press, 1961), 46.

66. Raymond Williams, "Base and Superstructure in Marxist Cultural Theory," *New Left Review* 82 (November/December 1973): 15–6.

67. Among such accounts, see Williams, *Marxism and Literature*, 21–44.

68. John Downing, "Audiences and Readers of Alternative Media; The Absent Lure of the Virtually Unknown," *Media, Culture & Society* 25, no. 5 (September 2003), 625–45.

69. Williams, *Marxism and Literature*, 94.

Part One

Market Formations

Introduction to Part One

The distinction between mainstream communications and alternative communications is an outcome of the emergence of capitalism, which was enabled by, while also enabling in turn, other key developments. An exploration of market formations of communication provides a way of historicizing this distinction and thus understanding better its social and political basis and implications.

The term "market formations" names the dominant way in which communications is practiced in capitalist societies—with Anglo America being the paragon example. However, claiming that they are dominant does not also mean that they serve uniformly and without contradiction.[1] As will be discussed, forms, modes, and practices of communications remain open with an ever-present progressive and democratic (as well as regressive and authoritarian) potential. The examination of market formations thus requires an historical recovery of the interaction between forms, modes, and practices of communications, and the pressures and imperatives of particular moments of capitalist development.

Accordingly, Part One comprises a critical reinterpretation of existing historical accounts of communications. The chapters recover more fundamental bases of communications in Anglo-American capitalist societies and explore ways in which these bases were present within different, even opposing, social formations and uses.[2] These common bases are what I call in the Introduction "cultures of truth," which is a shorthand way of referring to complexes of particular forms, modes, and practices of representation taken to be authoritative.

The intention here is not to search for and establish the origin of a culture of truth, which is an antiquarian if not positivist goal instead of a critical one that recognizes the historical as a continuous outcome of the interaction between present concerns and past practices.[3] Rather, the intention is to recover significant points of challenge and transformation and their enabling relationships, all of which that have yet to receive sufficient attention.

Although Anglo-focused histories of journalism identify the emergence of capitalist forms and organization of communications institutions in early modern English society of the seventeenth century, such accounts are only of limited help in addressing the intentions here. Such studies discuss technological innovation, the invention of news as a commodity, and the emergence of rationalized, commercialized, and professionalized organizations.[4] The more incisive among them seek not simply to narrate a whiggish story of progress, but to examination the kinds of social intentions and social imperatives that inaugurated the drive in the first place toward expanded technologies, the commodification of news, and the rationalization and professionalization of communications organizations.[5] As has been extensively studied, the formation of capitalism provided significant pressures toward such developments, but the relationships between such changes and the larger social and cultural process deserve greater clarity.

Attention to deeper issues and processes is suggested by the titles of recent studies outside communications history proper such as *A Culture of Fact*, *Objectivity in the Making*, and *A Social History of Truth*. Their conclusions go beyond familiar historicist claim that truths change over time, suggesting more fundamentally that the very conception of "Truth" itself is an ontological and epistemological category that is socially and historically produced.[6] To explore the process of the production of truth and its negation as a category of knowledge, recent scholarship in the sociology of knowledge and the history of science has, in the words of one such scholar, "promoted new studies of the 'ways in which language and science are mutually constructive'," as well as of "the intellectual, discursive, and socioeconomic conditions" of such practices.[7]

To address the practice of the construction of truth and its negation as a cultural category of knowledge and what part this practice plays in the formation of communications practices and institutions which in turn remake "truth," Part One focuses on the *longue durée* to better understand the mutually constitutive relations between cultures of truth and their conditions.[8] While analytically separable, cultures of truth and their conditions must always be seen as inseparable and the complex of which they are a part irreducible.

Although varied contributions to the emergence and development of capitalism have been made throughout many countries and eras, the account here will focus on the emergence of capitalist forms of communication within early modern England. Two reasons justify this choice. First, the British civil wars of the mid-1600s were the first of the western revolutions that ushered in constitutionalist forms of national government—a definitional precondition for modern representative democracies. Second, England underwent the dramatic transformations of capitalism earlier than did other western countries. Thus, it makes sense to explore the emergence of market formations of communications and their relationship to democratic practices in England and its most prominent renegade colony, what became the United States.

Chapter 1, "Providentialism and Rationalist Empiricism in Early Modern England," identifies and characterizes Providentialism as a medieval-derived culture of truth built upon faith in divine authority and perpetuated through prophecy within the distributed, colloquial authorship of manuscripting. In contrast, rationalist empiricism as a culture of truth was an Enlightenment means of struggling against what it regarded as the arbitrariness of Providentialism. Based in empiricism and perpetuated through polemic (which valorized an individual's point of view) and reportage (which valorized an organization's systematic collection of data), rationalist empiricism as a culture of truth provided a rationale and working procedure for experimentation and technical innovation that, in turn, assisted the rise of capitalism.

The chapter also recounts how both Providentialism and rationalist empiricism were reformed within the context of early capitalist expansion. Where Providentialism was incorporated into the market through such modes as popular commercial literature, it also was a means of refusing the coming rationalist, capitalist social order. In its amenability to commercialization (individual writers as a bankable commodity) and to bureaucratization and elite control (the codification and professionalization of information gathering as a claim to value), rationalist empiricism fit more clearly within the imperatives of capitalist expansion in ways that will be discussed.

Chapter 2, "The Emergence of Broadcasting and the Rationalization of Participation," addresses the intensification of a capitalist inflection of Providentialism and rationalist empiricism within the expanding and industrializing social order. Capitalist transformation in such ways as accelerating industrial production, growing consumer economies, and technical innovations of transportation and reproduction continued to remake Providentialism and rationalist empiricism through an emerging division between makers and users of cultural products within the capitalist marketplace. The simmering oppositional formation that had relied upon Providentialism for a position of resignation or separatism emerged with vigor in the 1700s as Protestant evangelism, enabling the formation of evangelical publishing as a hybrid incorporation of capitalist organization and Providential intention, and the emergence of a new mode of communications known in the 1840s as broadcasting, a pioneering formulation that has served since as the basis for market communications practice.

Notes

1. An example of a study that confronts the contradictions of television in a commercial system is Douglas Kellner, *Television and the Crisis of Democracy* (Boulder, Colo.: Westview, 1990).

2. Such an intention is mean as a supplement and, in some cases, a critique of a number of reigning frameworks of historical explanation of communications. Such frameworks are usefully characterized in James Curran, "Media and the Making of British Society, c. 1700–2000," *Media History* 8, no. 2 (December 2002): 135–54. A feminist critique of this essay is Maria DiCenzo, "Feminist Media and History: a Response to James Curran," *Media History* 10, no. 1 (April 2004): 43–49.

3. A useful, general statement is Hanno Hardt and Bonnie Brennen, "Introduction: Communication and the Question of History," *Communication Theory* 3, no. 2 (May 1993): 130–6.

4. Representative works include James Sutherland, *The Restoration Newspaper and Its Development* (Cambridge: Cambridge University Press, 1986); and C. John Sommerville, *The News Revolution in England; Cultural Dynamics of Daily Information* (Oxford: Oxford University Press, 1996).

5. Curran, "Media."

6. Barbara J. Shapiro, *A Culture of Fact; England 1550–1720* (Ithaca, N.Y.: Cornell University Press, 2000); Julie Robin Solomon, *Objectivity in the Making; Francis Bacon and the Politics of Inquiry* (Baltimore, Md.: Johns Hopkins University Press, 1998); Steven Shapin, *A Social History of Truth: Civility and Science in Seventeenth-Century England* (Chicago, Ill.: University of Chicago Press, 1994).

7. Solomon, *Objectivity*, 8.

8. The phrase *longue durée* was relevant to the work of Annales School historians such as Fernand Braudel, Marc Bloch, Lucien Febvre, Emmanuel Le Roy Ladurie, and Pierre Goubert, who emphasized the need to address historical analysis in swaths of time and process much broader than the biographical or the event. For a discussion of the Annales School, see Traian Stoianovich, *French Historical Method: The Annales Paradigm* (Ithaca, N.Y.: Cornell University Press, 1976). A discussion of it as one of many different approaches to cultural history can be found in Lynn Hunt, "Introduction: History, Culture, and Text," in *The New Cultural History*, ed. Lynn Hunt (Berkeley, Calif.: University of California Press, 1989), 1–22. While the perspective of the present study differs from the generally structuralist approach taken by Annales historians, it nevertheless also recognizes the value of addressing longer spans of time and deeper changes that take place within it.

Chapter 1

Providentialism and Rationalist Empiricism In Early Modern England

"Abhor inaccuracies, carelessness, bias or distortions." Make every effort to be sure "news content is accurate, [and] free from bias." Always "test the accuracy of information from all sources and exercise care to avoid inadvertent error [and] deliberate distortion." Unfailingly "distinguish between advocacy and news reporting." Although these and other such statements come from journalism codes of ethics from separate organizations, they collectively express the same deep faith in the necessity of basing authoritative, truthful accounts on facts, as well as the ability to discern the truthfulness of accounts by evaluating the degree to which they are based on facts.[1] Such a truism also provides a key rationale and design for alternative-media practice centered around the detection and publication of hidden bias in media products, with bias seen as the tainting of ostensibly authoritative accounts and conclusions drawn from them with wishful thinking, purposive slanting for personal gain, or ideological indoctrination for political advantage.[2] It also provides the rationale for a different kind of alternative-media practice focused on making available the story and information hidden by or omitted from mainstream media accounts.[3]

The authority of facts within mainstream communications as well as alternative communications, underscores the status of facts as a shared building block of media practice. The "authentic" expression of an individual writer, the positing of witnessed events and their transparent presentation as showing "what really happened," the critical cases of media watchdoggery and studies of bias that take issue with particular representations, and efforts to counterbalance selective accounts with the "full story" all rely on the same claim: that, indeed, the authority of representation is directly a result of fidelity to the original, with fidelity assured to the degree that such accounts are based on facts.

Despite its status as a commonsense truism (but also—and more importantly—because of it), such a claim and the considerable authority it wields requires critical attention. This chapter attends to this claim critically and historically by discussing key processes of an immense social reorganization, which began to coalesce in England during a roughly defined early-modern period.[4] The time span of interest can be characterized in a shorthand way for the purposes here as a) the high point of royal rule characterized in particular by the reigns of Henry VIII and Elizabeth I during the 1500s; b) following the increasing challenge to royal authority that waxed during the first half of the 1600s, the period of civil war and republican rule during 1642–1660; c) the reassertion of an increasingly decaying royal rule, which culminated in 1688 as the assertion of Parliamentary prerogative over a now constitutional rather than Divine monarchy. Despite unavoidably simplifying the immense complexity of this time span, this shorthand schematic nevertheless provides a way of locating key points of development and challenge that are relevant to the broader argument here.

Practiced within the context of this social reorganization were two characteristically western cultures of truth, which will be referred to here as Providentialism and rationalist empiricism. Although claims of authority of knowledge undoubtedly date from the beginnings of human societies, such claims are particularly relevant to the practice of modern communications systems, gaining importance as they did in England beginning in the sixteenth century and from within the "capitalist world system in embryonic form," which first emerged in vague outlines during the fifteenth century "with the beginnings of the mass expropriation of the peasantry" and the resulting necessity of engaging in wage work and consumer purchases at scales that soon dwarfed earlier times.[5] The general argument to be made here is that the rise of capitalism and secularization helped bring about a limited pluralization of society and of cultures of truth, creating new possibilities for democratic communications while at the same time new and renewed means of social control.

Providentialism

The emergence of key cultures of truth within capitalism can best be grasped by focusing first on early modern England in the sixteenth century during the reign of the Tudors—particularly Henry VIII and Elizabeth I. The reign of the Tudor monarchs marked a high point of the social concentration of power in Anglo society, consolidated as it was within the head of state who, by the rule of Henry VIII, was also the head of the church.[6] This had not always been the case, of course. Prior to the Protestant Reformation and its manifestation in England, and despite the authority of the king as divinely originating and thus an all-encompassing confluence of the earthly and the religious, authority had been split between the crown as the secular head of state and the pope as the religious head of the church. The installation with Parliament's help of Henry VIII himself as the head of the church as well as of the state made him into "an ecclesiastical emperor on earth second only to Christ."[7]

Despite such a concentration of power, exercising rule in early modern England still required legitimation, "a process in which subjects were intimately involved, one which they learned to manipulate, to criticize, and even to change."[8] One reason why was that, despite the merging of political and religious power, other sources of legitimacy remained outside royal control. For example, although the monarch enacted the common-law legal system of precedent and custom, the authority of this system was tradition and thus to a great degree outside rulers' capriciousness.[9] Even the claim of the divine right to rule "both supported and subverted authority, often simultaneously" in that a variety of groups at one time or another all claimed the divine right.[10] The many rebellions and riots during the Tudor reigns of Henry and of Elizabeth alone suggest a far from quiescent populace.[11] To consolidate power despite such complexity, royal efforts at legitimation most often stressed the value of maintaining social order, and maintaining order meant supporting the existing authority against challenge. Hindle suggests that such a formulation was typical not only of Henry, but of "most European regimes," in that the lack of social order was often among "the most widespread and deeply felt grievances amongst the population as a whole."[12]

Within such a context of centralization and need for legitimation, Providentialism characterized a key culture of truth in that claims of authority and claims that challenged it both relied upon the same Providential basis. As many scholars have noted, the commonplace framework of making sense of worldly events and trends in early modern England was religious and Christian. As Smuts summarizes, Providentialism labels a view in which "God had created the social order and appointed each individual in his place within it."[13] And yet, the social order and worldly events were not experienced simply as unconnected and unexplainable manifestations of divine will. Rather than being a simple blind faith, Providentialism was an interpretive framework—not a replacement—for worldly knowledge that linked action and result, cause and effect. As Hindle notes, people at the time were well aware of secular causes of scarcity, famine, and other forms of hardship, but what concerned them more were "the long-term metaphysical causes, especially the providential judgments of God." Within such a framework, "the immediate preventative measures for dearth were repentance and prayer" instead of changes in the political and economic landscape.[14]

Providentialism operated in contradictory ways, fitting both dominant justifications for the existing state of affairs as well as dissenting challenges to it. One form of Providentialism emphasized the value of existing, dominant hierarchies for maintaining order. This view was institutionalized in such forms as monopolistic trading companies, trade guilds, oligarchic town corporations and closed parish vestries, all of which restricted decisionmaking to a small coterie.[15]

However, Providentialism could also be the basis for erasing historical time and precedent with a synchronic moral struggle that required ongoing vigilance and action, which often became a basis for challenges to "royal authority whenever it failed in its sacred mission."[16] In doing so, Providential claims to authority both bolstered and downplayed human involvement. On one hand, claims based on the will of God or the spiritual struggle of Good against Evil relied upon divine sources of authority outside human abilities to control. On the other hand, claims based on an individual's ability to directly know God and His will or to personally interpret worldly events in terms of their Providential import elevated the role of human abilities. Each of these capabilities was used to justify as well as to challenge the dominant as well as the oppositional.

Within the contradictions of Providentialism, prophecy was a key cultural form. Although only one of a variety of forms "in which public affairs were presented and commented upon in later medieval England," prophecy played a crucial political role as a cultural form of public critique from the early twelfth century through the late seventeenth century, while persisting far beyond that period into our own within a developing English Protestant literary tradition.[17]

Just as appeals to Providential authority both valorized and discounted specific human abilities, the authority of prophecies emerged from the skill and insightfulness of the prophet as well as from the past authorities cited in the prophecy. John Bale, "the most influential English Protestant author of his time" according to King, felt that "a rhetorically sensitive reading could unveil hidden truths," thus placing emphasis on the skill of the person.[18] In addition to the prophet's skill, prophecies' claim to authority could also be based on the prominence of recalled saints, legendary figures, and biblical characters who had "no personal interests in the difficulties of the present," and whose prophetic utterances "carried with them the authority of the past."[19] This dispersed authority, spread as it was between the prophecy and the prophets both present and past, was embodied in a dialogic form of ongoing social composition between the prophecy and its interpretation that explained its allusions. While a typically unchanging prophecy was couched in generalized imagery and elliptical references, its constantly updated interpretation related the prophecy to people and events of the day in detailed and specific ways. Thus, the prophecy and its interpretation together continually recalled the past while extending it into the present and future.

The reign of Henry VIII is a particularly clear period in which to see how prophecy was used both by challengers and by the dominant, marked as the period was by such exceptionalities as the break with the pope and the string of royal remarriages. Prophecy has been used for political challenge in other times and places, such as "the prophecies of the twelfth-century mystic Joachim of Fiore [which] had fuelled not only imperial campaigns against the papacy's power but also popular rebellions aimed at establishing egalitarian commonwealths."[20] The distinctive characteristic of prophecy both then and in the Tudor era was its availability to "lower orders" of society. In the England of Henry VIII, critiques that took the form of prophecy engaged a wide variety of people who opposed royal authority for varying reasons, including such commoners as tenant farmers, apprentice saddlers, innkeepers, provincial lawyers, and servants, as well as "very ordinary women," all of whom were questioned and prosecuted by the crown "for their involvement in politically oriented comment, action, and reaction." Prophecy and its prosecution aligned such outsiders with similarly discontented people who had been "an intimate part of the political and religious establishment," including "parish vicars, local parsons, priests, and chaplains, as well as of religious institutions, represented by priors, abbots, canons, monks, and friars who suddenly found themselves cut away from their traditional place in the social order."[21]

A great potential for popular participation carried with it potential dangers for the crown. Jansen writes how "both the times and the prophecies had become more dangerous" to Henry's authority, for prophecies were repeated to others and thus had a chance to circulate widely, elaborate, and help substantiate challenge.[22] Records of numerous examinations of "nuisance, sedition, and treason" recorded in the state papers of the day show that "political prophecies play a considerable part" in such trials.[23] A popular example that underscores this concern was "The Prophecies of Rhymer, Bede, and Merlin," which "foresees the coming of a new king who will, in his grace and strength and wisdom, return England to the blessings of Rome" while also predicting "revolution from within [the country] as well as intervention from without."[24] While dozens of Tudor-Stuart laws dealt with language regulation, official acts such as the Treason Acts in 1534, a 1542 act that specifically outlawed prophecies, and the "'seditious words and rumors' statutes" of 1580–1581 that prohibited among other acts "'prophesying, witchcraft, conjuration, or other like unlawful means" to predict the queen's death all suggest the degree to which prophecy as a threat was taken seriously.[25]

And yet, in addition to outlawing prophecy through fines, imprisonment and corporeal punishment, Tudor monarchs embraced for a time the prophetic form, responding to critics of the crown kind for kind. While it may at first seem strange for a monarch at the apex of social power to engage in such seemingly compromising activities (why not just imprison or execute everyone?), recent, revisionist scholarship has substantially modified conceptions of the degree of control exercised by the Tudor monarchs, suggesting among other things that "censorship tended to be a haphazard affair, less a matter of systematic repression than intermittent crackdowns in response to such local contingencies as an ambassador's protest, a foreign-policy crisis, a conflict between court factions, or the need to placate a political ally."[26]

Henry VIII responded to prophetic challenges by both circulating royally penned prophecies that bolstered his authority and by refuting the interpretation of existing prophecies that foretold his fall.[27] For example, the crown paid to have counter-prophecies composed and circulated, such as those by Richard Morison who, in 1539, introduced a counter-prophecy in Henry's favor, in which a terrible eagle is defeated ultimately by a noble lion.[28] Henry also paid for the preparation and circulation of contrary interpretations of widely circulating prophecies, such as for a reinterpretation by Thomas Gibson of a popular prophecy that showed how all the key allusions "*really* referred to the King, who was himself the predicted savior of the people and the conqueror who would win the Holy Cross."[29] Despite trying this for a time, the government realized, as Fletcher argues, "that this was too dangerous a tactic to follow, because prophecies could equally well be a justification for opposition." As a result, "determined efforts at suppression" uniformly became the preferred method.[30] Generally available criminal penalties ranged from fines and imprisonment, to branding, whipping, and execution.[31]

Prophecies' distributed, colloquial appropriation and recomposition were enabled by their reproduction within a hybrid oral, manuscript, and print culture. While royal concerns about printing are well-known, manuscripts were no less capable of challenging authority. For example, Justice notes how an "insurgent literacy" aided by handwritten manuscripts existed in England as early as the fourteenth century.[32] And, as Keeble argues, censorship of manuscripts existed as early as 1408, when "all manuscripts intended for copying were required first to be submitted to ecclesiastical censors for scrutiny."[33] Even during the two centuries after the introduction of the printing press in England, monarchs or Parliament sought to control scribal production.[34]

Such a degree of importance was not due only to the challenge posed by the actual manuscripts themselves but also due to the need to control the hybrid, distributed, social means of cultural reproduction which characterized it. One can call this process of reproduction "manuscripting" in order to emphasize an affinity with the process of nonprofessional scribal production as well as its embodiment as a generally available activity rather than simply a concrete artifact.[35] The distinction between activity and artifact is essential. While it would be naïve to extend to all social classes the ability of skillful reading and the means of engaging in or buying scribally produced manuscripts, the activity of manuscripting was indeed commonly available.

Manuscripting underscores the innumerable ways in which knowledge conveyed in scribal and printed materials was fused with that from oral culture, thus rendering efforts to distinguish between the original "source" and those of its reproduction beside the point. Like the readers in early modern France that Chartier explores, in the England by the sixteenth and seventeenth centuries, no steady, clear dividing line demarcated the oral from the scribal or the printed, and no firm relationship can be drawn between a specific utterance (excepting legal, administrative, and commercial information meant to remain proprietary) and its ownership by a specific class.[36] This point becomes increasingly important the more one seeks to assess the relationship between printed works and lower classes who are often judged to be outside the influence of such works. While recognizing that basic literacy in pre-industrial England was slowly extending beyond the aristocratic elite, in the case of manuscripting and in some important ways, the extension hardly matters.[37] As Fox notes, "centuries of manuscript transmission, sermon exempla, minstrelsy, and drama had seen to it that the folklore of the village contained elements which came down from learned and literate culture." At the same time, however, printed texts of "all manner of information and entertainment" were produced as broadsides that were cheap to buy and that were posted publicly, sometimes enshrining material picked up from the oral realm as much as feeding back into it.[38] Printed sources from books to broadsides posted on alehouse walls were quoted and reproduced through speech as well as in handwritten form, with the ability to do so increasingly held by people outside the aristocratic elite.[39] Manuscripts were quoted and reproduced in printed books at the same time portions of books were recirculated as manuscripts, discussed on the street, or even rendered in more idiosyncratic forms as wall-writing.[40] Thus, as Fox argues, "even those people who could not read the handwritten or printed word for themselves traded in forms which were derived from such sources."[41]

Prophecies benefited immensely from the hybridity and distributed composition of manuscripting. On one hand, as Fox notes, "prophecy was a supremely oral vehicle for political comment and what passed for news, something told and retold in alehouses and around firesides, easily retained and accessible to all." Yet, their use as a means of building opposition to royal power was often initiated by "clerics, courtiers, or gentlemen" who sought to "invent or recycle semi-magical predictions about the downfall of the monarch." Such uses were not original as much as they were supplemental to and even dependent on the oral. As Fox summarizes, "clergy and gentry who formed 'scribal communities' for the swapping and transcribing of texts passed this seditious material amongst themselves on paper [. . .] to augment and reinforce songs and sayings which would have passed even more quickly from mouth to mouth."[42] The 1537 case of John Dobson provides an example of how a "long chain of scribal copying [. . .] clearly went hand-in-hand with their oral circulation."[43]

More generally, prophecies and forms of their use and reproduction were already present and resonant within a long-standing culture of truth based in Providentialism, one that, as Love points out, "knew no boundaries of class or distinction," and that were a part of courtly as well as common life.[44] Even a member of the court of Elizabeth such as the Earl of Leicester sought to protect the prophesying movement in the 1570s and 1580s by offering patronage support to radical preachers such as Thomas Cartwright, John Knewstub, and Humphrey Fenn.[45] The importance of prophecy within a broad swath of English society persisted well into the 1600s, and popular to the extent that, as Hill notes, it had become "almost a new profession."[46] As an example of the aristocratic resonance of prophecy during the period of civil war, republic rule, and provisional royal restoration in the mid-1600s (close to 100 years after the time of Henry VIII), Love recounts the case of Sir William Haward, a courtier in 1641–1689 who was knighted in 1643, and whose personal collection of manuscripts included an extract in verse from Nostradamus as well as prophecies citing Merlin, and others such as "the long Fifth-Monarchist prophetic allegory, *The Panther*, copied in April 1681."[47] Prophecies that were popular among the decidedly anti-royal Fifth-Monarchist radicals included those attributed to Merlin, Mother Shipton, Nostradamus, Paracelsus, and astrologers.[48]

Given such patterns of use, one can conclude as Kingsley does that prophecy "remained a common genre in which to couch popular protest, particularly complaints about religious, social and economic inequities."[49] As a means of challenge but also as a means of support, Providentialism emerged in early modern England as a form and as a practice through which positions were established, justified, and challenged.

Capitalism and Secularization

After the death of Elizabeth in 1603, the comparative stability of the Tudor reigns gave way to a century of factional challenge and dissidence, marked most deeply by civil war and the uneven, but continued ascendance of Parliament as an increasingly prominent source of political authority. Underlying the struggles were two immense social, economic, and political processes of accelerating economic expansion and secularization. What emerged through these changes was a limited pluralization of power that had far-reaching implications for the forms and practices of cultures of truth and of democratic communications today. As Kingsley argues, "shifts—both perceived and real—in the political and economic structure of English print culture contributed to a need for the reconsideration and reconstruction of the very idea of authority—both political and textual," thus transforming the use of prophecy in the post-Civil-War era as well as opening possibilities for different cultures of truth, in ways that will be discussed.[50]

One facet of the social reorganization of the early modern era was an accelerating economic expansion (despite many often severe economic depressions and increasing structural imbalances). This was and is no uniform, steady, teleological change, in view of the billions of people today who live in close-to medieval conditions. Yet, despite this unevenness, one can trace beginnings in the expansion of empire dating from the second half of the 1500s and the reign of Elizabeth I with the intimations of an imperial mercantilist system. By the mid-1600s and the establishment of the Protectorate headed by Cromwell, the state had increased its involvement in the imperial economy to become "a proactive authority in commercial matters," with a good portion of this involvement related to matters of colonization and trade beyond the confines of Europe which, as Loades argues, was where "the most spectacular changes took place."[51]

While one could characterize this economic expansion in terms of an increase in volume, what was more novel and important was the corresponding, fitful, and often contentious reorganization of society from mercantilist to capitalist, as the presence of extensive criticism of the emergent capitalist order at the time suggests.[52] Hawkes summarizes the decisive shift in character marked by the emergence of capitalism as a "system of production for exchange rather than for use," which the internationalization of the system of finance and production helped bring about and support.[53] Additional developments Withington identifies include "the creation of a national market, the final establishment of agrarian capitalism, overseas expansion and colonialism, and the emphatic enlargement of public infrastructures."[54]

This expansion and reorganization helped make possible a limited pluralization of social, political, and economic power. This is not to say that power extended to all tiers of society—quite the contrary. During the reigns of Elizabeth and James I prior to civil war, English society had already undergone "'a massive redistribution of income in favor of the landed class'," with such inequalities "characterized not only by social mobility, but also by social polarization and by a greater range of economic differentiation."[55] From this, Hindle concludes that such increases in "inequalities of wealth and status ensured that social relations were politicized at every level in early modern England."[56]

In addition to this increased polarization, monarchical power came to be increasingly diluted, not only through popular critique and action but also through the emergence of rival sources of power. Traditional means of generating wealth such as taxation came to be supplemented then overtaken by increasingly national then international commercial and financial sources, with a concomitant increase in the importance of commercial centers of power based in the rising class of merchants. This development was part of a broader economic and political rationalization of the national economy, which manifested itself in an increasingly national policy that embodied classical political economy and its particularly orthodox brand of market economics.[57] As Stevenson summarizes, the merchants who gained economic power during the reign of Elizabeth were increasingly separated not only from lower classes by wealth, but also from the landed gentry, "who had been brought up to think that manipulation of the economy was demeaning and that trade kept a man from joining the social elite, no matter how rich he might be."[58]

The rise of commercial and financial centers of power had as much political as social and economic implications. Royal monopolies, patronage, licensing, and chartering had been key ways of providing financial support for exploration in exchange for a share of revenues and for political support. Such controls were gradually superceded as commercial centers of finance and investment grew, diversified, and internationalized, aided in part by the changing composition of

Parliament, which was itself increasingly intertwined in the emergent capitalist enterprise.[59] Thus, in the process of building an empire, economic expansion also helped validate a new class of financiers and merchants who were increasingly unbound by class rules of custom and traditional decorum and deference, but upon whom the crown was increasingly dependent.

Yet, monarchs often promoted instead of resisted such developments. As imperatives of empire expanded, capital that was generated through economic expansion from international trade came to be increasingly essential for the continued expansion of the empire. But promoting economic expansion ultimately reduced the crown's political power. As Loades argues, this political change in royal status had proceeded to the point where, by civil war and the rise of Cromwell, "a new type of executive power" had emerged "based not on any form of Divine mandate, nor on widespread popular consent, but upon a coalition of interest groups, particularly the army and the City of London," a process that distributed power outside a single head of state but that limited it as well to elite circles.[60] Royal power would from this point be subject to increasing Parliamentary oversight.

The second simultaneous process within this social, political and economic expansion and reorganization was the acceleration of secularization, which emerged with the fragmentation of institutionalized Christianity brought about by the Reformation, the establishment of the Church of England, and the growing bureaucratic and commercial value of humanism. Indeed, given its importance as both a justification for the emergent order and a means of its perpetuation, it is no overstatement to claim as Achinstein does that the trend of "the withdrawal of God-centered arguments from politics" is justifiably seen as "a central story about modernity."[61] Among its many other implications, a key social and political effect of secularization was the decline in the degree to which institutionalized religious authority coincided with state authority. This supplemented the corresponding limited pluralization of power due to economic expansion that was also underway, due to the emergence of rival, yet elite centers of social power—in the case of secularization, the steady disengagement (at least constitutionally) of the institutions of secular governance from institutions of religious orthodoxy.

Of course, it would be glib to push a claim about secularization too far, as the preceding discussion of the importance of Providentialism alone makes clear. As Harris argues, an historical truism always worthy of challenge is the assumption that the collapse of royalism with the Civil War, the end of the Puritan-inspired Protectorate with the death of Cromwell, and the restoration of a limited royalty with the reign of Charles II marked "an age when politics was becoming increasingly secularized."[62] While habits consistent with secularization can be seen among Elizabethans, who had become in significant ways already and increasingly concerned with "the human and less with the cosmic, and exalted the utilitarian virtues of honest, thrift, and hard work in all walks of life," the authority of Providentialism persisted during the Restoration and beyond.[63] However, despite the unevenness of secularization then and the continued presence today of various combinations of religious-based governance, the general trend of the secularization of state authority in the West is clear, marked in Anglo experience by the waxing of parliamentary power throughout the 1600s. Similar to the earlier claim about the emergence and expansion of capitalism, the point here is not to assert the teleological decline of the importance of organized religion in society. It of course continued (and continues) to play an immense and in some ways increasingly institutionalized political role in such forms today as Christian fundamentalist funding and influence in U.S. national politics as well as the institutionalized theocratic governments of the Middle East. Rather, the important point related to the social reorganization of society at this time is that institutions of religion became increasingly distinct from institutions of an increasingly secular state.

Rationalist Empiricism

The uneven and contested pace of economic expansion and secularization manifested itself in part as a general, pragmatic, growing emphasis on worldly, practical knowledge made valuable by the recognition of the increased problems of governance and administration due to growth and expansion. Within such a context, and along with the persistence of Providentialism as a key culture of truth, rationalist empiricism emerged as a distinct culture of truth, increasingly important with the ascendance in 1603 of James I amidst population expansion, economic crisis, and rising social inequality. Aided by the declining relevance to monarchs and merchants of Aristotelian–Thomist scholasticism and its emphasis on ontological inquiry, the emergence of rationalist empiricism formalized and institutionalized open-ended empirical inquiry, supported an unbounded economic expansion, and persuasively explained the value of doing so within an increasingly commercial, bureaucratized social order.[64]

The phrase "rationalist empiricism" places equal weight on the explanatory framework and on the process by which material for reflection is obtained. Inferred by "rational" is an explanation that consists of consistent cause-and-effect relationships, with such relationships deduced through concentrated mental effort. Inferred by "empiricism" is the necessity of systematically collecting evidence either from the natural world or from the results of experimentation.[65] Solomon specifies each of these facets in her detection within rationalist empiricism of a Neoplatonist strain, which set the will and intention of the knower above the object of the known, and a nominalist empiricist strain, which regarded the world and its objects rather than the knower as the source of knowledge.[66] The phrase together expresses how Jacobean rationalist–empiricist exemplar Francis Bacon and others saw as inseparable the skilled, systematic deduction by naturalist philosophers and the testing such knowledge in the real world by artisans and mechanics. Where Bacon saw rationalists simply as spiders "'spinning webs out of themselves'," and brute empiricists as ants that simply "'gather and consume'," the bee exhibited for Bacon the benefits of combining both in, like empiricists, gathering "'its material from flowers of garden and field'" and, like rationalists, also digesting and transforming this material "'by a faculty of its own'."[67]

Similar to Providentialism, rationalist empiricism was and is highly contradictory. On one hand, it valorizes the ability of humans to know and transform their immediate natural world but, on the other hand, asserted that the most valuable knowledge was located in the world's data unsullied (as it was seen to be) by human intention. Both the rationalist and the empiricist facets together lent authority to rationalist–empiricist claims of truth, whether such claims rested on realms external to and unaffected by human intention or on the insight and skill of the inquirer. Due to these fundamental contradictions, rationalist empiricism—like Providentialism—fit royal justifications for maintaining existing social hierarchies while also providing a justification for their challenge.

As alluded to above, Francis Bacon as Lord Chancellor in the court of James I paradigmatically embodies rationalist empiricism and its values of scientific progress, economic expansion and social order. Forged within the rising emphasis on applying knowledge to useful ends, Bacon reconceived the process and goals of scientific inquiry.[68] Where, in Providentialism, truth was defined as the knowledge of God's will and was deduced as well as intuited through an individual grasping of the relation between the absolute and divine and the specific and worldly, Bacon proposed an inductive, bureaucratized process whereby "a veritable army of assistants" would gather data gained from observations and experiments consistent with Bacon's highly standardized procedures, with the goal of systematizing the gathering and distillation of knowledge useful to the crown and to industry. And, where prophecy was highly distributed and dialogic particularly within the fragmented social hierarchies of the civil-war years of the 1640s, Bacon sought to extend participation in the process of scientific inquiry but at the same time limit who would interpret this data to "a small cadre of natural philosophers sifting through the masses of data collected by less skilled practitioners and using the data to derive inductively the axioms and laws—too subtle for others to see—of the new philosophy."[69] Rank empiricism, Bacon felt, best fit the skills and temperament of lower orders of society, who could tolerate and perhaps even find pleasure in tedious collecting, cataloguing, and classifying. By contrast, Bacon "reserves the constructivist mentality for the king's scientific elite" who would discern the principles and natural laws that would benefit the crown.[70] In this way, Bacon's project "was to be collaborative, but segregated," not only in these social terms, but also in terms of limiting "the dissemination of both his method and its results" to "the intellectual elite."[71]

The goals of scientific inquiry were similarly reconceived, becoming "as much about *using* and *controlling* nature as it was about understanding it [emphasis original]," which, as Ash argues, "were really the same thing" for Bacon and his contemporaries.[72] Despite an emphasis on producing knowledge that would have a practical use in the world, Bacon argued that the value of discoveries thus made was not simply in their practical utility, but in the degree to which they enabled a clearer understanding of nature in a never-ending cycle of experimentation and formulation of natural law. This was in stark contrast to what Bacon saw as the medieval Scholastic process of never-ending discussion and disputation without the production of any new knowledge, and what Bacon regarded as the "unmethodical empiricism of the alchemists" unguided by any developing framework of understanding.[73]

By virtue of his moves to bureaucratize and rationalize the production of knowledge, Bacon's natural philosophy was not simply an system of abstract thought but the underpinnings of a new means of governance and economic expansion, one whose influence in Hill's estimation "was spread wide in England after 1640."[74] The perpetual political problem for monarchs had been how to encourage discovery and increase the production of knowledge to aid economic expansion, but at the same time manage such expansion so it did not enable the formation of a new power base and thus help create a serious political challenge.[75] Bacon's social and political innovation put forth in his natural philosophy was to harness the activities of merchants, mechanics and inventors in service of the crown, producing wealth through technological innovation and economic expansion instead of consuming it through a system of patronage, patents, and licensing. Such a program sought not simply to balance social control with unleashed expansion, but to use economic expansion as a means of achieving social control. The scientific enterprise for Bacon, as Solomon summarizes, was "a standardized process of observation, collection, experimentation, and production that would serve as a tool of royal governance."[76] As Ash summarizes the implication, "Bacon self-consciously envisioned his reformed philosophy as a fundamental part of an imperial monarch."[77] Bacon's work inspired and structured the late-seventeenth-century founding of the British Royal Society and its focus on natural-philosophic investigation, inspired Auguste Comte and his social science in the nineteenth century, as well as the popular conception of how knowledge is generated. Since then and continuing to today, the model of scientific inquiry codified by Bacon exemplifies a deeply entrenched as well as increasingly critiqued process of knowledge production and use.

Within the contradictions of rationalist empiricism, a diversity of cultural forms embodied claims to truth. Forms such as the essay and the polemic highlighted individual skills of perspicacity and rationalist argument, while reportage emphasized instead an organization's capabilities of extensive reach, comprehensive recovery and empirical veracity. Despite these differences, both individualist and organizational forms worked within rationalist empiricism to validate claims about the world, and did so within a context of the increasing commercialization of communications organizations and industries within a similiarly commercializing society. Because the fluidity and hybridity of forms and uses evades analysis based on firm categorization, a more useful approach here is to note specific examples within a range of forms, modes, and uses from scribal and private, to print and public, although even these polarities inevitably fall short of taking full account of this diversity.

One group of cultural forms emphasized the abilities of individuals, thus valorizing the rationalist pole of rationalist empiricism. Forms in this grouping were typically associated with rational argumentation and included the essay, polemic, and satire. These have important commonalities with classical traditions of rhetoric and argumentation as well as with the more contemporary practice of individual reflection and commentary formulated by Montaigne, whose work, as Chan argues, had "a wide influence in seventeenth-century England."[78] Similarly individualist forms as the sermon and homily coincided to some degree with rationalist empiricism despite their religious purposes, particularly when composed through formal, logical reasoning. Although individually composed, these and other such individualist forms were in the larger sense not simply an isolated, individual viewpoint or opinion, but instead contributed to a larger social process by expressing concurrence with or counter-positions, responses, or proposals to already existing positions.

Despite discussions, stories, arguments, and ridicule taking place in the course of everyday life among all levels of society, essays, polemics, and satire—when meeting social expectations embodied in the form— required learned skills of formal argumentation and language-use beyond the reach of most people. At the very least, they required the time and ability (whether learned or self-taught) to pursue them at length. Groups that this pressure increasingly favored included aristocrats, gentry and the wealthy who had the financial wherewithal to engage in such writing as a pastime or sideline to their primary vocation; people engaged in the emerging vocation of writing for pay and who were able to make a living from it; and clergy whose livelihood consisted of generally conceived literary work in the course of their religious duties.

Within the emergent capitalist society, the individuality that essays and polemics emphasized came to be viewed as a possible source of value, in that individual authors and their personal skill—if and when popular—could become a bankable commodity. The commodification of individual authors within a rational–empiricist celebration of individual worldly knowledge was fitful and uneven, however. Within the burgeoning commercial culture of an early modern England still awash in Providentialism, it is little surprise that many best-selling books were religious, such as those of William Perkins, which, as Eales notes, had become by the early 1600s "such a guaranteed [commercial] success that his executors embarked on an ambitious program of posthumous publication which continued for over three decades after Perkins' death in 1602."[79] And the Protestant writings of John Bale, which illustrate "variations in a pattern of apocalyptic conflict," made him "the most successful professional writer of the mid-1600s."[80] In addition to best-sellers that worked within a Providentialist rather than rational-empiricist culture of truth, for a time the distributed authorship characteristic of manuscripting co-existed with individual authoring, with both for a time commercially viable. Existing side-by-side with books as well as pamphlets that were authored and signed by popular writers such as Robert Greene and Thomas Nashe were pamphlets authored anonymously or pseudonymously or that, in displaying a distributed nature of authorship consistent with manuscripting, were "constructed like patchwork entirely out of other men's writings."[81]

Despite this unevenness, Clark notes that Elizabethan pamphleteers too helped usher in an era where "writing for money, whether verse, prose or drama, was their [popular pamphlet writers'] main occupation, and whatever else they did to make a living a secondary one."[82] Later varieties of professionalization included individual writers such as Marchamont Nedham, who wrote in support of both sides in the civil war, foreshadows for Sommerville the "cynical attributes" of professional journalists whose work is not based on individual principles, but solely on the requirements of making money by doing whatever makes a quick buck.[83] The emergence of the commercial periodical later in the century solidified the marketable value of individual writers as a collective commercial enterprise.[84]

By contrast to these individualizing variations that stress rational argumentation and personal observation, a second variation through which rationalist empiricism worked can be collectively referred to as reportage. By contrast to the individualizing form, reportage embodies an empiricist emphasis on the significance of isolated, concrete, and discrete events and developments. In spite of this fragmentation, however, reports not only relied implicitly upon knowledge of already existing issues, trends, disputes, and bodies of knowledge (and, as such, were in some ways as much a part as essays and polemics were of a larger social process), they often included an implied or sometimes explicit scientific or moral lesson to be learned.

Reportage was pioneered by handwritten letters of news, one of many examples of the hybridity of oral, scribal, and print modes. As Schneider notes, the composition and distribution of personally reported "intelligence" from afar was an important facet of a broad epistolary culture that was engaged in by higher classes for personal reasons, as well as by diplomats and other such officials, who reported in a less overtly personal way events and thoughts relevant to their official duties.[85] Forerunners of reportage included the diplomatic report and private handwritten newsletters, with more closely related precursors including town criers' announcements and broadsides in addition to news pamphlets and manuscript newsletters.[86] Through this proto-news matrix both personal and administrative, epistolary culture laid the basis for reportage, not only in its intention of conveying intelligence from distant places, but also in its use of increasingly formulaic claims that, in the same mold as the empiricist strictures of Bacon and others, were displayed as standards of proof in order to establish credibility.

Hybridization

Neither Providentialism nor rationalist empiricism were practiced in any pure form. Cultures of truth during the emergence and initial expansion of capitalism were very unstable, with many overlaps and hybridizations for at least two reasons. Not only did Providentialism and rationalist empiricism both exhibit internal contradictions as well as implicit linkages between them, they were also both embedded within the consolidating, commercial print trade, which encouraged experimentation to find increasingly profitable combinations within the forms and possibilities available. For example, with the continued ascendance of print and decline of popular uses of manuscripts, critical forms that had first appeared in manuscripts—such as parodies and what Love calls the "scribal lampoon"—were simply picked up and transferred to printed sources.[87]

Forms and uses were hybridized because these cultures of truth as communicative practices co-existed in time and space, manifesting themselves in combinations hybrid and unstable if not outright contradictory. For example, despite the formation of rationalist empiricism, popular interest persisted in prodigies, prophecies, astrology, and the apocalyptic millennium. Residual faith in pre-Christian beliefs existed alongside those that were Christian and rationalist. As Hill summarizes, it was a long time before Protestantism and empiricism affected ordinary—and not so ordinary—men and women.[88] Economic depressions were viewed by kings James and Charles not as the results of human actions, but as acts of God.[89] As Shapiro recounts, "great storms, earthquakes, comets, and monstrous births continued to provide moral instruction throughout the seventeenth century" through the framework of Providentialism.[90] The Bible was "the most commonly owned book in the seventeenth century, in England and New England, and could even be found in households where nobody could read," thus exerting an almost talismanic power for many.[91] Thomas Scott, Protestant minister and pamphlet writer of the 1620s, unproblematically blended Providentialism and rationalist humanism, with the ease of their merging thus suggesting "that the values he promulgates hold together [for him] naturally."[92] And, in regards to the seemingly critical stance of rationalist empiricism toward Providential explanations, and much as present-day apologists of "intelligent design" fashion an amalgam of science and religion, the "new science" of the Restoration in Spurr's judgment "sustained rather than undermined Providentialism: God now became the first cause, the architect and supervisor of the cosmos, and divine providence provided the essential link through which the spiritual and material worlds could interact."[93]

Like hybrids of Providentialism and rational empiricism, while news composition at the time relied upon discourses of fact to establish their veracity, reportage was also used to recount wonders, strange events, and marvels of the day, frequently embedding them in a "religious framework and used to illustrate God's Providence or provide timely warnings to the sinful."[94] Indeed, in some ways, news of strange or exceptional natural phenomena blurred divisions between reportage, scientific information, and Providential sign, emphasizing how cultures of truth were an active stance taken in relation to writing and the world, not simply a formal, textual feature. "Contemporary wonders and marvels were news to the periodical publisher, providential events with moral implications to the preacher and theologian, and, to the virtuoso, the subject of reports offered to the Royal Society."[95] Not only were such marvels widely reported, to the degree that such reports conformed to Bacon's guidelines by including details of time, place, and circumstance; by identifying and evaluating witnesses and their testimony; and by rejecting hearsay, such news items were occasionally used as authoritative reports by naturalist philosophers of the day.[96] In addition, polemic and news were melded in such ways as what Raymond calls "anti-mercuries" of the mid-seventeenth century, which "offered news at the same time as they dismissed the value of news and criticized those who appeared to be interested in it."[97]

Even the Baconian program of institutionalized scientific inquiry rested upon Providential justifications. While the goal of his program was to discern natural laws and put that knowledge to use for the benefit of society, Bacon argued that people were not capable of actually creating anything new, as this was an ability only attributable to God. As he put it in *Of the Interpretation of Nature*, Bacon asserts that, while "human beings have a God-given right to pursue natural knowledge to ameliorate the human condition," not only the right but the very natural laws that govern the apparent world derive from Him.[98] To assert otherwise was seen as a form of idolatry.[99] Even claims that a plainness of writing style was more appropriate to scientific and philosophical discourse rested on a Providential basis. As Stark suggests, "arguments against obscure styles are arguments against the assumptions of a magical worldview" and in favor of a rationalist one informed by clear hierarchies, divisions, and distinctions, thus fitting both religious as well as secular justifications for existing authority.[100]

Within such hybridization, entertainment and news reportage could be as much a means of moral and spiritual instruction as were sermons, while sermons could involve not simply spiritual issues, but instead be the "main means by which a broader section of the populace was incorporated into political debate."[101] Popular literature such as Elizabethan pamphlets and cheap murder pamphlets provided moral instruction as well as sermons did by presenting "an austere vision of the power of divine providence over all second order causes in the world and of the inexorability of divine justice."[102] News pamphlets of the Elizabethan era sought to show how every event could be seen to illustrate God's "providential control of human affairs and his careful and constant warning of the inevitable consequences of sinful living."[103]

Commodification and Professionalization

As remade within a burgeoning commercial market, reportage soon became much more than an individual account of events and situations at a distance, whether providentially or secularly interpreted. Similar to essays and polemics and the commodification of individual expression in an emerging professionalized market, reportage gradually became a commodified organizational product that presented a commercially viable, authoritative gathering of "intelligence." By borrowing in many ways from conventions of epistolary culture, commercial news reportage intended for general publics was in many ways letter-writing stripped of its social form. Long-standing conventions of letter-writing "rooted in duties of administration, clientage, and kinship, or in the intimacies of friendship," were drawn upon in the early stages of reportage to underwrite their authority. Epistolary forms were incorporated directly into reportage in such ways as the reprinting of letters from correspondents—whether or not written intentionally for publication—that "engaged polemical debate, religious controversy, scholarly exposition, [and] news reportage" while they also drew upon the authenticating conventions of manuscript and personal correspondence.[104] Printed letters presented as news also employed textual conventions of letter-writing including "salutations, dates, signatures, first-person discourse, and the textualization of transmission processes." However, with the continued commercialization of news as an organizational product first in professional and institutional manuscript newsletters and later in printed news, such features appeared less and less, and later were entirely absent.[105] To become more commercially attractive, reportage was made increasingly impersonal, thus boosting the veracity (and commercial value) of what was recounted through an assertion that the writer's particular individual identity was irrelevant if not damaging to the claim of disinterested, empirical validity of the observations recounted.

To enhance further its value within a commercializing rationalist–empiricist culture of truth, reportage also began to explicitly call attention to the use of increasingly codified procedures (again, recall Bacon's efforts) typical of what Shapiro has called "discourses of fact." As an increasingly organizational practice of establishing authority within the rationalist–empiricist culture of truth, discourses of fact in the English context had first been articulated in the legal sphere, slowly becoming more prevalent in other practices of representation such as history, chorography, travel reporting, and, of course, reportage. Even at the earliest stages of the emerging news genres, conventions that established claims to truth included an emphasis on credible witnesses, impartiality, a clear separation of the fictional from the factual, a distinction between reporting facts and speculating based on those facts, and an emphasis on particulars and details using plain and unadorned prose.[106] When viewed as a group, argues Shapiro, "news genres [. . .] played a role in transforming 'fact' from a category limited to human actions and deeds into one that comprehended both human and natural phenomena"—an epistemological claim that clearly identifies the debt news genres have and the contribution they make to the broader culture of truth of rationalist empiricism.[107]

By the end of the seventeenth century, and whether reportage, or essay or polemic, the economic advantages of print production and distribution made print instead of manuscripting the preferred mode of reproduction within rationalist empiricism, despite early and severe efforts to control it. Although the court occasional produced official news publications to "present orthodox and approved views and to prevent or correct any tendencies to heterodox thinking on significant current events" that had been circulated in anonymous pamphlets, suppression was the preferred means, beginning with the Tudors and often severe if uneven systems of control.[108] Licensing sought to limit legal printing to those loyal to the crown, and unlicensed printing warranted severe penalties and even execution.[109] Efforts under Elizabeth to control printed materials included chartering self-regulating monopolies of printers, with the penalty for noncompliance the loss of lucrative, exclusive rights to print categories of materials and the perpetuation of licensing as investigated by the Star Chamber.[110]

However, with the passing of the Tudor era and the expanding power of the merchant class came regulatory changes. As Harris notes, the first commercial English newsbooks and corantos appeared in the 1620s, patterned upon successful, similar Dutch publications.[111] By 1641 and the Civil War, the licensing system formulated under the Tudors had collapsed, and a dozen weekly newsbooks and eight weeklies were available in London alone soon after in the two decades of effective lack of regulation.[112] During the protectorate, Cromwell reinstated a state monopoly on news, which by 1662 and the Restoration became once again a system of licensing, presumably as an attempt to promote what one historian calls retrospectively "a Puritan cultural revolution" as well as to help prevent a counter-Restoration from forming.[113] By 1688 and the Glorious Revolution, the practice of censorship and control and the political will for it in the face of commercial expansion had diminished to the point where, in 1695, the system of licensing had ceased.[114]

The end of the seventeenth century marks what Harris calls "a watershed in the history of the English press." Within the context of accelerating commercial expansion, government concerns about the increasing availability of radical information, and commercial printers' concerns about the increasing difficulty in making printing pay in an overcrowded market, efforts to reinstate earlier forms of press controls were ineffective, and were replaced beginning in 1712 with efforts to tax all publications and their advertisements rather than overtly suppress specific publications.[115]

The century's end is a watershed as well because, as Sutherland claims, "many of the essential features of the English periodical [commercial] press (waiting, to be sure, for development) were already there in embryo or plainly visible."[116] Not only had essays and polemics become the staple of the commercial periodical press, reportage was practiced within a gradual specialization taking place in the production process, with the emergence of roles of editors, printers, and publishers and, later, the writer remade into the anonymous worker in a generic role inside an organization, which was a marked contrast to distributed means of composition typical of manuscripting.[117] Raymond notes how newsbooks and pamphlets gradually came to be produced not by one person, but collaboratively by editors, printers, and publishers who were often also booksellers, all of whom were related in primarily economic ways.[118] Sutherland also notes an example of increasingly specialized production of the *Impartial Protestant Mercury* of the 1680s, in which the publisher arranged for, received, and distributed copies of the paper while the writers handled the material.[119]

In the context of increasing specialization, there existed many varieties of professionalization. In addition to professional writers who could make a living from their own individual talents and resulting personal reputation, much more common in the seventeenth century were people employed by a publisher for other reasons, or who provided accounts to newspapers as an unpaid sideline, hobby, or out of a sense of duty. For example, Shaaber suggests that the nearest seventeenth-century equivalent to the modern working journalist was "translators who worked for the publishers." Secretaries and recorders were also professional newswriters of a sort, in that they created official accounts that were copied and reprinted elsewhere. Amateurs played a similar role by supplying eyewitness accounts of events either witnessed or personally engaged in. Reprinted personal letters also constituted an unpaid source of many items.[120] And, although paid, the work of "hack" authors of Grub Street enabled a generally meager subsistence—more a way of life than a way of making a living, as Pinkus puts it.[121]

Grub Street writers, however, were one indication of the gradual emergence of a second sense of professionalization, which emphasizes how the named author of personal reputation and skill was replaced by the anonymous worker in a generic role inside an organization, foreshadowing such developments in the nineteenth century and later, which are characterized by Barnhurst and Nerone as a change from a craft-oriented practice of a printer and a few assistants to an industrial- and later a corporate-oriented practice of large organizations.[122] Sutherland notes the increasing employment of what were known by the late-seventeenth and the eighteenth centuries as "news-gatherers or collectors of news," who were anonymous workers retained on a regular or freelance basis. While they obtained some items by being by chance in the right place at the right time (such as at the location of a house fire or a street accident), most items were obtained by overhearing coffee-house conversations and writing them up for pay as short pieces for publication in a newspaper.[123]

In general, the most important effect of the rationalist-empiricist social reorganization of communications within the emergent culture of capitalism was the reformulation of participation, suggested by Bacon's proposal for a rationalized distribution of duties, one that corresponds to the commercially driven polarization of participation into reporters and readers. Whether the product of an individual author or the product of an organization, cultural forms of rationalist empiricism meshed in important ways with the emergent market economy by becoming more professionalized. Although, as Pinkus notes, "writers had attempted to earn a living from their writing as early as the Elizabethan period," it was "in the last part of the seventeenth century, when the professional writer became a familiar part of society."[124] Hammond concurs, noting that "between, roughly, 1690 and 1720, wholly new forms of literary production and thoroughly hybridized literary kinds emerged or were in their gestation period (newspapers, literary periodicals, the novel) that provided new opportunities for workers in the writing industry."[125] By the 1690s, with the growing importance in public life of newspapers, periodicals, and review publications, "we can speak of the beginnings of a mass market for literature [broadly conceived]."[126] One can also speak of the increasing articulation of rationalist empiricism with the emerging capitalist society, along with the accelerating presence of increasingly institutionalized forms of participation as, for a minority of professionals, making and, for the rest of the population, buying, reading, and commenting on.

Toward Modernity

The slow emergence of capitalism and secularization that marked the uneven social reorganization of early modern English society underwrote as well as depended upon Providentialism and rationalist empiricism, both of which were used to authorize, support and justify such developments as well as to help generate the very formation in which they arose. In this context, forms and practices of communications were similarly remade, with long-standing ones reduced in importance or transferred variably to others while new ones found resources in the emerging and expanding social order. The pressures and imperatives of commercialization manifested themselves in an increasingly institutionalized distinction between makers of communications and those who used (or did not use) them. Both Providentialism and rationalist empiricism were inflected in ways that fit such a division.

A crucial feature of this emergent landscape was new ways to think about, address, practice, and manage new kinds of social collectivities. Concomitant with the rise of the Enlightenment individual was the paradoxical emergence of movement politics, whose outline can be seen in the political ferment of English civil war in the 1640s, and of the peculiarly modern political actor prefigured early on by the elite conception of the "mob," but which came to be enshrined within modern constitutional republics as the public.[127] Along with their emergence and recognition were transformed uses of Providentialism and rationalist empiricism in ways amenable to the accelerating capitalist and secular reorganization of society signaled by such developments as political presses, popular commercial presses, advertising and marketing, and evangelical publishing, all of which came to be reconceived within the entirely original communicative mode of broadcasting.

Notes

1. Associated Press, "The Associated Press Statement of News Values," 1 December 2005, <http://www.asne.org/index.cfm?ID=387> (14 July 2006); American Society of Newspaper Editors, "ASNE Statement of Principles," 28 August 2002, <http://www.asne.org/kiosk/archive/principl.htm> (14 July 2006); Society of Professional Journalists, "SPJ Code of Ethics," <http://www.spj.org/ethicscode.asp> (14 July 2006).

2. The study of bias is itself worthy of booklength treatment. Recent representative studies include Benjamin Radford, *Media Mythmakers: How Journalists, Activists, and Advertisers Mislead Us* (Amherst, N.Y.: Prometheus Books, 2003); Eric Alterman, *What Liberal Media?: The Truth About Bias and the News* (New York: Basic Books, 2003); Bernard Goldberg, *Bias: A CBS Insider Exposes How the Media Distort the News* (Washington, D.C.: Regnery Publishers, 2002); Barry Gunther, *Measuring Bias on Television* (Luton: University of Luton Press, 1997); and Jeff Cohen and Norman Solomon, *Through the Media Looking Glass: Decoding Bias and Blather in the News* (Monroe, Maine: Common Courage Press, 1995).

3. A recent example includes documentaries about the publicity campaign prior to launching military action in Iraq in 2003. A partial list of such documentaries includes *Uncovered; The War on Iraq* (produced and directed by Robert Greenwald, Cinema Libre Studio, 2004); *WMD: Weapons of Mass Deception* (directed by Danny Schecter, Cinema Libre Studio, 2005); and *Iraq for Sale: The War Profiteers* (produced and directed by Robert Greenwald, Brave New Films, 2006).

4. Mielants argues for the Western European specificity of initial capitalist consolidation and reorganization. See Eric Mielants, *The Origins of Capitalism and the "Rise of the West"* (Philadelphia, Penn.: Temple University Press, 2007).

5. David Hawkes, *Idols of the Marketplace; Idolatry and Commodity Fetishism in English Literature, 1580–1680* (New York: Palgrave, 2001), 8. See also Neil McKendrick, *The Birth of Consumer Society; The Commercialization of Eighteenth-Century England* (Bloomington, Ind.: Indiana University Press, 1982).

6. For a useful summary, see John N. King, *English Reformation Literature; The Tudor Origins of the Protestant Tradition* (Princeton, N.J.: Princeton University Press, 1982), 19–32.

7. J.P.D. Cooper, *Propaganda and the Tudor State; Political Culture in the West Country* (Oxford: Clarendon Press, 2003), 214.

8. Steve Hindle, *The State and Social Change in Early Modern England, c. 1550–1640* (New York: St. Martin's, 2000), 237.

9. R. Malcolm Smuts, *Culture and Power in England 1585–1685* (New York: St. Martin's Press, 1999), 17.

10. Smuts, *Culture and Power*, 32.

11. Hindle, *State and Social Change*, 48–54; Anthony Fletcher and Diarmaid MacCulloch, *Tudor Rebellions*, 4th ed. (London: Longman, 1997).

12. Hindle, *State and Social Change*, 237.

13. Smuts, *Culture and Power*, 28.

14. Hindle, *State and Social Change*, 55.

15. Smuts, *Culture and Power*, 30.

16. Smuts, *Culture and Power*, 31.

17. King, *English Reformation Literature*; and Lesley A. Coote, *Prophecy and Public Affairs in Later Medieval England* (Rochester, N.Y.: Boydell Press, 2000), 41. For a discussion of the relationship between prophecy and conversion narratives, see Nigel Smith, *Perfection Proclaimed; Language and Literature in English Radical Religion 1640–1660* (Oxford: Clarendon Press, 1989), 23–72.

18. King, *English Reformation Literature*, 56; David Norbrook, *Poetry and Politics in the English Renaissance*, rev. ed. (Oxford: Oxford University Press, 2002), 33.

19. Sharon L. Jansen, *Political Protest and Prophecy Under Henry VIII* (Rochester, N.Y.: Boydell Press, 1991), 15. The list of sources is from Coote, *Prophecy*, 38–39.

20. Norbrook, *Poetry and Politics*, 35.

21. Norbrook, *Poetry and Politics*, 28; and Jansen, *Political Protest*, 19, 149. See also Ethan A. Shagan, "Rumors and Popular Politics in the Reign of Henry VIII," in *The Politics of the Excluded, c. 1500–1850*, ed. Tim Harris (New York: Palgrave, 2001), 41–3

22. Jansen, *Political Protest*, 26.

23. Jansen, *Political Protest*, 24.

24. Jansen, *Political Protest*, 63–67.

25. Jansen, *Political Protest*, 36, 60–61; and Debora Shuger, *Censorship and Cultural Sensibility: the Regulation of Language in Tudor-Stuart England* (Philadelphia, Pa.: University of Pennsylvania Press, 2006), 74.

26. Shuger, *Censorship*, 2.

27. Coote, *Prophecy*, 235–36.

28. Jansen, *Political Protest*, 59.

29. Jansen, *Political Protest*, 112.

30. Fletcher and MacCulloch, *Tudor Rebellions*, 12.

31. Hindle, *State and Social Change*.

32. Steven Justice, *Writing and Rebellion; England in 1381* (Berkeley, Calif.: University of California Press, 1994).

33. N.H. Keeble, *The Literary Culture of Nonconformity in Later Seventeenth-Century England* (Athens, Ga.: University of Georgia Press, 1987), 93.

34. Fredrick Seaton Siebert, *Freedom of the Press in England 1476–1776; The Rise and Decline of Government Controls* (Urbana, Ill.: University of Illinois Press, 1952).

35. For more about aristocratic uses of manuscripts, see Peter Beal, *In Praise of Scribes; Manuscripts and Their Makers in Seventeenth-Century England* (Oxford: Clarendon Press, 1998); and Harold Love, *Scribal Publication in Seventeenth-Century England* (Oxford: Clarendon Press, 1993).

36. Roger Chartier, *The Cultural Uses of Print in Early Modern France*, trans. Lydia G. Cochrane (Princeton: Princeton University Press, 1987), 6–7. Love points out the suitability of manuscripts for material meant to remain proprietary. See Love, *Scribal Publication*, 177–78.

37. Laura Caroline Stevenson, *Praise and Paradox; Merchants and Craftsmen in Elizabethan Popular Literature*, (Cambridge: Cambridge University Press, 1984), 57. The primary work she draws upon is David Cressy, *Literacy and the Social Order: Reading and Writing in Tudor and Stuart England* (Cambridge: Cambridge University Press, 1980).

38. Adam Fox, *Oral and Literate Culture in England 1500–1700* (Oxford: Oxford University Press, 2000), 9.

39. Stevenson, *Praise and Paradox*, 55.

40. Juliet Fleming, *Graffiti and the Writing Arts of Early Modern England* (Philadelphia, Pa.: University of Pennsylvania Press, 2001); Elizabeth L. Eisenstein, *The Printing Revolution in Early Modern Europe*, reprint ed. (Cambridge: Cambridge University Press, 1984), 20.

41. Fox, *Oral and Literate Culture*, 5–6.

42. Fox, *Oral and Literate Culture*, 366.

43. Fox, *Oral and Literate Culture*, 364.

44. Love, *Scribal Publication*, 211–16.

45. Jacqueline Eales, "A Road to Revolution: The Continuity of Puritanism, 1559–1642," in *The Culture of English Puritanism, 1560–1700*, ed. Christopher Durston and Jacqueline Eales (London: Macmillan, 1996), 189.

46. Christopher Hill, *The World Turned Upside Down; Radical Ideas During the English Revolution* (New York: Viking, 1972), 72–3.

47. Love, *Scribal Publication*, 216.

48. Hill, *World*, 72.

49. Margery A. Kingsley, *Transforming the Word; Prophecy, Poetry, and Politics in England, 1650–1742* (Newark, N.J.: University of Delaware Press, 2001), 16.

50. Kingsley, *Transforming the Word*, 18.

51. David Loades, *England's Maritime Empire: Seapower, Commerce, and Policy, 1490–1690* (New York: Longman, 2000), 215–9.

52. Hawkes, *Idols*.

53. Hawkes, *Idols*, 15.

54. Phil Withington, *The Politics of Commonwealth; Citizens and Freemen in Early Modern England* (Cambridge: Cambridge University Press, 2005), 5.

55. Quoted in Hindle, *State and Social Change*, 46. See also Keith Wrightson, *English Society 1580–1680* (London: Routledge, 2003), 231–32; and Patricia Fumerton, *Unsettled; The Culture of Mobility and the Working Poor in Early Modern England* (Chicago, Ill.: University of Chicago Press, 2006).

56. Hindle, *State and Social Change*, 233.

57. Douglas Hay and Nicholas Rogers, *Eighteenth Century English Society; Shuttles and Swords* (Oxford: Oxford University Press, 1997), 98–102.

58. Stevenson, *Praise and Paradox*, 38.

59. Julie Robin Solomon, *Objectivity in the Making; Francis Bacon and the Politics of Inquiry* (Baltimore, Md.: Johns Hopkins University Press, 1998), 65–9.

60. Loades, *England's Maritime Empire*, 248.

61. Sharon Achinstein, *Literature and Dissent in Milton's England* (Cambridge: Cambridge University Press, 2003), 5.

62. Tim Harris, "Introduction: Revising the Restoration," in *The Politics of Religion in Restoration England*, ed. Tim Harris, Paul Seaward, and Mark Goldie (Oxford: Basil Blackwell, 1990), 2. See also Smuts, *Culture and Power*, 28–9.

63. Sandra Clark, *The Elizabethan Pamphleteers; Popular Moralistic Pamphlets 1580–1640* (Rutherford, N.J.: Fairleigh Dickinson University Press, 1983), 35.

64. For a useful summary discussion of Aristotelian-Thomist scholasticism, see Solomon, *Objectivity*, 27–33.

65. The rationale for this label derives from Solomon's passing adjectival use of the phrase "rational-empirical." See Solomon, *Objectivity*, 56.

66. Solomon, *Objectivity*, 46–7.

67. Francis Bacon, "Thoughts and Conclusions," in *The Philosophy of Francis Bacon*, trans. Benjamin Farrington (Chicago, Ill.: University of Chicago Press, 1966). Quoted in Solomon, *Objectivity*, 55.

68. For summaries of humanism during Bacon's lifetime, see Solomon, *Objectivity*, 163–7; and Eric H. Ash, *Power, Knowledge, and Expertise in Elizabethan England* (Baltimore, Md.: Johns Hopkins University Press, 2004), 193–4.

69. Ash, *Power*, 186–92.

70. Solomon, *Objectivity*, 58.

71. Ash, *Power*, 205–6.

72. Ash, *Power*, 189.

73. Ash, *Power*, 196.

74. Hill, *World*, 232.

75. Solomon, *Objectivity*, 59.

76. Solomon, *Objectivity*, xv–xvi.

77. Ash, *Power*, 212.

78. Mary Chan, "From Essayist to Author: Roger North and Evolving Narrative Forms," *Seventeenth Century* 15, no. 2 (Autumn 2000): 268.

79. Eales, "Road to Revolution," 196.

80. King, *English Reformation Literature*, 111.

81. Clark, *Elizabethan Pamphleteers*, 33.

82. Clark, *Elizabethan Pamphleteers*, 17–27.

83. C. John Sommerville, *The News Revolution in England; Cultural Dynamics of Daily Information* (New York: Oxford University Press, 1996), 39.

84. Helen Berry, *Gender, Society and Print Culture in Late-Stuart England; the Cultural World of the* Athenian Mercury (Burlington, Vt.: Ashgate, 2003), 17–8.

85. Gary Schneider, *The Culture of Epistolarity; Vernacular Letters and Letter Writing in Early Modern England, 1500–1700* (Newark, Del.: University of Delaware Press, 2005); Barbara J. Shapiro, *A Culture of Fact; England 1550–1720* (Ithaca, N.Y.: Cornell University Press, 2000), 77–8, 87.

86. Shapiro, *Culture of Fact*, 86.

87. Love, *Scribal Publication*, 293–95.

88. Hill, *World*, 71.

89. Loades, *England's Maritime Empire*, 215.

90. Shapiro, *Culture of Fact*, 89.

91. David Cressy, "Books as Totems in Seventeenth-Century England and New England," *Journal of Library History, Philosophy and Comparative Librarianship* 21, no. 1 (Winter 1986): 93.

92. David Colclough, *Freedom of Speech in Early Stuart England* (Cambridge: Cambridge University Press, 2005), 103.

93. John Spurr, "'Virtue, Religion and Government': The Anglican Uses of Providence," in *The Politics of Religion in Restoration England*, ed. Tim Harris, Paul Seaward, and Mark Goldie (Oxford: Basil Blackwell, 1990), 31.

94. Shapiro, *Culture of Fact*, 88.

95. Shapiro, *Culture of Fact*, 99.

96. Shapiro, *Culture of Fact*, 99–103.

97. Joad Raymond, *The Invention of the Newspaper; English Newsbooks 1641–1649* (Oxford: Clarendon Press, 1996), 226.

98. Solomon, *Objectivity*, 62.

99. Hawkes, *Idols*.

100. Ryan J. Stark, "From Mysticism to Skepticism: Stylistic Reform in Seventeenth-century British Philosophy and Rhetoric," *Philosophy and Rhetoric* 34, no. 4 (2001): 323.

101. Raymond, *Invention of the Newspaper*, 192.

102. Clark, *Elizabethan Pamphleteers*, 35; Peter Lake, "Deeds against Nature: Cheap Print, Protestantism and Murder in Early Seventeenth-Century England," in *Culture and Politics in Early Stuart England*, ed. Kevin Sharpe and Peter Lake (Stanford, Calif.: Stanford University Press, 1993), 272.

103. Clark, *Elizabethan Pamphleteers*, 89.

104. Schneider, *Culture of Epistolarity*, 183–4.

105. Schneider, *Culture of Epistolarity*, 143–52.

106. Shapiro, *Culture of Fact*, 86, 94. See also Clark, *Elizabethan Pamphleteers*, 99; and Raymond, *Invention of the Newspaper*, 130–3. On the dynamic relation between fact and fiction, see Lennard Davis, *Factual Fictions: the Origins of the English Novel* (New York: Columbia University Press, 1983).

107. Shapiro, *Culture of Fact*, 4.

108. Clark, *Elizabethan Pamphleteers*, 86. For a revisionist perspective on Tudor-era censorship, see Shuger, *Censorship.*

109. Siebert, *Freedom*; Martin Conboy, *Journalism: A Critical History* (London: Sage, 2004), 10–11.

110. Seibert, *Freedom*; Conboy, *Journalism*, 13.

111. Michael Harris, "The Structure, Ownership and Control of the Press, 1620–1780," in *Newspaper History from the Seventeenth Century to the Present Day*, ed. George Boyce, James Curran, and Pauline Wingate (London: Constable, 1978), 82–97. See also Raymond, *Invention of the Newspaper*.

112. Shuger, *Censorship*, 12.

113. Christopher Durston, "Puritan Rule and the Failure of Cultural Revolution, 1645–1660," in *The Culture of English Puritanism, 1560–1700*, ed. Christopher Durston and Jacqueline Eales (London: Macmillan, 1996), 210–33.

114. Harris, "Structure," 83.

115. Harris, "Structure," 84.

116. James Sutherland, *The Restoration Newspaper and Its Development* (Cambridge: Cambridge University Press, 1986), 232.

117. Raymond, *Invention of the Newspaper*, 21; Sutherland, *Restoration Newspaper*, 220–1.

118. Raymond, *Invention of the Newspaper*, 21. See also Brean S. Hammond, *Professional Imaginative Writing in England, 1670–1740; "Hackney for Bread"* (Oxford: Clarendon Press, 1997), 27. For an overview of the organization of the London pamphlet market, see Dagmar Freist, *Governed by Opinion; Politics, Religion and the Dynamics of*

Communication in Stuart London 1637–1645 (London: Tauris Academic Studies, 1997), 87–124.

119. Sutherland, *Restoration Newspaper*, 210.

120. Matthias A. Shaaber, *Some Forerunners of the Newspaper in England, 1476–1622* (Philadelphia, Pa.: University of Pennsylvania Press, 1929), 225–50.

121. Philip Pinkus, *Grub St. Stripped Bare* (London: Constable, 1968), 13–15.

122. William Solomon, "The Site of Newsroom Labor; The Division of Editorial Practices," in *Newsworkers; Toward a History of the Rank and File*, ed. Hanno Hardt and Bonnie Brennen (Minneapolis, Minn.: University of Minnesota Press, 1995), 110–34; Catherine McKercher, *Newsworkers Unite; Labor, Convergence, and North American Newspapers* (Lanham, Md.: Rowman & Littlefield, 2002); Kevin Barnhurst and John Nerone, *The Form of News; A History* (New York: Guilford, 2001).

123. Sutherland, *Restoration Newspaper*, 220–21.

124. Pinkus, *Grub St.*, 13.

125. Hammond, *Professional*, 9.

126. Hammond, *Professional*, 32.

127. Tim Harris, "The Leveller Legacy: From the Restoration to the Exclusion Crisis," in *The Putney Debates of 1647; The Army, the Levellers and the English State*, ed. Michael Mendle (Cambridge: Cambridge University Press, 2001), 232.

Chapter 2

The Emergence of Broadcasting and the Rationalization of Participation

The problem with alternative media today, argued activist and alternative-media contributor Michael Albert in a post-9/11 commentary, is not a lack of power, but lack of reach. "Our current radical media are powerful and effective for small communities of already committed folks and for local consciousness raising," he writes. "But they have little or no national mass appeal." Because "the Left has no national presence able to affect consciousness, policy, and program on a national scale," he argues that establishing a national radical mass-media system should be a paramount goal. His "overarching vision of radical media" includes a 150- to 300-member radio network that would carry "a good portion of simultaneously aired shows, including national radical call-in talk shows"; a twenty to fifty member regional newspaper network that would contain in addition to local material "the same national reporting, visionary pieces, [and] cultural pieces"; and "an array of national magazines and journals" presumably with the same mix of material specific to its locale along with nationally syndicated material.[1] Although "the odds of this happening seem slim," Albert concludes that "this level of radical mass media [. . .] could fulfill at least one essential prerequisite for building national political organization, for having national electoral impact, and for arousing national movement activism in crises."[2]

What is telling about this commentary—written as it is by a prolific, committed writer currently in the thick of alternative-media practice in the United States—is the form in which communications in the service of oppositional politics is imagined. The crucial facets of this vision are a significant uniformity (ensured by the national scale) and simultaneity of apprehension (ensured by the simultaneity of delivery and availability), with both made possible through the work of a centralized organization. As contemporary-sounding as such a vision is, such claims echo not only the holy grail of commercialized media systems and the advertisers that support them, but also the hopes and practices of evangelical activists of eighteenth-century England and nineteenth-century America, thus suggesting not only the persistence of the vision but also its resonance for purposes as distinct as mass marketing, evangelism, and secular political activism (which, in many ways, have emerged as not quite so distinct). Such a commonality between quite different intentions recalls the ways that Providentialism and rationalist empiricism as cultures of truth were central to otherwise deeply opposing formations in early modern England, despite the deep rifts between Catholics and Protestants, royalists and parliamentarians, and the popular and the aristocratic (of course, with infinite variations and relationships both within and between them).

Whether a topic of discussion among corporate boardroom executives or progressive political activists today, or evangelical antislavery social activists as early as 1840, this form of activist communications is referred to as "broadcasting." The use of this term as early as the 1840s in conjunction with discussions of communications—more than 60 years before the ostensible technical beginnings of such a system with the emergence of wireless voice transmission, and close to 80 years prior to the beginnings of its institutionalization—suggests the need to recognize broadcasting as much more than a technical description of signal transmission.[3] Recent work by Peters and Scannell suggests as much by addressing broadcasting as a particular form of social relationships as well as—and perhaps more fundamentally than—technologies and institutions.[4]

This chapter addresses this seeming incongruity by addressing the rise of broadcasting as a new mode of communications enabled through accelerating processes of rationalization, one that further entrenched the practice of Providentialism and rationalist empiricism within capitalist societies. As such, however, the focus in this chapter shifts somewhat from the prior one. While the continued remaking of forms—prophetic, polemic, empirical—within the emergent industrial capitalism, mass marketing, and mass politics is an inquiry well worth conducting, here I want to focus on a different but equally crucial transformation in the practice of communication in market societies—this time, by attending not to forms so much as to a novel mode of communications.

As has already been discussed, the rationalization of communications had proceeded with the commercialization of publishing and the remaking of Providentialism and rationalist empiricism to fit its imperatives. This chapter will discuss further the rationalization of communications by reference to political communications in mid-1600s England, and to the practice of evangelical Protestantism, which had emerged in a powerful way by the end of the 1700s as Wesleyan Methodism. Enabled by its rise was the formation of national bible-tract societies in the England and, in the United States, not-for-profit corporations, both of which were dedicated to evangelical publishing and the purpose of fomenting the wholesale religious conversion of society. Such organizations intertwined political, commercial, and evangelical intentions and practices, and, by doing so within the context of an expanding industrial capitalism, brought about the communicative mode of such intentions that came to be known by the mid-1800s and beyond as broadcasting.

Contradictions of Capitalism

The uneven but steady coalescence of capitalism in early modern England brought about, on one hand, clearly liberatory possibilities, with one being the limited pluralization of authority. Along with other pressures, the emergent capitalist system had made possible the formation of concentrations of wealth and influence outside the court, enough so in England, in the span of 100 years between the reign of Elizabeth and the Glorious Revolution, for parliament by the end of the seventeenth century to supercede monarchical authority and its claim of legitimacy based in divine right. By 1714 and the ascendance of George I to the throne, England had emerged as a constitutional monarchy, with Parliament as the state's supreme political authority. Instead of an absolute monarch who ruled over subjects (the norm at the start of the seventeenth century), by the end of the same century state authority had become based in secular, political factions (later parties) leading their respective adherents with a still powerful but now accountable monarch.

While the overall trend is clear, such labels mask a more complex situation. As accounts of court intrigue make clear, monarchical authority was far from singular and uniform, and Parliamentary authority even less so. Thus, a more accurate characterization of this development is not the clean replacement of monarchy by parliamentarism, but rather a process by which authority itself became increasingly factionalized—and publicly so—and subject to debate within an enabling, shared constitutional framework that, through limitations on suffrage, legitimized the perpetuation of elite rule, defined increasingly in terms of property as much as by lineage and tradition. Through its pluralization and fragmentation, authority had also become historicized (again, in a limited way) by being located no longer solely in a specific person, heritage or lineage but in a specific position as legitimized by parliamentary process. While stances within this framework could and did conflict, the constitutional framework itself comprised the implicit, accepted horizon of disagreement and debate.

A second liberatory development was the emergence of the public as a political actor, which validated popular participation in the social and political process. With its emergence, challenge to reigning authority became increasingly a legitimate although highly contested part of the political process. Despite the fact that the vast majority of people did not possess significant voting rights, building popular support for elite positions in an increasingly (but still limited) pluralized polity—and engaging the public in processes of legitimation—had become by virtue of the civil wars in mid-seventeenth century a political necessity.[5]

However, the liberatory promise of these developments was not uniform nor without limitation, as such democratizing developments engendered their own controlling responses. In few other realms was this as clear as in the rise of the public as a political actor. The limiting contradiction of the emergent bourgeois public sphere of the eighteenth century has been noted by scholars such as Habermas, who argued that, while such a public sphere was in a very real sense an embodiment of democratic possibility, it also had (as Warner summarizes Habermas' argument) "always been structured by a set of ideals that were contradicted by its own organization and compromised by its own ideology."[6]

Such a limiting contradiction has been explored and qualified through historical work that characterizes the emergence of the public in social protest in early modern England. The pioneering historical studies by Rudé, Hobsbawm, and Thompson rehabilitated the legitimacy of popular political action as well as recovered the substantive, selective role of Protestantism in the later formation of working-class radicalism.[7] More recently, the work of Harris on what he calls "crowd politics" documents both the emergence of crowds as political actors as well as the ways in which "the masses themselves were divided" and, further, formed not through some organic process but at least in part a product of "a close interaction of elite encouragement."[8] The Habermasian critique, supported by its recognition of the dissociation between, on the one hand, the ideals of open and rational debate and, on the other hand, the social and historical conditions that worked against such an ideal, highlights ways in which the struggle of expanding popular participation was conducted against not only legal limitation but also instrumental rationalization. The public had indeed emerged, but was regarded by elites as an external mob to defang, placate, or selectively mobilize in the service of elite purposes.

As discussed in the prior chapter, a key way this rationalization occurred was through the capitalist inflection of cultures of truth—in particular, the progressive formation of increasingly defined roles of readers and writers, brought about through commercialization within an increasing social and economic polarization. The demands of commercial print culture and the increasingly specialized roles and rationalized process demanded by it separated and solidified the experience of buyers and readers as distinct from that of makers and writers. In doing so, it also solidified boundaries between the two, which consisted most typically of the bricks of education, available time, class, access to capital, and professional skill. It is crucial to point out, however, that producing and solidifying boundaries between readers and writers does not mean that each was unrelated to the other or that there was any kind of linear, one-way relationship between them. Popular cultures were continually drawn upon, and the popular pulse continually taken in terms of rates of sales and commercial success. Writers then as now were also readers, and some people whose lives began as readers were able to become writers, too. Thus, instead of separate spheres, these are better seen as fluid positions within the larger formation of commercial culture.

At the same time, however, cultures of truth underwent selective, hybrid transformations within the context of emerging capitalism. Providentialism within manuscript culture had provided a significantly dialogic process, one that had also been significantly limited by the monopoly power of the court. However, as reformed within the burgeoning commercial print culture of early modern capitalism, by the end of the civil wars Providentialism had become accommodated to state power in the forms of, on the one hand, accommodation and acceptance and, on the other hand, a secularized and tolerated tradition of dissent.[9] Providentialism provided such institutions as the Church of England a form through which, as Spurr puts it, to "sustain virtue, religion, and government" while also providing a basis for accommodating dissent as constant reflection on "the 'controversy' between a wrathful God and an incorrigible nation."[10] Forms of rationalist empiricism were more easily accommodated to capitalism. Polemic had become accommodated to commercialization as individual opinion, and reportage as bankable information.

Rationalization of Political Communications

Related to the rationalization of popular culture were parallel processes taking place in what would be called today political communications. Working in conjunction with other developments already noted, the increasing bureaucratization of the state and the collapse of systems of licensing and censorship helped encourage "the emergence of increasingly bureaucratized, centralized and professionalized methods of producing [political] propaganda."[11] The scale and organization was in marked contrast to earlier "renaissance [era] operation[s]" in which communications to royal subjects was "the product of individual patrons and authors." Examples of such an operation can be seen prior to the English civil wars, and kings James's and Charles's use of official statements, declarations, and manifestos, including royal speeches reproduced and distributed in print.[12]

However, as Peacey summarizes, the need for an institutionalized system of organized political communications "became acute in the conditions of the 1640s, as rival political groups claimed authority, and sought to secure public support."[13] By the mid-seventeenth century, the system of organized political communications had become one in which "the resources of the executive and its bureaucracy were brought to bear on the world of publishing," a manifestation of "the transition from the renaissance monarchy to the modern state, and from aristocratic patronage to state's servants."[14]

This reformation of political communications made use of rationalist empiricism and its corresponding individualist form of polemic, but within an increasingly large and bureaucratized process. Individualist polemic—ostensible assertions of individual writers making reasoned arguments supported by evidence—embodied the rationalist–empiricist faith in humans to discover truth. Yet, such statements were the result of "politicians and writers working together in order to produce political and religious polemic."[15] Such working relationships soon became institutionalized to the point where specific, professional writers were paid by Parliament through salary or pension.[16]

They also put into practice increasingly sophisticated means of feigning public consensus. For example, tracts written by commoners were often in fact commissioned by members of Parliament for the purpose of presenting an impression of spontaneous support.[17] Thus, "while it was necessary to produce works which were transparently official, in order to establish orthodoxy, it was also important to inundate the market with books which could not easily be traced back to their official origins, in order to create the impression of spontaneous support for particular parties and positions."[18]

In sum, the need for political factions to secure public support enlarged the "growth of public debate [. . .] and the growth of a realm of contestation," but these moves toward a democratization of participation were dampened not by censorship but, paradoxically, by the very same selective enlargement of participation in communications. Peacey argues that the need to secure public support "may have provided a more subtle means of undermining opponents, and one which did not necessarily have the appearance of stifling debate." In this way, "propaganda, and the engagement with the public through a broad range of literary forms," which was in one sense a broadening of relationships structurally speaking, also "represents an attempt to restrict the boundaries of discussion and to control the terms of debate."[19] Although "civil-war politicians lacked a monopoly over the public domain of print," Peacey argues that "they arguably sought to use propaganda in order to ensure their dominant position within it."[20]

The narrowing of modes of participation noted earlier can be seen clearly within this developing system. Whether mobilizing support for factions and parties, securing allegiance and maintaining support, explaining and justifying policies and actions, encouraging compliance and participation, inoculating against propaganda of the opposition, and providing "'spectacles' for those who were viewing the events and personalities of the civil wars insufficiently well," all such practices rehearse the relationship of an elite working on supporters.[21] Whether used to explain policies, generate and maintain support, create an orthodox position, serve the interests of friends and patrons, or display their personal support for religious and political causes, these diverse efforts had in common the elite management of public involvement, except as the "target" of such efforts.[22] For the public here was largely seen not as co-equals with whom to collaborate, but as an irrational force to defang and placate. As Peacey concludes, "many, though not all, parliamentarians appear to have been willing to see the emergence of a broad and inclusive political nation, but [they] sought to use print in order to constrain the terms of debate within it."[23]

Evangelical Publishing

The narrowing of modes of participation through rationalization and professionalization fit not only commercial needs for profitability and political needs for appropriate popular mobilization. It also fit with the intentions of evangelical Protestants and their efforts to sustain their denominational existence as well as more generally to fulfill their spiritual duties and responsibilities to themselves and to God. Understanding the intertwinement of evangelical intentions with the market provides yet a third way—along with commercialization and the professionalization of political communications—in which the pressures and imperatives of an accelerating capitalism were refashioning cultures of truth.

The evangelical conception of communications was intimately tied to a belief in the primacy of the Word. Brown recounts the core of this belief by noting that, for evangelicals,

> the Word existed before the world's creation and spoke the world into being. In the incarnation, the Word became flesh, Jesus Christ, in order to redeem creation from its fall to sin. Shortly before Jesus' crucifixion, he promised to send his disciples the Holy Spirit—from the same root as "inspired" or "God-breathed" word—to guide Christians into all truth.[24]

What inspired "a passion for preaching, writing, and reading" were key Reformation doctrines of *sola scriptura* (that religious truth was found exclusively in the Bible), the priesthood of believers (that the word of God can only be directly apprehended by individual believers without the intersession of clerical authority), preparation for grace (that it was necessary for people to prepare themselves to receive divine grace if such a fate was part of God's preordained plan), and sanctification (that persistent study and personal spiritual reflection throughout one's life was necessary to prepare for receiving grace).[25] Thus, the act of publicizing carried not only an element of worldly self-preservation, but a spiritual if not divine weight and importance.

With the advent of the printing press, Protestant reformers soon saw that the extension of preaching, writing, and reading is publishing. As Nord recounts, "Protestant reformers of the sixteenth century not only reaffirmed the textual basis of Christianity; they preached the virtue of Bible reading by ordinary people in their own vernacular languages, and they idealized a powerful new technology for reproducing texts: the printing press."[26] Although Protestant publishing had been crucial since Luther, evangelical publishing in the Anglo context emerged from Puritan and dissenting traditions of the seventeenth century. By the time of the establishment of the constitutional monarchy, dissenting writing and publishing had been legalized by the Toleration Act of 1689, "which legally recognized loyal but nonconformist religious groups, namely Presbyterians, Congregationalists, Baptists, and subsequently Quakers."[27]

By the late 1700s, however, the importance of evangelical publishing had grown commensurate with increases in the size and mobility of populations both in England and in its expanding colonies, as well as with the increased volume and perceived crassness of popular commercial culture. The establishment of a particularly aggressive form of evangelicalism can be seen in the Methodism of John Wesley, "High Tory in politics, sacerdotal in his approach to organization." As Thompson characterizes it, Wesleyan Methodism was a Protestant sect that was indebted to the "poor man's Dissent" but, based on Wesley's authoritarian iteration, was no longer "a religion *of* the poor" but, paternally and severely, "a religion *for* the poor [emphasis original]."[28] Combined with the emergence of new legal and financial means of support—within England, the national tract society, and, within the newly emerged United States, the not-for-profit corporation—the context, rationale, and means now existed for a previously unattainable level and centrality of evangelism.

With the emergence and spread of Wesleyan Methodism and other overtly evangelical denominations as devoted to spreading the Word as they were highly organized, significant effort was placed in creating secular institutions that had a godly role and purpose. The U.S. context was particularly amenable to such efforts. The distinctive context of American evangelism (including comparatively greater religious diversity of immigrants, and the Establishment Clause of the First Amendment that separated church and state at the federal level) encouraged the formation of many denominations that gained influence by, as Congregationalist spokeperson Lyman Beecher noted in his autobiography of the time regarding the spread of Congregationalism, "voluntary efforts, societies, missions, and revivals"—constituting what Brown calls an "entrepreneurial milieu of antebellum America."[29]

The characterization of evangelical publishing as entrepreneurial underscores the degree to which success at the propagation of one's denomination was increasingly tied to as well as indistinguishable from success in sales. In such a competitive milieu, by the mid-1800s, a "widespread perception" was that "competitive failure in the print market spelled disaster in the religious marketplace as well."[30] While the imperative and the initial means were drawn from already existing organizations in England such as the Society for the Propagation of the Gospel in Foreign Parts (chartered in 1701), in the context of antebellum America, what were in 1815 "hundreds of missionary, Bible, and religious tract societies" converged later in the century into "several large national organizations, which led America into the modern era of mass publication and systematic distribution of printed material."[31]

The means for conducting a widespread program of print-assisted evangelism existed already in the tools of capitalist commercial expansion, in such ways as mechanized systems of printing and book production, national systems of transportation and distribution and, as already noted, emerging means of financing such ventures. Bible and tract societies played a central role in this redefinition, which in many ways created a key template for strategies, organizations, and technologies of modern mass media.[32] Nord argued that it was "evangelical Christian publicists in the Bible and tract societies [. . . who] first dreamed the dream of a genuinely mass medium—that is, they proposed to deliver the same printed message to *everyone* in America [emphasis original]."[33] American evangelicals owed much to older tract societies in England, which "inspired the Americans with their vision of what religious mass media should be."[34]

The evangelical embrace of the tools and techniques of mass marketing and commercial publishing, despite their perceived misuse by secular companies in the service of spreading sin and irreligion, represented no contradiction as long as the ends justified the means. As Brown summarizes, evangelicals "envisioned their secular [publishing] work as functionally sacred because it was useful in communicating the gospel."[35] The emergence of Protestant evangelism and its entwinement with capitalism together produced the forms and practices associated with the formation of a novel mode of communication called by the early nineteenth century "broadcasting."

Evangelism and Broadcasting

Citing the term "broadcasting" in a discussion of pre-electronic means of communications might seem quite out of place. While its pre-network sense is a single person sowing seeds for him/herself using individual muscle power, its sense today is the "sending [of] a wide range of sounds [. . .] from powerful transmitters to thousands of individual receivers."[36] The coincidence by which a term refers both to a pre-modern method of sowing seeds by hand and perhaps the most fitting means in capitalist societies of electromagnetic dissemination of visual/audio programming may seem initially curious but ultimately irrelevant.

Yet, when seen as an emergent mode of remaking social relations, the articulation of agricultural and media practice through "broadcasting" indicates more than an archaic, metaphorical equivalence between the spreading of seeds and the spreading of messages. Indeed, an examination of examining expressions regarding broadcasting from the contexts of agriculture in eighteenth-century England to its place in nineteenth-century evangelical publishing recovers how the first usage became incorporated into the second, and describes the conventional frameworks of understanding that made its incorporation possible and meaningful.

The time when evangelical publishing first appeared in England was also the time when "broadcasting" first appeared in the English language in the mid-1700s. Literally speaking, "broad-casting" meant walking through a field, scooping a palmful of seeds from a seed bag slung over one's shoulder, then throwing them through the air to the ground in the hopes they germinated and grew into crops to be later harvested. Arguments against broadcasting appeared commonly in English farming books, a centuries-old hybrid genre part self-help manual, empirical study, promotional advertisement, and political treatise that "were the classics, part of every gentleman's education of the day."[37] By contrast to this labor-intensive, imprecise, and inefficient method of sowing, critics argued that seed-drilling (the use of animal-drawn machines to place seeds in the ground at a precise location and depth) made it possible to cultivate fields continuously without requiring fields in intervening years to lie fallow, thus enabling farmers to extract greater economic value from the land they owned or rented.[38] This benefited not only the individual farmer, but the nation, recalling a linkage articulated by classical political economists of the day. By the early 1800s, such a linkage had the status of an axiom for one writer, who asserted that "every step that a man makes for his own good [also] promotes that of his country."[39] Working for the good of the country maintained an abstract nationalist pride, but it also sought to avoid poverty, hunger, and resulting widespread rioting, a practical policy objective not only for England, but for "all European governments in the eighteenth century."[40] To help the country as well as oneself, one had to farm using the most productive, efficient methods possible.

What was present even in this pre-communications usage was a clear social division between gentry farmers and laborers, between those who profited from the activity and those who labored for subsistence within it. Not unlike the political elite's aim in mid-1600s England to politically defang and placate the newly emerged public, critics of broadcasting expressed a similar desire to bring agriculture labor further under the logics of accelerating capitalist expansion and profitability. Gentry farmers aimed to become more productive, in part, by minimizing their dependence on and the cost of rural peasant labor. A writer at the end of the eighteenth century noted that, compared to "hand-hoeing [, which] requires a great many laborers," mechanized cultivation "requires but a few labourers [. . .] and therefore may be practiced in many places where the other cannot, for want of labourers."[41] And, as Fussell notes, another farmer "designed his seed-drill to combat the intransigence of his workers [. . .] after they had refused to do the planting in rows that he required."[42]

By contrast to gentry proponents of seed-drilling, agricultural laborers preferred broadcasting for practical reasons. Not only were seed-drills notoriously unreliable and difficult to fix, the mechanization of agriculture was gradually depriving rural laborers from a means of making a living.[43] Seen in a broader scope, the technical rationale for seed-drilling was part of the same accelerating economic and political rationalization that was remaking cultures of truth, with local customary law becoming increasingly superceded by a national policy that embodied classical political economy and its particularly orthodox brand of market economics.[44] Rationalization had its own costs, however. Decreasing the need for laborers meant increasing the burdens on local communities to take care of them when unable to find work, a trend contrary to the long-standing "social code of the countryside and the unwritten law that the onus of finding work should be spread among all employers in the parish."[45] Such a trend was not accepted lightly. The examples of Luddism and the agricultural riots of the eighteenth and nineteenth centuries suggest the depth of the working-class response to attempts to "control time, the wage, and levels of skill [. . .] both [through] the law and new machines [. . .]."[46] What the context of its emergence indicates is that "broadcasting" from the beginning was not a simple description of a technique. It marked a terrain constituted by politics, technology, and class, emerging just as it was seen as a thing of the past.

The link between the agricultural and the communications usage and practice was formed through evangelism. While the term "broadcasting" does not appear in the Bible, the practice to which broadcasting refers operates as a central trope to embody a claim about the process of finding Christ. Rather than privilege the power of the text by making a mechanical equivalence between messages and seeds, the parable of the sower argues that the effect of the Word is determined by who encounters it.[47] In Peters's summary of "the synoptic Gospels of Matthew, Mark and Luke [. . .], the Word is scattered uniformly, addressed to no one in particular, and open to its destiny."[48]

However, this liberal-sounding New Testament admonition deserves to be compared with one much sterner found in the Old Testament, which provides a clear imperative for a concentrated, controlled effort. The prophet Isaiah sees the task of communication quite differently. Just as rain and snow from heaven water the earth and cause plants to grow, bear fruit, and provide seed for more planting as well as for food to eat, the holy Word that comes through the mouths of prophets shall succeed just as necessarily in its purpose of conveying the Word of God. "[I]t shall not return to me empty, but it shall accomplish that which I purpose, and prosper in the thing for which I sent it."[49]

The emergence and evolution of evangelical practice of communications in nineteenth-century United States aligns much more clearly with the Old Testament conception. Evangelical Christian publishing of the nineteenth century forged the articulation of broadcasting with media in the context of vanguard, mass politics.[50] For nineteenth-century tract societies, broadcasting was a means of message dissemination seen as eminently suitable for reaching lower classes, from whom emanated "the rising tide of democracy and irreligion."[51] It is here that the class indexing performed by broadcasting increases in complexity, illustrating how meaning, significance, and class operations depended on historical context. Broadcasting in agriculture was disparaged by gentry and cultural elites as something suitable only for lower classes. Yet, with the ideological merger of it with evangelical activism and publishing, broadcasting was not a popular practice in opposition to elite preferences, but a popularizing practice to aid elite purposes. The shift is based in earlier agricultural and biblical usages, but remade to fit the new context, thus emerging from but also severed from the past.

While evangelicals' expressly ideological aim was to stem the tide of irreligion and democracy and thus restore the past, their practical contribution consisted of innovations in media reproduction and organization. They helped develop and refine stereotyping, steam-powered printing presses, and mechanical paper-making. They also pioneered mass means of financing and distribution in such ways as "the building of a genuine national organization" composed of "elaborate systems of local auxiliary societies and branch distributors" that were "subsidiary to a centralized, national organization."[52] The emergent media strategy, first put into practice by the New York City Tract Society in 1829 (fully four years prior to the *New York Sun* of Benjamin Day, typically identified in the United States as the initiator of mass media in the form of the penny press) was to "place in the hands of every city resident at least one tract, the same tract, each month." Such a strategy and practice exemplified for Nord "the ideology of modern mass media: to have everyone reading and talking about the same thing at the same time."[53]

The trajectory traced through an examination of discussions of broadcasting as found in evangelical publications of the day documents a complex process of mystification and naturalization—the transmogrification of a human muscle-powered activity into a non-human, invisible if not divine force of nature. Parallel with this is an increasing confluence between broadcasting in the agricultural sense and the then-new practice of mass media in the service of mass political organizing.[54]

Despite its great variability, this confluence can be mapped along some key dimensions. The first concerns the concreteness of the practice. On one hand, a metaphorical usage expressed the natural widespread distribution of non-material qualities or ideas. By contrast, a practical usage named the intentional human action of distributing media products, such as newspapers, books, pamphlets, and broadsides. Each position tended to place human agency differently. To the degree that the metaphorical usage removed humans except as implied recipients of this process, it mystified and reified the process and its effects. Thus, authors wrote of "the growth of those [desirable] affections whose seeds are sown at broadcast in the natural relations of life."[55] Horace Mann stated in a speech that "the way is open here in Kentucky for sowing the seeds of freedom broadcast."[56]And others argued that "the seeds of oppression had been scattered broadcast over this land."[57] By contrast, the practical usage placed people as the instigators and concrete means of distribution. Writers advocated to their followers that "such notices [for abolition] should be thrown into the tract form, and sown broadcast over the land."[58] The facts of "the brutal character of American slavery" are available for all to read because they "are sent broadcast over the British empire, and are published in nine-tenths of all the British newspapers."[59]

Yet, variations existed that render simplistic an attempt to rigidly equate the metaphorical with the mystified. Some items referred to the active distribution of media products, but as a natural process unaided by human intention or intervention, thus also mystifying their production and use. One statement noted how documents seemingly under their own power "would go broadcast all over the free States in this country."[60] On the contrary, other variations referred to an intentional human activity, but of the spreading of ideas that lacked any material embodiment. People would "go sowing broadcast the seeds of rich blessing," "sow also broadcast and with a liberal hand, the seed of truth," or have "sown broadcast the seeds of bitterness."[61]

Specifying this exploration was the increasing frequent discussion throughout the 1850s and 1860s regarding the role of media and broadcasting in fomenting social change. Such items express an increasingly vanguard relationship between the activist group and its target publics. The pace of such mentions accelerates throughout the 1850s to the point where broadcasting describes the political work done by media employed by vanguard groups. Whether metaphorical or practical, "broadcasting" in the service of the movement increasingly meant, for the instigators, an intentional, concrete practice and, for the audiences, reception and acceptance. A commentary in *The North Star* remarked on the "mighty and overwhelming" ability of the press to spread "truths broadcast in the world."[62] In a promotion for a tract, the "friends of Freedom will, it is hoped, send their orders immediately for this seasonable pamphlet, and cause it to be spread, broadcast, over the whole country."[63] Similarly, "anti-slavery books, tracts, and newspapers should be scattered broadcast over the land; [and] the question of abolition should be brought home to every man's hearthstone."[64]

As time went on, the certainty of effect increased. As one writer put it in 1856, no one can "doubt for a moment the utility of scattering broadcast over our land, such documents and speeches as will have a tendency to enlighten the public mind [. . .]."[65] Manuscripts, books, and tracts are seen to enable the spread, "broadcast, [of] the very truths which are there contained."[66] Others note that media distribute "the seeds of anti-slavery truth, which shall thus through your instrumentality be scattered broadcast over the land."[67] Readers were increasingly urged to "inform the masses. Circulate the documents broadcast through the land, until every honest farmer and mechanic had read and understood the principles of our party."[68] One organization suggested that "our publications should be sown broadcast in all our advances, and keep pace with our advancing missionary operations, as a sure means of permanent success."[69]

By the early 1860s, this complex of relationships and practices was even more explicitly and uniformly expressed. "Broadcasting" had come to mean a vanguard relation between advocacy groups and their targets, and a reified spread of sentiment as well as a practical tactic of political agitation. The key tactic of mass social activism was broadcasting, whatever the technology.[70] While some items admonished readers to promote abolition "in all practical ways" including "spreading broadcast the issues of that mighty engine, the public press," others preferred to privilege the means by advocating that "every true friend of liberty and humanity" should do everything possible to "scatter broadcast the seeds of Anti-Slavery truth."[71] By 1859, the irrelevance of the specific technology in favor of a recognition of the power of broadcasting was increasingly axiomatic. "Scatter information broadcast among the people," urged a writer in 1859. "The harvest will be sure."[72]

However, the power of media to broadcast new ideas was not always seen as beneficial. Those who opposed abolition felt the press "contributed to bring about a ferocious discontent, which needed only the insidious and inflammatory articles spread broadcast over the land by designing men to fan into an insurrection."[73] This class-bound association of broadcasting with non-elites emerged within broader debates about the moral effects of mass, commercialized popular culture. On one hand, "broadcasting" expressed an authentic popular culture as framed within and linked to the Protestant tradition of direct access to God and the popularization of religious practice. As one Protestant, evangelical interpretation of medieval society remarked, "the Word of Truth [at that time] was read only by the learned men and cloistered monks, and its benign influence was not shed broadcast over the earth as now."[74] However, the moral effect of other kinds of materials and sentiments broadcast was less laudable, noted one commentator. "For if these masses are not soon educated and Christianized, they will sow broadcast the worst forms of infidelity imported from the Old World, and by and by they will overturn the very foundations of our free institutions."[75] Growing concerns were expressed about the popular distribution of French novels "sown broadcast through our land [. . . , which] corrupt the imaginations of our young men and maidens, and [. . .] waste the time of more mature readers."[76] Similar claims appeared regarding how "the mails are extensively prostituted to immoral and vicious purposes, and that through this channel obscene books, circulars, &c. are sown broadcast throughout the country."[77]

The evangelical articulation not only extended "broadcasting" conceptually into media and politics, but newly defined its social nature and class status. "Broadcasting" described not only a way of distributing ideas, but one uniquely suited to vanguard intentions and uses. It—and not any specific media technology—was increasingly seen as the determinant, fundamental agent and cause of moral and political transformation. Unlike its agricultural iteration, evangelical broadcasting was not a practice of non-elites, but a practice of elites in political programs directed at the rest of society.

Yet, existing in the midst of this dominant articulation of broadcasting was an emergent articulation with democracy and, implicitly, with consumerism. Within it, the power of broadcasting was seen as something other than audience responses subsumed within the wishes of a vanguard. Some articles expressed an active, shaping response by the audience consistent with classical-liberal conceptions of the marketplace, such as in an account of a broadside author who had "sent so many thousands of 'lower law sermons' broadcast throughout the free States." However, he must, as the commentator argued, "meet the penalties of the popular will" should his argument not be received approvingly.[78] The fit of this political statement about the ideal workings of a democracy as the sovereignty of the people with an implicit, parallel claim about the sovereignty of consumers as conceived in liberal economic thought became only more apparent as the century closed and a new one dawned.

The Nature of Broadcasting at the Turn of the Century

In the contexts of the increasing pace of consolidation, centralization, and rationalization of the state, and the popular celebration of the creative capacities of Americans to exert control over their world (seen perhaps most broadly in the geopolitical consolidation of the continent under the banner of Manifest Destiny), the vanguard social relations underlying broadcasting that were pioneered by Bible tract societies became increasingly articulated through forms and organization of commercial media.[79] This emergent articulation became increasingly important to the point where, in 1907, inventor and entrepreneur Lee de Forest could conceive the general outlines of a broadcasting system in terms not of centralized political activism directed at the masses, but as equally centralized popular entertainment delivered to the masses for their individual pleasure. In a diary entry of that year, the evangelical aim of creating a social movement had become a secular wish to extend the political community beyond its geographic borders. In the entry, de Forest explained that "My present task is to distribute sweet melody broadcast over the city and sea so that in time even the mariner far out across the silent waves may hear the music of his homeland, sung from unseen sources."[80]

The conception of a broadcasting system as a means of achieving social solidarity via the centralized distribution of entertainment derives from longstanding millennial hopes and dreams regarding electronic communication and its almost magical power by which communities could be (re)constituted.[81] Furthering the mystification and naturalization of broadcasting was its articulation with the popular fascination with electronic technologies and the seeming disappearance of the materiality of the message.[82] Such a fascination was in part a response to sectional strife, significant labor activism, social and cultural dislocation related to urbanization, and the reconstitution of the United States as an industrial rather than agricultural society, and large-scale immigration and resulting debates regarding its impact on American society and identity.[83] Such a view appeared in Britain as well, which was experiencing many of the same pressures. A British writer in a popular science magazine of 1884 concurred that "everything which can knit a community together and which can cause a rapid interchange of sentiment and ideas, annihilate distance, isolation and prejudice, is of the greatest happiness to the greatest number [of people]."[84]

An overlap with its evangelical articulation expresses this community building in the form of broadcasting religious services. A laudatory item promoted one such effort in Old St. Patrick's Church in Pittsburgh during two weeks late in 1921, whose author estimated (perhaps optimistically) that one million people "every night [were] 'listening in' to these messages picked out of the air." The result was that "the whole world is brought to our very ears while we lounge at ease amid the comforts of our own drawing-rooms, libraries, or bed-rooms." By this means, the author hoped that "the seed of further conversions could be sown and scattered wherever human beings congregate."[85]

The emergent conception of human involvement in broadcasting also continued to change. Rather than being a human practice or spiritual manifestation, broadcasting was becoming not only a nonhuman force of nature to be tamed by human ingenuity but, for some commentators, the infinite, unimaginable energy flow itself in the natural world. Some early twentieth-century writers regarded broadcasting as the harnessing of the "æther" [*sic*], which was seen to constitute the enabling medium of wireless (described in a 1924 popularized account of broadcasting as a kind of material denser than the finest steel but undetectable by human senses).[86] Another British writer in 1923 remarked on the wonder of harnessing this force. Even more amazing than how "the electrical influence from the aerial emerges out of the unknown and manifests itself at some place, perhaps hundreds or even thousands of miles away" is the discovery that this force "is actually under control and can be made to carry messages."[87] Other contemporary commentators elevated broadcasting from the human harnessing of this force, to the force itself, such as one in 1924, which asserted that "nature has been 'broadcasting' since the earliest thunderstorm. With the first lightning flash, wireless waves were sent rippling across space, penetrating primæval [*sic*] forests, rocky caverns and the haunts of such animal life as then existed." By contrast, people have been broadcasting for little more than twenty-five years, "using harnessed forces of nature."[88]

This ambiguity regarding control (had humans tamed this force, or were they simply riding a tiger that could turn on them at any time?) found expression, alongside approval of such technologies, in concerns about electronic media as a demagogic means of subverting or erasing traditional social hierarchies.[89] Concerns about the social and political consequences of the telephone had appeared less than a year after Alexander Graham Bell's initial public demonstration. A March 15, 1877, cartoon in the *New York Daily Graphic* "illustrated 'the terrors of the telephone' by showing an announcer at a telephone-like device mesmerizing masses of people listening simultaneously throughout the world."[90]

Within this overall transformation from human to heavenly to natural force, a parallel reifying trajectory can also be identified. A preoccupation with the wonder of the technology and a celebration of individuals and their technical achievements soon became the broader celebration of the teleological emergence of consumerist "'media' for the millions."[91] As explained in a promotional book published by the Radio Corporation of America (R.C.A.), the problem to tackle after point-to-point telephony was how to devise "a system whereby one voice, one speech, one concert of music might be heard instantaneously and simultaneously in millions of homes," with the answer being broadcasting, seen as the next evolutionary stage in the development of communication technology, with the need met by the newly formed National Broadcasting Company.[92]

Yet the emergence of broadcasting systems was not the natural fruition of intrinsic properties of communications technology, as R.C.A. asserted. By contrast, its emergence was directed by conceptions of what broadcasting should be, thus picking up on the vanguard service pioneered by evangelicals and print. As Briggs notes (corresponding to Nord's claim about evangelical publishing), the idea of "everyone everywhere sharing the same 'programmes' at the same time preceded the introduction of the required technology."[93] In addition to evangelical statements, conceptions of broadcasting were publicly worked out in such forms as "cartoons, articles, and even popular songs [that] lauded the potential wonders of the telephone, including the distance dissemination to mass audiences of music, information, drama, and education."[94] News and information had been broadcast to ships telegraphically for almost two decades before the first commercial broadcast, and propaganda by the Germans and others during World War One were also noted as early instances of broadcasting, albeit with telegraphy rather than telephony.[95]

Other attempts to realize broadcasting using prewireless point-to-point electronic technologies took place in 1877, when Edison stumbled on a means of using the telephone as a public receiver. After further experimentation and refinement, "by placing the [modified] telephone in the centre of the room [,] persons seated around could hear [transmitted music] with perfect distinctness."[96] Conceptual experimentation included French engineer Clément Ader's patent application for "Improvement of Telephone Equipment in Theaters," in which he proposed to put telephones on the stage so subscribers could listen to the performance at home.[97] De Forest's own contribution occurred in 1906 through hanging a microphone on the stage of the Metropolitan Opera House in New York City and sending the signal to receiving sets installed in several public locations throughout the city for passersby to encounter and listen to.[98] Yet another example occurred later in 1906, with Canadian Reginald Fessenden's Christmas Eve transmission to ships of a program consisting of "a person speaking, then a woman singing, then someone reading a poem, followed by a violin solo, and verses [read] from the Bible."[99]

In this complex of the agricultural, the vanguard, and the consumerist contexts, broadcasting emerged as a synthesis during the 1910s in the activities of physics and engineering faculty and their students at Midwestern U.S. universities. As part of learning more about electronic theory, transmission facilities set up in such departments distributed to farmers "reports from time to time on weather conditions, crops, producer prices, Department of Agriculture advisories, and other things that an isolated farmer might otherwise have to wait days to learn."[100] A parallel effort took place in the then-rural Santa Clara Valley of California, where Charles Herrold developed a wireless information-distribution service during the 1910s that provided "programming music and talk" for "amateur men" in the area. Hilliard and Keith note that this effort "evolved out of his role as the headmaster of a wireless trade school"—with the school providing the institutional basis, rationale, and technical facilities, but the social uses and relations deriving from that of a headmaster with his students.[101]

Increasingly in the United States, however, the template for organizing broadcasting was commercial, with a broadcasting system becoming a primarily consumerist pastime that fit the basic redefinition of social relations in an accelerating capitalist economy.[102] A consumerist, often paternal vanguardism as opposed to political vanguardism drew on formulas developed by mass-circulation newspapers, popular magazines, popular books, and live popular entertainment. A broadcaster "must be both showman and concert manager, newspaper editor and music-hall director"—all of which derive from earlier commercial forms of popular entertainment.[103] A contemporaneous assessment asserted that "broadcasting can only reach its ultimate goal through centralization of efforts."[104] A non-Anglo example noted in U.S. press accounts of such centralization was the "telephone newspaper" *Telephon Hírmondó*, whose customers "receive[d] the news as they would ordinary telephone messages." By applying long-standing ideas of organizing live commercial entertainment (such as scheduling the day and time of performances as akin to programming) and commercial forms of media production (rationalizing, professionalizing, and centralizing its practice), the telephone newspaper also helped pioneer what became the media practice and organization of broadcasting.[105] Even amateurs' early experimentation worked through a commercialized intention of pleasing an audience. As one author put it, early amateurs' means of assessing their success consisted of "timid [inferred] requests" of their audience: "'Did you like it? Would you like more of it?'"[106]

The emergent mass-consumerist articulation of broadcasting pushed the mystification of broadcasting further than the evangelical, in that it appeared to be a form of direct delivery to individual households and a direct connection between performance and its experience. This was perhaps the most powerful social and cultural effect, which helped set terms of participation, although not without detractors.[107] Unlike newspapers and magazines, which in their presentation and narrative structure were overtly professionally gathered and produced, broadcasting relied on the aesthetic evaporation of the organizational means by which the message was propagated, prompting the then-current and already noted term to describe audiences' use as a direct, unmediated "listening-in."[108]

In these ways, the move from evangelical to consumerist turned on its head the power relationship implicit in vanguard forms. Instead of placing the public in the service of elites, the public was ostensibly being served by a coterie of scientists, inventors, visionaries, and, later, entrepreneurs and national corporations. Yet the very separation of makers from listener-consumers was inherited from the evangelical form, aided by rationalizing organization and the general shift from distributed, pragmatic, individual use to centralized, entertainment, "mass" use. Such a development was institutionalized with the merging of governmental and industrial interests and the creation in 1919 of R.C.A., which, in Hilmes's estimation, constituted the conditions by which "American broadcasting would never again be as accessible to individuals."[109]

Providentialism, Rationalist Empiricism, and Broadcasting

Providentialism and rationalist empiricism were key cultures of truth that were rationalized through capitalist expansion and social reorganization. While the pluralization of power in society (more so in the United States than in England, although still limited in both compared to today), and the rise of the public as a political actor marked the slow process of democratization, the public itself had placed upon it clear boundaries on the legitimate role it was to play in the emerging social order. These political boundaries corresponded with commercial boundaries, which were comprised of a slowly solidifying divide between a producing minority and a buying majority, each actively contributing but each typically and increasingly in its own defined and institutionalized way.

The crucible of nineteenth-century evangelism melded more tightly the realms of the political, the commercial, and the religious while it enhanced the social divisions increasingly typical of each between a producing and directing minority and a receiving majority. Paraphrasing Thompson's point about Wesleyan Methodism, with the emergence of evangelical broadcasting in the nineteenth century, communications of this kind and scale was a cultural process not *of* the people, but paternalistically *for* the people.[110]

The further elaboration of Providentialism and rationalist empiricism as cultures of truth and of broadcasting as a dominant if not quintessential formation of communication in market societies scarcely needs recounting here, having received as they have voluminous attention in a variety of areas as journalism history, television studies, film studies, literary history, art history, media history, and so on. The emphasis here by way of preliminary summary is that this complex formation underlies a dominant communications practice in Anglo-American societies within the context of capitalism. Whether commercial ventures that seek a large and/or lucrative audience for advertisers, or alternative-media practitioners that seek a large or active audience for mobilization, the intentions of each are not simply neutral or technical matters. Whether offering commentaries on current events or comprehensive reports, the forms and practices used to do so are hardly without social implication, nor able to be understood or practiced—however implicitly and contradictorily—outside of their historical contexts. When broadcasting either commentaries or reports, the social implications of such a practice not only reproduce rationalized social divisions, by doing so they enact a social order that limits democratic participation by enshrining it. Within such a formation, serving the public becomes a replacement for enabling it.

Notes

1. Michael Albert, "Mass Media: Theirs and Ours," *Z Magazine*, <http://zena.secureforum.com/Znet/ZMag/articles/oldalbert6.htm> (19 Jan. 2006); "Michael Albert," *Z Magazine*, <http://zena.secureforum.com/Znet/ZMag/fralbert.htm> (22 Aug. 2006).

2. Albert, "Mass Media." Similar statements for the need for a national alternative/progressive media organization include Jessica Clark and Tracy Van Slyke, "Welcome to the Media Revolution; How Today's Media Makers are Shaping Tomorrow's News," *In These Times*, 28 June 2006, <http://www.inthesetimes.com/site/main/article/2687/> (31 Aug. 2006); and "National Progressive Media: Who's Left?," *Media File*, <http://www.media-alliance.org/article.php?story= 20031109003346211> (31 Aug. 2006).

3. One early example of such a usage of broadcasting is "Counter Statement," *The Colored American*, 16 May 1840. Key historical accounts of broadcasting include Erik Barnouw, *A History of Broadcasting in the United States*, 3 vols. (New York: Oxford University Press, 1966–70); Raymond Williams, *Television; Technology and Cultural Form* (New York: Schocken, 1975); Susan Douglas, *Inventing American Broadcasting, 1899–1922* (Baltimore, Md.: Johns Hopkins University Press, 1987); Robert Britt Horwitz, *The Irony of Regulatory Reform; The Deregulation of American Telecommunications* (New York: Oxford University Press, 1989); Robert W. McChesney, *Telecommunications, Mass Media, and Democracy: The Battle for the Control of U.S. Broadcasting, 1928–1935* (New York: Oxford University Press, 1993); and Susan Smulyan, *Selling Radio: The Commercialization of American Broadcasting, 1920–1934* (Washington, D.C.: Smithsonian Institution Press, 1994).

4. John Durham Peters, *Speaking into the Air; A History of the Idea of Communication* (Chicago, Ill.: University of Chicago Press, 1999); Paddy Scannell, "Introduction: The Relevance of Talk," in *Broadcast Talk*, ed. Paddy Scannell (Newbury Park, Calif.: Sage, 1991), 1–13. See also Thomas Streeter, *Selling the Air; A Critique of the Policy of Commercial Broadcasting in the United States* (Chicago, Ill.: University of Chicago Press, 1996).

5. Gary S. De Krey, *A Fractured Society: The Politics of London in the First Age of Party 1688–1715* (New York: Oxford University Press, 1985).

6. Jürgen Habermas, *The Structural Transformation of the Public Sphere: An Inquiry into a Category of Bourgeois Society*, trans. Thomas Burger (Cambridge, Mass.: MIT Press, 1989); Michael Warner, *Publics and Counterpublics* (New York: Zone Books, 2002), 46. For critical reflection on Habermas and this formulation, see Craig Calhoun, ed., *Habermas and the Public Sphere* (Cambridge, Mass.: MIT Press, 1992).

7. Examples include George Rudé, *The Crowd in History; a Study of Popular Disturbances in France and England, 1730–1848* (New York: Wiley, 1964); Eric Hobsbawm, *Social Bandits and Primitive Rebels; Studies in Archaic Forms of Social Move-*

ment in the 19th and 20th Centuries (Glencoe, Ill.: Free Press, 1959); and E.P. Thompson, *The Making of the English Working Class* (New York: Pantheon, 1963).

8. Tim Harris, *London Crowds in the Reign of Charles II; Propaganda and Politics from the Restoration Until the Exclusion Crisis* (Cambridge: Cambridge University Press, 1987), 218.

9. John Spurr, "From Puritanism to Dissent, 1660–1700," in *The Culture of English Puritanism, 1560–1700*, ed. Christopher Durston and Jacqueline Eales (Basingstoke: Macmillan, 1996), 234–65.

10. John Spurr, "'Virtue, Religion and Government': The Anglican Uses of Providence," in *The Politics of Religion in Restoration England*, ed. Tim Harris, Paul Seaward, and Mark Goldie (Oxford: Basil Blackwell, 1990), 30.

11. Jason Peacey, *Politicians and Pamphleteers; Propaganda During the English Civil Wars and Interregnum* (Burlington, Vt.: Ashgate, 2004), 305. See also Carole Levin, *Propaganda in the English Reformation: Heroic and Villainous Images of King John* (Lewiston, Maine: Edward Mellen Press, 1988); John P.D. Cooper, *Propaganda and the Tudor State: Political Culture in the Westcountry* (New York: Oxford University Press, 2003); John Patrick Montano, *Courting the Moderates: Ideology, Propaganda, and the Emergence of Party, 1660–1678* (Newark, N.J.: University of Delaware Press, 2002); Dagmar Freist, *Governed by Opinion; Politics, Religion and the Dynamics of Communication in Stuart London 1637–1645* (London: Tauris Academic Studies, 1997); and Tone Sundt Urstad, *Sir Robert Walpole's Poets: The Use of Literature as Pro-Government Propaganda, 1721–1742* (Newark, N.J.: University of Delaware Press, 1999).

12. Peacey, *Politicians*, 42.

13. Peacey, *Politicians*, 62.

14. Peacey, *Politicians*, 306.

15. Peacey, *Politicians*, 303.

16. Peacey, *Politicians*, 196–200.

17. Peacey, *Politicians*, 203–33.

18. Peacey, *Politicians*, 63.

19. Peacey, *Politicians*, 331.

20. Peacey, *Politicians*, 332.

21. The list is summarized from Peacey, *Politicians*, 323.

22. Paraphrased from Peacey, *Politicians*, 303.

23. Peacey, *Politicians*, 332.

24. Candy Gunther Brown, *The Word in the World; Evangelical Writing, Publishing, and Reading in America, 1789–1880* (Chapel Hill, N.C.: University of North Carolina Press, 2004), 2.

25. David Paul Nord, *Faith in Reading; Religious Publishing and the Birth of Mass Media in America* (Oxford: Oxford University Press, 2004), 14–17.

26. Nord, *Faith*, 13.

27. Brown, *Word*, 33. Accounts of the particularly prolific Quaker writing in England include Thomas N. Corns and David Lowenstein, eds., *The Emergence of Quaker Writing; Dissenting Literature in Seventeenth-Century England* (London: Frank Cass, 1995); and Kate Peters, *Print Culture and the Early Quakers* (Cambridge: Cambridge University Press, 2005). General accounts of dissenting literary practice of the seventeenth century in addition to works cited in Chapter 1 include Nigel Smith, *Literature and Revolution in England 1640–1660* (New Haven, Conn.: Yale University Press, 1994).

28. Thompson, *Making*, 37.

29. Brown, *Word*, 36.

30. Brown, *Word*, 59.

31. Nord, *Faith*, 5, 32.

32. David Paul Nord, "The Evangelical Origins of Mass Media in America, 1815–1835," *Journalism Monographs* no. 88 (May 1984), 23. A less sweeping account of evangelical inspiration of journalism is Doug Underwood, *From Yahweh to Yahoo!: The Religious Roots of the Secular Press* (Urbana, Ill.: University of Illinois Press, 2002).

33. Nord, "Evangelical Origins," 2.

34. Nord "Evangelical Origins," 4.

35. Brown, *Word*, 46.

36. John Cain, *The BBC: 70 Years of Broadcasting* (London: BBC, 1992), 7.

37. G.E. Fussell, *The Old English Farming Books; From Fitzberbert to Tull, 1523 to 1730* (London: Crosby Lockwood & Son, 1947), 1.

38. Asa Briggs, *A Social History of England*, 3rd ed. (London: Penguin, 1999), 192.

39. Adam Smith, *An Inquiry into the Nature and Causes of the Wealth of Nations*, ed. Edwin Cannan, 6th ed. (London: Methuen, 1961); William Amos, *The Theory and Practice of the Drill Husbandry* (1802), v–vi.

40. Douglas Hay and Nicholas Rogers, *Eighteenth Century English Society; Shuttles and Swords* (Oxford: Oxford University Press, 1997), 71.

41. John Anstruther, *Remarks on the Drill Husbandry* (London: T. Egerton, 1796), 133.

42. Fussell, *Old English Farming Books*, 20; G.E. Fussell, *The Farmer's Tools, 1500–1900; The History of British Farm Implements, Tools and Machinery Before the Tractor Came* (London: A. Melrose, 1952).

43. G.E. Mingay, *Land and Society in England 1750–1980* (London and New York: Longman, 1994), 41; G.E. Mingay, *A Social History of the English Countryside*, reprint ed. (London and New York: Routledge, 2002), 46-50; E.J.T. Collins, "The 'Machinery Question' in English Agriculture in the Nineteenth Century," in *Research in Economic History*, Suppl. 5, Pt. A: *Agrarian Organization in the Century of Industrialization: Europe, Russia and North America*, ed. George Grantham and Carol Leonard (London: JAI Press, 1989), 213.

44. Hay and Rogers, *English Society*, 98–102.

45. Collins, "Machinery Question," 206.

46. Collins, "Machinery Question," 203; Hay and Rogers, *English Society*, 124. See also Eric Hobsbawm and George Rudé, *Captain Swing* (London: Lawrence and Wishart, 1969); Barry Reay, *The Last Rising of the Agricultural Laborers; Rural Life and Protest in Nineteenth-Century England* (Oxford: Clarendon, 1990); and Lee Davison, Tim Hitchcock, Tim Keirn, and Robert B. Shoemaker, eds., *Stilling the Grumbling Hive; The Response to Social and Economic Problems in England, 1689–1750* (New York: St. Martin's Press, 1992).

47. See also V. George Shillington, *Jesus and His Parables: Interpreting the Parables of Jesus Today* (Edinburgh: T&T Clark, 1997).

48. Peters, *Speaking*, 35.

49. Isa. 55:5.

50. Richard P. McCormick, *The Second American Party System; Party Formation in the Jacksonian Era* (New York: W. W. Norton, 1973); Ronald P. Formisano, *The Birth of Mass Political Parties, Michigan, 1827–1861* (Princeton, N.J.: Princeton University Press, 1971); and Margaret Finnegan, *Selling Suffrage; Consumer Culture & Votes for Women* (New York: Columbia University Press, 1999).

51. Nord, "Evangelical Origins," 3.

52. Nord, "Evangelical Origins," 13–15. A greater exploration of these contributions can be found in Nord, *Faith*, 27–111.

53. Nord, "Evangelical Origins," 20.

54. The body of materials consulted for this section consists of publications available through Accessible Archives (http://www.accessible.com): *Godey's Lady's Book* (1830–1880), *The Pennsylvania Gazette* (1728–1800), *Freedom's Journal* (1827–1829), *The Colored American (Weekly Advocate)* (1837–1841), *The North Star* (1847–1851), *The National Era* (1847–1860), *Provincial Freeman* (1854–1857), *Frederick Douglass Paper* (1851–1859), and *The Christian Recorder* (1861–1902).

55. C.M. Sedgwick, "Scenes from Life in Town," *The Lady's Book*, April 1843.

56. "Speech of the Hon. Horace Mann," *The National Era*, 21 September 1848.

57. "Selections: Address of H.W. Johnson," *The North Star*, 21 August 1848.

58. Untitled, *The National Era*, 15 April 1847, 4.

59. "Monthly Illustrations of Slavery," *The North Star*, 25 August 1848.

60. "Counter Statement."

61. "Review of The Young Lady's Friend," *The Lady's Book*, April 1838; "Spring!," *The North Star*, 14 April 1848; "Michigan Politics," *The National Era*, 25 October 1849.

62. "The Press," *The North Star*, 30 June 1848.

63. "Fugitive Slave Bill—Third Edition," *The National Era*, 21 November 1850.

64. "Resolutions Adopted by the Late Anti-Slavery Convention in Cincinnati," *The National Era*, 15 May 1851.

65. "A Circular to the Friends of the Republican Movement," *The National Era*, 17 January 1856.

66. "'Facts for the People'—A Want Supplied," *The National Era*, 21 April 1853.

67. "To the Members of the 'Rochester Anti-Slavery Sewing Society'," *Frederick Douglass Paper*, 27 August 1852.

68. "Extracts from Our Correspondence," *The National Era*, 5 January 1854.

69. "Have We Made Any Advancement?," *The Christian Recorder*, 4 June 1864.

70. "An Address to the Anti-Slavery Christians of the United States," *The National Era*, 24 June 1852; "The Annual Report of the New York State Temperance Society," *Frederick Douglass Paper*, 2 September 1853; "Facts for the People," *The National Era*, 18 August 1853; "The Present and the Future," *The National Era*, 31 March 1853.

71. "The Present and the Future"; "Extracts from Our Correspondence," *The National Era*, 11 August 1853.

72. "Speech of Gov. Chase at Cincinnati," *The National Era*, 10 November 1859.

73. Charles E. Blumenthal, "Develour; A Sequel to 'The Niebelungen'," *Godey's Lady's Book*, January 1851.

74. "Clerical Humanity," *The National Era*, 10 July 1851.

75. R.D. Harper, "The Church—Her Danger and Duty," *The Christian Recorder*, 7 June 1862.

76. "Editors' Table; Novels in French and English," *Godey's Lady's Book*, March 1863.

77. "Domestic Items," *The Christian Recorder*, 30 May 1863.

78. "Congressional Proceedings," *The National Era*, 8 January 1852.

79. Stephen Skowronek, *Building a New American State: The Expansion of National Administrative Capacities, 1877–1920* (New York: Cambridge University Press, 1982); Richard L. McCormick, *The Party Period and Public Policy: American Politics from the Age of Jackson to the Progressive Era* (New York: Oxford University Press, 1986); Daniel Schiller, *Theorizing Communication: A History* (New York: Oxford University Press, 1996), 3–38.

80. Quoted in Lee de Forest, *Father of Radio: The Autobiography of Lee de Forest* (Chicago: Wilcox and Follett, 1950), 225–26.

81. Carolyn Marvin, *When Old Technologies Were New: Thinking about Communications in the Late Nineteenth Century* (New York: Oxford University Press, 1987), 63–108; Daniel Czitrom, *Media and the American Mind: From Morse to McLuhan* (Chapel Hill, N.C.: University of North Carolina Press, 1982), 91–121.

82. In addition to Marvin, *Old Technologies*, and Czitrom, *Media and the American Mind*, see Catherine L. Covert, "'We May Hear Too Much': American Sensibility and the Response to Radio, 1919–1924," in *Mass Media between the Wars: Perceptions of Cultural Tension, 1918–1941*, ed. Catherine L. Covert and John D. Stevens (Syracuse, N.Y.: Syracuse University Press, 1984), 199–220; Mary Mander, "The Public Debate about Broadcasting in the Twenties: An Interpretive History," *Journal of Broadcasting* 28, no. 2 (Spring 1984): 167–85; Hugh Aitken, *The Continuous Wave: Technology and American Radio, 1900–1932* (Princeton: Princeton University Press, 1985); James Carey and John J. Quirk, "The Mythos of the Electronic Revolution," in James W. Carey, *Communication as Culture: Essays on Media and Society* (Boston, Mass.: Unwin Hyman, 1989), 113–41; Lisa Gitelman and Geoffrey B. Pingree, eds., *New Media, 1740–1915* (Cambridge, Mass.: MIT Press, 2003); David Thorburn and Henry Jenkins, eds., *Rethinking Media Change; The Aesthetics of Transition* (Cambridge, Mass.: MIT Press, 2003); and Martin Spinelli, "Democratic Rhetoric and Emergent Media: The Marketing of Participatory Community on Radio and the Internet," *International Journal of Cultural Studies* 3, no. 2 (August 2000): 268–78.

83. Raymond Wilson, *In Quest of Community: Social Philosophy in the United States, 1860–1920* (New York: John Wiley, 1968); Jean Quandt, *From the Small Town to the Great Community: The Social Thought of Progressive Intellectuals* (New Brunswick, N.J.: Rutgers University Press, 1970); Fred Matthews, *Quest for an American Sociology: Robert E. Park and the Chicago School* (Montreál, Quebéc: McGill-Queens University Press, 1977); Martin Bulmer, *The Chicago School of Sociology* (Chicago, Ill.: University of Chicago Press, 1984).

84. Asa Briggs, *The BBC: The First Fifty Years* (Oxford: Oxford University Press, 1985), 9.

85. Thomas F. Coakley, "Preaching the Gospel by Wireless," *Catholic World*, January 1922, 517–18. See also "The Gospel by Wireless," *The Literary Digest*, 4 February 1922, 32; "Preaching by Wireless," *Current Opinion*, May 1922, 649–50; and "Radio Sermons for Pastorless Churches," *The Literary Digest*, 17 June 1922, 27.

86. Arthur Richard Burrows, *The Story of Broadcasting* (London: Cassell and Company, 1924), ix. See also H. M. Dowsett, *Wireless Telephony and Broadcasting*, vol. 1 (London: Gresham Publishing, 1924), 179–201.

87. Ellison Hawkes, *The Romance and Reality of Radio* (London: T. C. & E. C. Jack, 1923), 1.

88. Burrows, *The Story of Broadcasting*, ix.

89. Example cited in Robert L. Hilliard and Michael Keith, *The Broadcast Century and Beyond: A Biography of American Broadcasting*, 3rd ed. (Boston, Mass.: Focal Press, 2001), 4.

90. Quoted in Hilliard and Keith, *Broadcast Century*, 4.

91. Briggs, *The BBC*, 9.

92. Radio Corporation of America, *The Place of Radio in World Communications* (New York: Radio Corporation of America, 1928).

93. Briggs, *The BBC*, 10.

94. Hilliard and Keith, *Broadcast Century*, 4.

95. Burrows, *Story of Broadcasting*, 17.

96. Hawkes, *Romance*, 175.

97. Hilliard and Keith, *Broadcast Century*, 4.
98. Hilliard and Keith, *Broadcast Century*, 12.
99. Hilliard and Keith, *Broadcast Century*, 7, 9.
100. Hilliard and Keith, *Broadcast Century*, 16.
101. Hilliard and Keith, *Broadcast Century*, 16, 20.
102. Raymond Williams, *Keywords: A Vocabulary of Culture and Society* (New York: Oxford University Press, 1976), 69.
103. Waldemar Kaempffert, "The Progress of Radio Broadcasting," *Review of Reviews*, September 1922, 305.
104. "About the Radio Round-Table: Opinions of Radio Leaders Regarding the Past, Present and Future of Broadcasting," *Scientific American*, December 1922, 379.
105. Briggs, *The BBC*, 8; "The Telephone Newspaper," *Scientific American*, October 1895, 267.
106. Kaempffert, "The Progress of Radio," 303.
107. For example, see "The Coming of the Wireless Telephone," *The Spectator*, 13 May 1922, 582–83.
108. Pierre Boucheron, "News and Music from the Air," *Scientific American*, December 1921, 104–5; "'Listening In,' Our New National Pastime," *Review of Reviews*, January 1923, 52; and Rex Palmer, "Future of Broadcasting as a Social Force," *English Review*, August 1923, 210–14. Claims that equated the fidelity of signal and skillfulness of the production with the degree of audience pleasure can be seen in "Organs of Sound That Do and Do Not Broadcast," *Current Opinion*, July 1922, 68–9.
109. Michele Hilmes, *Hollywood and Broadcasting; From Radio to Cable* (Urbana, Ill.: University of Illinois Press, 1990), 13–14.
110. Thompson, *Making*, 37.

Part Two

Struggling Against the Market

Introduction to Part Two

While the market was the crucial condition for the emergence of rationalist empiricism as the basis for corporate communications practice, it also enabled the rise of equally significant challenges to it. Suggesting that these responses are "against" the market does not mean that the market is neutralized, unimportant, or irrelevant, but that such efforts are in a negative, yet constitutive relation to it. This point should not be taken as an assertion for some binary, mechanical, inescapable dialectic but as the assertion of a dialogic struggle—critical responses that are products of the same enabling context as is the dominant—and thus a process of great range, variation and complexity.

In suggesting that the market exerts limits and pressures that in turn produce the conditions of resistance to it, a clearer sense is required of exactly how those limits and pressures are exerted. Granting the insufficiency of a claim that a handful of evil masterminds controls everything to its own personal benefit, how best to conceptualize the relationships between ostensibly oppositional projects and the more general formations they work within and through remains a key task.

One way such relationships have been conceptualized critically is through a dominant-ideology thesis. Such a position argues that projects have either been captured by or already operate within the priorities and structures of capitalism, a claim that for example informs much work on philanthropies.[1] Such an interpretation implicitly relies upon theories of dependent relationships between bureaucracies and the state.[2] The general difficulty with such a formulation is that it posits with mechanical certainty the inevitability of social control in a way that reduces the full variety of how control is exercised in modern societies as well as how it is challenged.[3]

A more nuanced variation of the dominant-ideology thesis is the argument that opposition projects are rationalized into manageable and non-threatening or only superficially threatening forms. In such a view, the dominant indeed responds to oppositional pressures, but leaves its basic premises unchallenged and deeper problems unaddressed.[4]

A further historicization of rationalization can be gained by seeing the process as one of setting limits and exerting pressures in a continual, variable struggle between different "sides" that themselves are provisional and often interwoven.[5] Such a view recognizes that opposing formations and practices are not simply eliminated or neutralized, but strategically incorporated and redirected. However, care must be taken not to use this incorporation thesis to set the existing dominant beyond the possibility of challenge or change by positing an already existing, essentialist dominant through which such actions are taken.[6] In contrast, by seeing rationalization as a contested struggle through which the sides themselves are produced, one can avoid both the idealism of undiluted celebrations of oppositional practice as well as the essentialism of taking the determinant context as inviolable and automatic.

Chapter 3, "Philanthropy, Professionalization, and Social-Reform Communications," discusses the challenge posed to market formations by philanthropic support. Such projects refuse the determining imperative of the market to produce commodities. But, in doing so, they also enshrine expertise and expert knowledge. While a number of projects engage key social issues, social relationships of professionalism rely on and reproduce a social division characteristic of market formations.

Chapter 4, "Community Media Projects and Their Containment Through the Mass-Culture Critique," anchors community media as a response to market formations in the historical formation of communities. By refusing the determining imperative of the market to massify audiences, community media projects also constitute a significant challenge. Yet, in doing so, such projects also too often essentialize bureaucratically defined localities, which encourages a corresponding fragmentation, isolation, and containment.

By organizing themselves largely within rationalist empiricism, philanthropic projects and community-media projects take issue with matters of organization rather than with the more substantial problems of form and representation. Such unaddressed problems include the empirical recovery of authoritative information, the collective rationalist mode of expert prescription, the individualist empirical mode of authentic voice or its transitive form of witnessing (documentary), and the collective empirical mode (reporting).

Chapter 5, "Modernism and the Aestheticization of Dissent," discusses a struggle against market formations that problematizes rationalist empiricism at the ontological and epistemological level. What one can call critical modernism refuses not only the determining relationships of the market of commodification and massification, but also of rationalist empiricism as the validating culture of truth. Modernist formations enable a deprofessionalized, popularized critique of the fragmentation and banalization of modern life and consumer society, with collage and montage as key forms of truth production outside the limitations of logical argument and naïve empiricism. The modernist production of truth is by contrast subjective, internal, but driven by ethical, normative considerations and seen to be rooted in a deeper reality of existence and experience. Such a recognition became enacted as a politics of spectacle and demonstration that, by the

1960s, entered general practice. But, in such a challenge, critical-modernist projects also encourage an aesthetic withdrawal from social life and from history.

The separation of these responses to market formations into individual discussions is done here for reasons only of clarity. As each discussion makes clear, each of these responses is interwoven with the others and with market formations in key ways. As only one example, since the 1960s, critical-modernist forms of media practice are often combined with other forms of truth production. General social criticism by the 1960s and more recently takes such varied forms as street marches, billboards and posters, and underground newspapers, all of which contain varying combinations of polemic arguments, naturalist exposés, and modernist critiques. Individual discussions in this part should therefore be taken as an analytically necessary way of organizing what in experience must be seen as an indissoluble complex of responses.

Notes

1. Such a perspective has emerged as a response to pluralist, voluntarist, and functionalist perspectives. A recent example can be found in Peter Frumkin, *On Being Nonprofit; A Conceptual and Policy Primer* (Cambridge, Mass.: Harvard University Press, 2002).

The dominant-ideology thesis is used with variations in works such as Edward H. Berman, *The Influence of the Carnegie, Ford, and Rockefeller Foundations on American Foreign Policy: The Ideology of Philanthropy* (Albany, N.Y.: State University of New York Press, 1983); Donald Fisher, "The Role of Philanthropic Foundations in the Reproduction and Production of Hegemony," *Sociology* 17 (1983): 206–23; and Robert F. Arnove, *Philanthropy and Cultural Imperialism: The Foundations at Home and Abroad* (Boston, Mass.: G.K. Hall, 1980). For summaries and overviews of such an approach and literature, see Judith Sealander, *Private Wealth and Public Life; Foundation Philanthropy and the Reshaping of American Social Policy from the Progressive Era to the New Deal* (Baltimore, Md.: Johns Hopkins University Press, 1997), 8–9; and Jeffrey D. Brison, *Rockefeller, Carnegie, and Canada: American Philanthropy and the Arts and Letters in Canada* (Montréal, Quebéc: McGill-Queen's University Press, 2005). For a response, see Barry D. Karl and Stanley N. Katz, "Foundations and Ruling Class Elites," *Daedalus* 116, no. 1 (1987): 1–40.

2. A fine summary discussion of capture theories as they relate to federal regulatory agencies is Robert Britt Horwitz, *The Irony of Regulatory Reform; The Deregulation of American Telecommunications* (New York: Oxford University Press, 1989), 22–45.

3. While many arguments explore this problem, useful ones for the purposes of the current study include Raymond Williams, "Base and Superstructure in Marxist Cultural Theory," *New Left Review* 82 (November/December 1973): 5–16; and Stuart Hall, "The Problem of Ideology: Marxism Without Guarantees," in *Marx, A Hundred Years On*, ed. Betty Matthews (London: Lawrence and Wishart, 1983), 57–85.

4. Useful works in relation to media and society include Max Horkheimer and Theodor Adorno, *Dialectic of Enlightenment*, transl. John Cumming (New York: Continuum, 1972); Jay Bernstein, "Critical Theory—The Very Idea (Reflections on Nihilism

and Domination)," in *The Frankfurt School: Critical Assessments*, vol. 1, ed. Jay Bernstein (London: Routledge, 1994), 3–38; Jeffrey T. Nealon and Caren Irr, eds., *Rethinking the Frankfurt School; Alternative Legacies of Cultural Critique* (Albany, N.Y.: State University of New York Press, 2002); Shane Gunster, *Capitalizing on Culture; Critical Theory for Cultural Studies* (Toronto: University of Toronto Press, 2004); and David Jenemann, *Adorno in America* (Minneapolis, Minn.: University of Minnesota Press, 2007).

Useful studies that approach regulation as a conflictual, historical process include Horwitz, *The Irony of Regulatory Reform*; and Thomas Streeter, *Selling the Air; A Critique of the Policy of Commercial Broadcasting in the United States* (Chicago, Ill.: University of Chicago Press, 1996).

5. Williams, "Base and Superstructure." Openly acknowledged by Williams as a crucial source for such an argument is the work of Antonio Gramsci. See Antonio Gramsci, *Prison Notebooks* (New York: Columbia University Press, 1991).

6. In addition to Williams, "Base and Superstructure," and Hall, "Marxism," a more recent exploration of the dilemma of such a position is John Holloway, *Change the World Without Taking Power* (London: Pluto Press, 2002).

Chapter 3

Philanthropy, Professionalization, and Social-Reform Communications

In a missive posted in March 2006, Trevor Davis recounts how he started the website Crisis Pictures "by accident." Upon posting on a weblog a series of pictures documenting the carnage in Fallujah, Iraq during the November 2004 U.S.-led major military offensive, the extent of response inspired him to expand the effort. "When a few thousand people wrote me in the space of a week—and ever since—when so many people identified, I became convinced that there was indeed a God, [that] people are good, but [that all] we lack [is simply] the basic information to internalize this."[1] Davis based his web site on a simple strategy: "I could just buy all these pictures that we weren't seeing that told these stories, and publish them every single day."[2]

Davis's subsequent effort to develop Crisis Pictures depicts well the pressures felt when adopting the imperatives of broadcasting. Davis's posting just a handful of pictures had already been a significant act. However, due to his aim to broadcast, the project soon overwhelmed him, for the imperatives of broadcasting created a paradox in which building a larger audience meant exhausting the means of doing so. In the case of Crisis Pictures, success quite literally meant failure. Not only had Davis "never done any [computer] programming or design in my life," he had "spent the last six months programming this elaborate content manager to manage thousands of pictures to tell stories in a way that goes beyond 'another bad day in the Third World'." He even went so far as to incorporate Crisis Pictures as a nonprofit organization. Despite these attempts to make the project manageable, he admitted in his message that "I don't think I can do this anymore without some significant help." By March 2006, the situation had reached the point where "it's Crisis Pictures or me. I am out of money and out of energy." Unlike many nonprofit organizations, which conclude similar notices with solicitations for donations, Davis' note ended with a personal apology, directed to people to whom he felt he had made a promise that he was now unable to keep. "Please understand, I really did my best. If I can find a way to keep going, I will. Otherwise, I'm sorry."[3]

A second commentary that also appeared in 2006 can be productively read as a response to situations such as the one Davis recounts. The article, which appeared in the June 2006 issue of the progressive U.S. publication *In These Times*, proposes a blueprint for what its two authors describe as a long-overdue progressive effort to "reshape the national media landscape."[4] Tired of continually being outgunned by "right-wing funders and ideologues [who] have over the past three decades created their own successful cadre of media and messaging organizations, from think tanks to magazines to radio and television outlets," the authors propose the creation of "a sustainable progressive media infrastructure that can inform and influence public opinion, encourage grassroots action and create political change."[5]

Such an infrastructure, argue the authors, cannot be a serendipitous, spontaneous convergence of separate projects, but a progressive media consortium, described as "a vibrant network of progressive media outlets, reporters, artists, grassroots organizers, politicians, researchers, celebrities, activists, think tanks and citizens working in concert," planned in detail, and intentionally built. Means for doing so include not only seeking increased funding from nonprofit foundations, but also considering "private-sector business models." Such models include commercial media companies' techniques of "amplifying" stories through multiple means of distribution ("a strategy known as 'vertical integration'"), a form of cross-promotion that can "reach new audiences and build momentum around an issue." Rapid-response book publishing paired with grassroots promotion helps build purchasing audiences quickly, and the authors laud various "multi-media masters" such as Farai Chideya, Al Franken, Janeane Garofalo, Juan Cole, and Al Gore for their entrepreneurial ability in spreading their "ideas through progressive and commercial print media, radio, online outlets, books, television appearances and speaking engagements." Groups of "independent journalists and analysts" need to be prepped and promoted to the mainstream media, perhaps through a "national progressive booking agency." National syndication could make available the same content to media organizations across the country. Think tanks need to more fully support "reporters, pundits and bloggers on the front lines of political debate."[6]

Not all progressive activists agree with this plan, the article authors note. Some critics claim that "adopting commercial forms and tactics for their media projects is antithetical to their mission of empowering the disenfranchised and critiquing capitalist culture." The two authors quote Micah Sifry, executive editor of the Personal Democracy Forum, to dismiss such concerns. "The old notion of creating an 'alternative' media in opposition to the 'mainstream'," notes Sifry, "has become meaningless. At a time when anyone can find any article or report from almost any news outlet in the world directly and instantaneously, it makes little sense to marginalize ourselves as 'alternative'."[7] The article's authors agree, concluding that, in order to gain national impact and avoid remaining small scale and irrelevant, "it's time for progressive media makers to [collectively] seize the moment."[8]

When viewed together, these two commentaries place long-standing issues and debates about the organization and practice of democratic communications in high relief. On one side is a refusal to replicate the strategies and tactics of the very mainstream that is seen to have helped produce the society in such dire need of change in the first place. Such efforts see nonprofit organization, volunteer work, and projects inspired by individual if not heroic acts of altruism as the hallmarks of such a refusal. As a recent commentary puts it, basing decisions on whether they pay off financially means "falling prey to the same ills as their corrupt mainstream counterparts." Although many online alternative publications eke out an existence through donations and limited advertising, and rely on freelancers and editors who work often for free, they do so because they "believe in their cause and are willing to engage in it without monetary compensation," with the benefit of "freedom from financial manipulation," and thus "avoiding the pitfall of selling out to the elite."[9]

Indeed, to the extent that democratic communications presupposes a struggle against the market, organizing as a nonprofit or volunteer group seems to be an ideal way to operate outside of market relationships. In contrast to producing capital by producing commodities, grants from philanthropic foundations provide capital from donors (occasionally from benefactors who are immensely wealthy). Such financial support is further removed from capitalist relationships because, in contrast to the imperative of amassing and concentrating capital, the purpose of a philanthropic foundation is to dispense it according to criteria other than whether doing so helps the philanthropy become profitable.

On the other side, however, is the intention of furthering social change by whatever means that work, no matter if such means are inspired by the social-reform nonprofit tradition, the commercial mainstream, or the radical Right.[10] If success means commercial instead of nonprofit, paid instead of volunteer, expert vanguard instead of broad-based collective, and initiated by political intention instead of individual inspiration, so be it. As this view asserts, just as the days of traditional media forms appear to be ending, the days of traditional "single-proprietor" progressive media organizations seem to be as well.[11] To become more effective, so goes this view, it's time to get big and out-do the radical Right and the mainstream at their own game.

Although presented as a novel situation, such a debate draws upon and reproduces long-standing received traditions. Despite their differences, the positions exemplified by Crisis Pictures and by the proposed progressive media consortium agree that, while noncommercial modes of organization have been justified for some time as necessary responses to what are seen as commercialized projects' problems of the corruption of content or limits placed on perspectives and information, haphazardly operated small groups of volunteers are seen as unable to sustain the necessary level and continuity of funding.[12] But, while one side sees tenuous support and informal organization as the necessary cost of not selling out, the other side sees it as a central problem to solve.

An area of even greater implicit agreement concerns the kinds of materials that should be produced in the first place. Both positions implicitly share a deep faith that reasoned arguments and transparent representations of "reality" are the most significant and effective means of authoritative social criticism and political mobilization. Informing Crisis Pictures from the start is the belief that simply seeing the carnage as depicted in photographs triggers a deep, lasting change not only in individuals who viewed them, but perhaps also in the level of social support for such actions and policies. Such an implicit belief is raised to the level of namesake by activist organizations with names such as "Seeing Is Believing" and "Witness."[13] And the selection of personnel (the array of analysts, experts, journalists, and documentary makers) that the progressive media consortium hopes one day to support suggests an emphasis on accounts of and commentary on the real.

In these two ways (traditions of organization and traditions of practice), recovering the embeddedness of philanthropically supported, nonprofit projects and responses to them within larger social formations is crucial for understanding their implications for democratic communications. To the extent that alternative-media organizations are staffed by volunteers, they draw upon traditions of voluntary association. To the extent that they are also nonprofit and seek as well to bring about social change, they draw upon traditions of noncommercial, social-reform organizations. Projects that reject such premises for their ostensible opposite (staffing by paid professionals, organization as for-profit) are related negatively to these same traditions. And, to the extent they advocate interpretive commentary and reporting, they draw upon traditions located in the culture of truth of rationalist empiricism.

Together, these traditions constitute a crucial complex, the implications of which have yet to receive their due attention. While the noncommercial and the commercial are seen today as opposites, a cultural and historical analysis of their emergence and the pressures that shape them suggest otherwise. Philanthropic practices are neither insulated from the market, neutrally benevolent, nor randomly organized. While an influential definition of philanthropies notes their purpose as to "maintain or aid socially useful activities and purposes," exactly in what ways "socially useful" is defined and practiced emerges from larger social formations, from which specific formulations of "socially useful" derive and to which they contribute.[14] Although the imperatives under which philanthropies and nonprofits operate release them from some of the direct pressures of capitalism, they are subjected to other pressures that just as surely shape their activities.

While philanthropic support aids many projects that otherwise would not be able to continue, this chapter argues that communications projects organized as social-reform or social-service organizations reproduce forms of professionalization, which work against democratic practice. Professionalism constitutes a long-standing topic in communications research in such forms as being a strategic ritual engaged in by newsworkers to manage the demands of their job; one half of the structural contradiction between the ostensible autonomy of professions and the deskilling of the workforce; an ideology that seeks to ensure that news coverage will pose minimal challenge to the dominant; or a characteristic development in modern societies.[15] What is of most interest here is not so much professionalism as the process of professionalization, defined as the social production of expertise and the resulting social and cultural division between experts and lay people, only one example of a broad range of similarly undemocratic relations between the powerful few and less powerful many, with analogs in politics (while pluralists talk about leaders and followers, more radical accounts talk about vanguards and the rank-and-file), bourgeois economics (producers and consumers), traditional medicine (doctors and patients), traditional education (teachers and students), and so on.

Despite the comparative inequality of each side, what is essential for an account of the social implications of professionalization is to leave behind a simplistic account of the dominated many by the preformed, already existing, and independently dominating few. Instead, one must address the division between two "sides" as a relationship and as a formation. Instead of initiating a dependency, leading is co-constituted by following, just as producing is by consuming.[16] Similarly, leaders without followers, a vanguard without a rank-and-file, and producers without consumers are not only impotent but nonsensical. However, such points should not be taken to their pluralist extreme. Such relationships are practiced within conditions that enable a focus and concentration of power in particular ways and in favor of particular interests, although not without challenge.[17]

Hybridizing Communications

Professionalization as the social production of experts and lay publics relied on a complex commercial-noncommercial, private-public hybridization of communications practice within an expanding market society. Anglo traditions of philanthropy took shape in early modern England with the emergence of a general system of charity (as opposed to informal, local efforts). With the emergence of such a system, charitable benevolence became politicized—and thus hybridized—by virtue of being institutionalized attempts to preserve the political order.[18] The Elizabethan system of general charity sought to defuse social discontent brought about by the "massive relocation of people from long-established rural communities" to urban areas, which had commenced as a response to the "transition to early modern capitalism."[19] This political intention was institutionalized in the Poor Law of 1601, which legally required individual church parishes to care for people who were unable to care for themselves. If work for them in the parish could not be found, or if they were physically unable to work and no relatives could take them in, parish taxpayers were required to support them.[20]

By the eighteenth century, charitable efforts had become as commercial as they were political and benevolent. Today, collaborations or combinations of charitable, political, and commercial seem novel, due largely to a market-capitalist orthodoxy which argues in favor of the purity of private, commercial initiative in contradistinction to public largesse, particularly since the 1980s and the rise of neoliberal conservatism.[21] However, hybridity is always an historical judgment constituted out of a relationship between the familiar present and the alien past. In their time, what would be seen today as hybrid public/private organizations were the commonplace norm, one that was suited particularly well for addressing large-scale intentions and needs. Variations and examples of the day that McCarthy identifies range from the commonwealth ideal ("the notion that public and private interests were blended under the sanction of a charter") to the method and organization of English revivalist George Whitefield's American tour during the spread of Protestant evangelism in the Great Awakening of the 1730s.[22]

Not only were such hybrid combinations the norm, they were seen as legitimate and necessary, particularly in the newly established United States and particularly for religious organizations, where the constitutional disestablishment of religion removed any possibility of publicly funding religious practice. For evangelical movements, the constitutional mandate against government support of religion encouraged the ecclesiastical to join forces with the entrepreneurial. Religious leaders in the early United States thus "began to compete for members, lacing the country with parishes and networks for fund-raising, proselytizing, and reform." Together with the cultural imperatives of Protestantism (in McCarthy's estimation, "the single most important factor in the growth and elaboration of American philanthropy during the early national period"), the hybridization of commercial means with spiritual ends "provided an enduring institutional framework for social mobilization," whether secular or religious.[23]

The hybridization of the charitable, the political, the commercial, and the spiritual included an emphasis on communications in the service of social reform. As discussed in the prior chapter, such projects were embodied first in Bible tract societies and evangelical publishers. Instead of simply meeting the minimal physical needs of its charges, charitable action starting with the Society for Propagating the Gospel among the Indians and Others in North America sought the broader political goal of incorporating native peoples and isolated settlers into the Kingdom of God. Despite its not being wholly original (having been modeled on the English Society for the Propagation of the Gospel in Foreign Parts, which had been chartered in England in 1701) nor very successful, the importance of the Society in Nord's estimation was in how it "energized a broad movement of missionary and publication work in New England" as well as pioneered "legal, administrative, and organizational arrangements that would become commonplace in the voluntary associational milieu of the new republic."[24]

Such arrangements included those involving communications. As noted in the prior chapter, nonprofit "mass communication" (in the form of evangelical articulations of broadcasting) pioneered modes of organization and forms of technology that soon became commonplace in many areas such as political-party organization, labor activism, abolition, suffrage and women's rights, temperance, moral reform, prison reform, education, and many other instances that together comprised the complex range of "charitable, educational, welfare, advocacy, and religious nonprofit organizations" during the first half of the nineteenth century.[25] The importance of communications to hybridized charitable-political-commercial organizations was magnified during the nineteenth-century explosion of evangelical publishing that followed the lead of the Methodist Book Concern, established in 1789.[26]

In such efforts, commercial means were seen as neutral techniques and thus, at minimum, uncorrupting of noncommercial ends and, at most, essential. An example of communications efforts with a complex relationship with commercialism included the suffrage movement and the case of *The Woman's Journal*, which was founded in 1870 as both a "political polemic" and a "joint-stock corporation." In Finnegan's estimation, capitalist enterprise in the form of the *Journal* was "not only a tool in the battle for enfranchisement, but also an intricate component of the movement itself."[27] Shore makes a similar point about socialist publishers in the United States in the nineteenth and early twentieth centuries, explaining attempts to rely on the capitalist practice of selling advertising space despite opposition to capitalism as the only possible adaptation to the rise of consumer culture and the dearth of other means of adequately supporting large-scale publications.[28]

Social Divisions, Social Reform, and Philanthropy

Social-reform communications became organized through a means derived from evangelism and commercialism, and it also adopted a similar scale of operation and organization. The hybridizing of social-reform efforts was a way of exploring new kinds of social organization that addressed problems and intentions of an entirely unparalleled scale. Through the nineteenth century, an awareness increased of the insufficiency of inherited forms of collective action. Managing or changing society during an era of immense industrial and urban expansion, upheaval, and struggle required means far beyond the capabilities of heretofore small-scale voluntary associations or governments. Drawing on countless precursors variably social, religious, political, and domestic, the form of effective collective action increasingly seen as best suited to confronting such problems and intentions was the large-scale, centralized organization, the most prominent examples of which were the religious denomination; the national political party; mechanics' institutes, associations, trade societies, and unions; and the corporation.[29] As a particularly influential form, corporations early on were formed as part of public-private "special development projects, chartered by legislatures to benefit the community as a whole." Contrary to the present-day legal status of corporations as an individual operating in its own interest, "early corporations embodied a clear public interest."[30] The hybrid public-private nature of corporations began to be remade through judicial action designed to foster competition by federalizing "all commercial law and by denying states the right to act on economic activity" in all but limited ways. As Horwitz argues, due to such changes, "the court unbundled the earlier link between corporations and the public interest," thus laying the legal groundwork for corporations to be "legally recognized as private entities distinct from the public purpose."[31] It is at this point that the division between public and private in the United States became clearly institutionalized.

Unhooking corporations from their public-service obligations paradoxically led to their increasing size, centralization, and influence, which was a concrete form of a "more general process of change" that Trachtenberg calls "the incorporation of America."[32] While "the corporate form reaches back to classical times," with communal or religious purposes embodied by groups such as boroughs, guilds, monasteries, and bishoprics as well as public-private purposes such as forms of joint-stock companies, Trachtenberg argues that the viability of the corporation "for large-scale commercial and industrial enterprises did not become evident until the rapid expansion of trade and manufacturing in Europe and American in the nineteenth century." Beyond matters solely of "the expansion of an industrial capitalist system across the continent," Trachtenberg suggests that incorporation was also a cultural process, "the reorganization of perceptions as well as of enterprise and institutions" whose key components included "a changed, more tightly structured society with new hierarchies of control, and also changed conceptions of that society." Such an organization was based on "the legally established authority of a small group of directors and managers to act in the name of a larger, amorphous body of otherwise unrelated stockholders," thus providing "capitalists with a more flexible and far-reaching instrument than earlier forms of ownership such as simple partnerships and family businesses."[33]

Such pressures in favor of incorporation in this more general sense were also insistent for charitable organizations in the United States and their projects of social reform. These organizations had consisted of a patchwork combination of benevolent societies, social-reform groups such as evangelical efforts, and a politicized benefit program underwritten by the federal government. But, by the late nineteenth century, this haphazard amalgam of programs was generally acknowledged as a failure.[34]

Addressing this failure meant reorganizing charity into larger, corporatized forms that had seem to have proven themselves in business and so many other areas.[35] Such a reorganization was as social as it was economic, with social divisions manifesting themselves in the transformation of voluntaristic social reform into an increasingly professionalized and rationalized social service. Such a transformation was a response to class concerns about mass society and the effects of popular culture in a context of an accelerating social dislocation brought about by the expansion of capitalism, the increasing concentration of wage labor in the cities, and waves of immigration.

By the turn of the twentieth century, reformers were persistently apprehensive about the potential for social upheaval, and they sought means of harmonizing differences within the current framework, thus fitting what many historians judge as the reformist rather than the radical character of what became Progressive-era activism.[36] Charitable organizations of the nineteenth century reflected the class basis of such concerns, peopled as they were by generally middle- and upper-class reformers who hoped to inculcate the masses with proper, middle-class American values.[37] The long-standing class basis of charity is captured well by Lubove in his characterization of "efficient charity" in the late-nineteenth century as that which "involved the direct influence of successful, educated, and cultured representatives of the middle class upon the dependent individual or family."[38]

The gradual incorporation of charity and its reformist as opposed to radical orientation enabled the emergence by the early twentieth century of professional social service as a form of public intervention. Reformist projects then and now seek to accommodate, fix, or otherwise help their "clients" fit into existing conditions or to help the existing order work better, instead of challenging the very basis and legitimacy of these conditions and existing order. By contrast to being "the practice and the prerogative of the many" and a means by which "men and women, rich and poor, black and white publicly contested for authority and power during the nation's youth," McCarthy argues that this voluntarism and inclusiveness that characterized nineteenth-century philanthropy was gradually replaced by the twentieth-century view of philanthropy as lavish giving by the fabulously wealthy in professionally administered programs.[39] While one may argue whether such a characterization inflates the degree of inclusiveness and pervasiveness of philanthropic practice in the nineteenth century, what is less arguable is the transformation of the philanthropic enterprise to one increasingly corporatized and thus more exclusive, professionalized, and rationalized.

The incorporation and specific class basis of charitable work directly challenged evangelical programs, which played so crucial a role in the conception, emergence and development of broadcasting and mass communication. The basis of the challenge was a philosophy organized around the Social Gospel, which refers to a range of positions that were inspired by the rise of a humanist, rationalist iteration of Protestant Christianity as an alternative to fundamentalism and evangelism, as well as by the European revolutions of 1848, the Paris Commune of 1870, and the Chartist movement in England.[40] The Social Gospel differed from traditional alms or programs of benevolent support by combining a Christian responsibility for altruistic help with an understanding of social as opposed to individual causes of problems that required this help.[41] As a result, reform movements inspired by the Social Gospel worked within the hybrid religious-secular associational tradition while also relating the need for moral improvement to more comprehensive programs of social reform.

This movement was more precisely a set of movements, which differed in some key areas, an important one being that between a more "conservative, individually oriented strain of social Christianity" and more "radical expressions of Christian socialism." However, most adherents in the United States of these originally English positions were "more concerned with the sentimental aspects of socialism—cooperation and brotherhood—than with the more radical aspects—collective ownership of land and capital."[42] Even the more radical ones saw socialism as "primarily a religious movement" that sought to address "neglected aspects of the church's social responsibility."[43]

A key means by which social-reform movements were becoming increasingly professionalized was through an increasing reliance on social science. By contrast to evangelical programs which saw sin as an inherent stain that could only be reconciled by the grace of God, sin was for those working under the Social Gospel a matter not only of personal responsibility but of social equality and justice, all of which needed to be addressed not simply by the individual action of accepting God, but through social action underwritten, as Agnew characterizes it, by "a 'bridge theology' that linked historical Christianity with modern social science."[44] This bridge theology underscored the already existing emphasis on expert direction of such efforts, with the emergence of "scientific philanthropy" as professional projects of social engineering solidifying the imperative for expert, rationalized programs.[45]

The work of social reform thus emerged as a means of holistically and professionally applying moral certitudes and scientific methods to social problems. It was seen generally as the scientific study of problems of urbanization and modern society, to be solved through rationalization, expert direction, and greater efficiency. In this professionalized social work, social engineering was seen as the province of skilled experts who drew from newly created social science as well as from examples of its application in business and manufacture. Whether seeking to Americanize immigrants or to boost assembly line productivity, social engineering was "dedicated to the use of new instruments being forged by the newly emerging social order itself, tools already being used with stunning success in business and industry."[46]

A key social result of the application of such techniques was the production of a clear distinction between experts and their "clients."[47] As Kirschner summarizes, many Progressive-era reformers pictured "the professional as they pictured a scientist—rational, objective, dispassionate, and generally disengaged socially" from the problems to which they attended. Reformers "saw themselves as a kind of priesthood defined by their monopoly of knowledge and expertise required to steer society" toward harmony.[48] Saveth concurs by discussing reformers' intention to "make 'deserving poor' of the 'undeserving poor' or, optimally, to make of them what James Bryce characterized [at the time] as 'well conducted men of small means.'" The political intention behind the reform of manners, habits and temperament was to immunize the deserving poor against what reformers saw as the ever-present threats of "venal politicians and the appeals of demagoguery."[49] Yet, such intentions were hardly a simple case of brute domination. Most reformers were in Kirschner's estimation "sincerely touched by the plight of the poor," yet many also "feared the wrath of the wretched or recoiled from the cultural norms of the foreign, and wanted to improve them by altering their behavior." Kirschner continues by pointing out that, while social control was clearly reformers' aim, "most of them saw no conflict between welfare and control. They saw these as complementary methods in a seamless quest to reintegrate a fragmented society."[50] Not everyone shared this view, however. By the 1920s, while some reformers sought to make "manipulation itself more sophisticated, more professional," thus trying to "reshape democracy so that the new professional elitism could function as part of the system," others "worried that democracy and professionalism were intrinsically incompatible and contemplated ways to relieve the tension between them without seriously modifying either [one]."[51]

But from where were the resources to come to pay for such comprehensive programs of social reform? Government in these days of minimal taxation could hardly be the answer. And, because the source of problems was seen as unfettered capitalism, the source of funds could not be capitalism, at least directly. Due in part to the drawbacks of each of these possibilities, newly emerging philanthropic foundations were an increasingly favored means of funding social reform outside of what were seen in the wake of Populist and Progressive challenges to "the Trusts" as irretrievably compromising capitalist relations. Philanthropic support seemed to avoid conflicts of interest that might arise or at least be suspected between companies' interests and the causes they supported financially.

While the first privately funded philanthropic foundations such as the Peabody Education Fund (1867) and the John F. Slater Fund for the Education of Freedmen in the South (1882) addressed narrowly focused social needs in the wake of the War Between the States, the establishment of the Carnegie Institution of Washington (1902) pioneered the philanthropy as a source of social-reform funding that was meant to be insulated from capitalist agendas or pressures, probably doing more, as Kiger argues, "toward setting up the foundation pattern of giving in the United States" than any other effort before or since.[52] The highly influential pattern it set emphasized attention to "major social problems" with the resulting task, as Karl and Katz summarize it, to use "scientific understanding, and [to promote] the application of that understanding to virtually every field of inquiry."[53] Carnegie's action was paralleled by organizations funded by John D. Rockefeller such as the Rockefeller Institute for Medical Research (1901) and the Rockefeller Foundation (1913), and was soon matched by "other rich men and women" who, as Bremner summarizes, "sought practical, socially useful ways of disposing of surplus wealth."[54] Given this pattern established early on, it is not surprising that support for the establishment of professional social work came from the Russell Sage Foundation, which "viewed the promotion and professionalization of social work as one of its chief 'progressive' projects." Its vision of social work was clearly reformist in the sense of helping to "rationalize social reform and channel it into the development of social welfare services rather than a more fundamental redistribution of wealth or power."[55]

Despite all the differences between voluntaristic charity and professionalized, philanthropically funded programs of social service and social reform, what remained common between them is the ceding of power to a leading minority, made possible through the rationalization and professionalization of reformism.[56] Such an emphasis has clear implications for democratic practice to the degree it reproduces now what it did in the early twentieth century, which was a stress on "the expertise of the professional and a service ideal which underscored the unequal relationship between the parties to the service transaction."[57]

Reportage and Documentary

While many nonprofit projects of democratic communications work within the tradition of philanthropically funded professionalized social service, their practice does not consist of case studies and counseling, but of producing communications in the hopes of enabling social change. To better understand the connection between traditions of professionalized social reform and nonprofit projects of democratic communications, a contextual analysis of the organizational form needs to be complemented by a cultural analysis of communicative practice.

As noted in Chapter 1, Anglo-American social movements of all kinds relied upon the enabling cultural forms of prophecy and polemic. From Puritanism, to Putney, the Declaration of Independence, the Federalist Papers and to reform movements of the nineteenth century—all were based on moral, normative arguments by individual writers who worked within the hybridized traditions of Providentialism and the rationalist tradition of logical argumentation.[58]

Such cultural practices relied for their authority not only on the status but also on the skill of the individual writer, and were widespread in the political press, movement speeches (which were often reprinted), as well as in pamphlets and newspapers produced by a wide variety of movements, associations, and societies.[59]

However, such individualist forms were of little use for those who hoped to grasp objectively (and, later, scientifically) the "true" nature of the problems and issues in a society undergoing unprecedentedly rapid industrial, economic, and political expansion along with equally unprecedented social change. Programs of comprehensive, wide-ranging social reform needed not only the will, resources, and expert direction, but a means of grasping and documenting the large-scale changes and problems beyond individual sensory apprehension and normative assertion. To address such intentions, reportage and the newly fashioned rationalist-empiricist means of statistical profiling relied upon, encouraged, supported, and valorized the knowledgeable professional. Rationally argued policy statements persisted as a key part of modern philanthropic foundations' projects, but they drew for their authority on the statistical report and the documentary, both of which embodied the empiricist rather than the rationalist pole of rationalist empiricism.

As practiced through and by virtue of philanthropic foundations, reportage and documentary had and have their characteristic social implications. Where the research report is a collective enterprise that relies for its authority on the scale, scope, and quality of data upon which it is based, the documentary is more often seen as an individualist enterprise that relies for its authority on the seemingly incontrovertible evidence of having been there to witness, itself relying in turn on an understanding of representation as a mimetic instead of a productive practice.[60] Statistics increasingly became a means of comprehensive representation, with testimony and documentary a means of specific understanding, with both composing a dual-focused complex of simply two different ways of viewing the same problem. As such, they were (and, in bureaucratic uses, remain today) deeply mimetic.[61]

Precise, scientific, and authoritative techniques of summary and abstraction had been initiated in such forms as population censuses, but were enhanced and expanded by the development of social statistics and the bureaucratic infrastructure needed for their gathering, processing, and analysis. In the United States, this effort is marked by such events as the establishment of the Statistics of Income Division of the Department of the Treasury in 1866, the Economic Research Service and the National Agricultural Statistical Service of the Department of Agriculture in 1867, the National Center for Educational Statistics of the Department of Education also in 1867, and the Bureau of Labor Statistics of the Department of Labor in 1884.[62] Such techniques went beyond gathering and cataloguing information. Consistent in many ways to Francis Bacon's seventeenth-century vision of a division of labor contributing to scientific understanding, social improvement, and the development and expert management of capitalism, the rise of governmental statistical bureaus enabled an authoritative means of embodying in a particular way an immense range of experience while also envisioning equally large-scale interventions.[63] Such federal efforts inspired social surveys carried out by social reformers who were also trained in social science. For example, in 1897 the Federation of Churches and Christian Workers in New York City began to conduct collective, systematic censuses of specific neighborhoods that "gave the Federation a tremendously valuable and significant fund of authoritative sociological data" for use in its various reform programs.[64]

However, programs of scientific philanthropy also required the contrary but complementary development of evocative and equally authoritative forms of personalization and contact, which were met through the development of naturalist, episodic recounting of individuals' acts, to become what was known by the early twentieth century as documentary. Where reportage refers to the report as expert-driven policy research as well as the practice of news reporting, documentary is often posed as a closely aligned, visual analog to the print practice of reportage although, with the emergence of radio news and video news, such a distinction is not entirely consistent.

Underwriting both reportage and documentary is the professional practice of objectivity, itself inherited from rationalist empiricism and its norm of scientific disinterestedness.[65] However, instead of a set of ideas or a professional creed, objectivity can be understood more fundamentally as a practice of objectification. It is the activity of turning the world—including the clients of social reform—into separate objects of contemplation and study, thus dividing and separating the world from the constitutive relationships of its apprehension and human making.

Upon its emergence in the 1920s, documentary referred to a kind of factual account grouped with "scientific, cultural, and sociological film"—all of which were implicitly opposed to fiction and entertainment film. Along with this distinction was a second one concerning epistemological value, in that a documentary film "must be a transcript of real life, a bit of what actually happened."[66] This second distinction regarding the epistemological value of documentary draws upon the correspondence between the documentary film and the document, which was seen by the nineteenth century as an object and account possessing evidentiary value with which to inform reasoned judgments, thus recalling the emergence of conventions of factual representation in sixteenth- and seventeenth-century England as described by Shapiro and others and its place within the culture of truth of rationalist empiricism.[67] However, documentary in its visual form was much more than functionally equivalent to a commercial receipt or a transcript of a legislative debate. Tagg discusses how the earliest documentary photography "transformed the flat rhetoric of evidence into an emotionalized drama of experience that worked to effect an imaginary identification of viewer and image."[68]

By the late nineteenth century, the faith in representation as mimetic—as a transparent depiction of truth—was thus well-entrenched, with documentary a dominant mode of social reform due to its embodiment of such rationalist-empiricist notions of truth. Whether for dominant, reformist, or radical intentions, documentary grew to be a cultural form of substantiating claims for as well as against change.[69] For example, reviewers of journalist Jacob Riis's groundbreaking book published in 1890 titled *How the Other Half Lives*—an effort to document what Riis calls in the book "the bald facts with which we have to deal in New York"—emphasized the empirical basis of the book's veracity.[70] The reviewer notes that the book is the result of "one of a practical turn of mind who describes exactly what he sees," with Riis lauded for going "into the slums with his camera and flash light, and in his illustrations he presents what the photograph has produced on the plate" without sentimental wishful thinking.[71] Such unflinching efforts to document conditions can also be seen in the early work of the Chicago School of sociology as the recovery of life histories within specific conditions, its application in the settlement-house movement, and the reigning aesthetic of naturalism, all of which provided a key basis for judging documentary as politically serious work.[72] By the 1930s, the federal government was increasingly involved not only as a collector and complier of statistics but also as a producer and distributor of documentaries, using its capacity to further not only patrician programs of social reform but also to document its programs' effectiveness for skeptical legislators.[73]

It was through such cultural forms and the practices which constitute them that philanthropies and their expert programs of social reform, reporting, and documentary emerged a significant cultural formation. It was and continues to be both an embodiment and a practice of deep faith in rational evaluation and empiricism and the resulting expert ability to guide the development of administrative solutions and their application to problems of social reform.

Social-Reform Communications Today

A way of exploring the practice of social-reform communications today as a professionalized project is through an analysis of its institutional imperatives. Despite the variety of combinations, the limits and pressures of rationalist empiricism that shape philanthropic foundations and the social-service communications projects they support professionalize the practice. The ways in which such a system and set of priorities is not only fully institutionalized but self-perpetuating can be seen in an analysis of national patterns of philanthropic spending. What must be noted first about philanthropic support for communications projects is the lack of grantmaking for such projects.[74] Of grants given in

2005 (the most recent year for which information is available) by 1,154 larger U.S. foundations, one percent ($157.9 million) went to communications projects.[75] As small as this is, it actually represents a modest increase from 2004, in which less than one percent ($132.4 million) had gone to communications projects.[76] To place the amount given to communications projects in 2005 in some kind of comparative perspective, the greatest proportion of grants (12.2 percent, $2 billion) went to support general and rehabilitative health. Ten times the amount given for communications projects went to higher education (9.6 percent, $1.6 billion), and four times as much (4.2 percent, $682 million) to museums, which was the category receiving the highest value of grants in the "Arts and Culture" category in which media and communications were also placed.[77]

What is also clear is the emphasis on supporting expert and professional organizations and practices. Of grant money that went to the fifty top recipients of grants for media and communications in 2005 (the most recent year for which this more detailed breakdown is available), the vast majority of funds—sixty percent ($110 million)—went to major public broadcasters, either urban stations such as WGBH (Boston), WETA (Washington D.C.), and KCET (southern California); statewide public-broadcasting organizations such as Minnesota Public Radio and Iowa Public Television; production organizations and program providers such as National Public Radio, Firelight Media, and Public Radio International; and umbrella funding organizations such as the PBS Foundation and the Educational Broadcasting Corporation. Fifteen percent of this grant money ($28.5 million) went to organizations and trade groups that either supported or addressed existing professional communications industries, such as the International Center for Journalists, the Advertising Council, and the Reporters Committee for Freedom of the Press. Organizations that promoted the arts through advocacy, and providing resources and support received 12 percent of foundation funding ($22.2 million). Three percent ($6.2 million) went to university programs.[78]

Of the remaining nine percent of funds ($17.1 million) for the top fifty recipients in media and communications in 2005, most went to a combination of seed-granting organizations and specific publishing projects. Only Youth Radio (an effort that provides free media training to economically disadvantaged young people, yet with the intention of offering professional training), Recording for the Blind and Dyslexic, and the Braille Institute of America addressed to any degree the concerns of clients who are initially outside the orbit of industry and professionalism. To support these programs for extending participation, in 2005 they together received $4.5 million—a significant amount to be sure, but 2.5 percent of the funding of top fifty grant receivers in media and communications. By contrast, Educational Broadcasting Corporation (the top recipient of the group of fifty) received $13.3 million, which is 7 percent of the funding provided to top-fifty organizations that year.[79] In sum, close to 98 percent of foundation funding to the top-fifty recipients for 2005 promoted, assisted, or provided programming for existing industry and professional groups and networks.

Not only are organizations themselves changed through the process of seeking grants, the grant-funding system as a whole perpetuates itself by channeling philanthropic support to social-reform communications projects that already have the appropriate professional, reformist credentials, or that begin to take on such a form as a result of the need to gain support. As Dowie argues, the more important power exercised by foundations is not necessarily the amount of money controlled (hardly rivaling that controlled by for-profit companies or by governments), but the way in which they structure, define, lead, guide, direct, and manage the priorities of the entire nonprofit sector—what Dowie calls the "leverage" gained by making a grant conditional on obtaining additional grants or other forms of support. Such a tactic has long been a means of forcing grantees to "establish a wide base of foundation support."[80] Appropriate credentials are most often earned through already having gained other grants, which itself is premised on an accepted means of validation by credible peers. A sense of decision-making priorities is suggested in a guide for for-profit companies who wish to set up a partnership with a nonprofit organization for marketing purposes. The authors of the guide emphasize the need to "*pick a partner that is credible in its field* [emphasis original]." Suggestions for assessing credibility range from whether it has had "substantial accomplishments in its field," whether it is "governed by recognized experts," whether it has already received "grants from reputable sources," and whether the majority of its funds are used for its programs—all of which favor organizations that exhibit the appropriately professional features and that have the resources to pursue such funding or who have already secured it.[81]

A second way of exploring the practice of social-reform, professionalized communications projects is through an analysis of cultural form. Such an analysis suggests that the conception that social-reform communications projects have of cultural practice draws centrally upon rationalist empiricism and the validation of expertise and often elaborate technical skills. While improvements in usability and comparative decreases in cost of means of communications lower somewhat the threshold of minimal competency and availability, social-reform communications nevertheless aspires to professional levels of accomplishment with its corresponding exclusionary emphasis. For example, such an aspiration forms the basis of a foundation's evaluation of video from nonprofit, social-reform organization Witness.org. "There is no trembling hand, amateurish, home-movie quality about Witness films," the evaluator writes. "Its product is compelling, professional, and creative with a proven capacity to deliver concrete change."[82]

As one form common to social-reform nonprofit organizations, research reports reproduce the disengagement of professional academic work, often shading into technocratic recommendations for optimizing organizational longevity and effectiveness. One recent report stressed the need for media activists to "build a field," described as the expert identification of "promising groups and/or individuals," assessing the efficiency and effectiveness of their organizations, then embarking on expert-directed "investment" in a "range of leadership and organizational strengthening activities."[83] Such needs in the United States are increasingly being responded to as well as promoted by the emergence of trade publications directed at professional activists, such as *Philanthropic World*, *Philanthropy Magazine*, and *Chronicle of Philanthropy*.

Reports of meetings regarding public media are similarly professionalizing, as seen in a 2005 meeting of 70 professional media makers, policy experts, and professional activists. The report suggests that emphasis within a system of public media be placed on "finding quality media that is of interest and/or serves a particular need." The intention of increasing audience size and the resulting impact on what are called "targeted audiences and the general public" is to be addressed in the form of "the aggregation of content by 'trusted sources' and strategic marketing and promotion," all of which base themselves in the vanguard practice of broadcasting.[84]

Research reports are directed at and meant for professional activists and their financial supporters, embodying the authority of disengaged experts and through it expert claims about the world. For example, reports from a variety of academics and activists submitted as a part of the program "Necessary Knowledge for a Democratic Public Sphere" organized by the Social Science Research Council and funded by the Ford Foundation define the needs, range, and nature of a "robust knowledge infrastructure for public interest media." However, based on the list of who was allowed to contribute, its recommendations excluded contributions at least in this stage from the subjects of this effort—namely, members of the public themselves.[85] Manuals and self-help guides are a popularized form of research report, with exhaustive explanations and itemizations reduced to expert directives, advice, and recommended generalized procedures.[86] For example, a series of publications from the Office of Communication of the United Church of Christ seek to "help educate the public about the role media plays [*sic*] in shaping current events," with a media empowerment manual providing "information about some of the many media justice organizing, policy and education drives going on around the United States" as well as how to "recruit people, how to organize your group, run meetings," "researching your media," and "several resource guides on actions you can take."[87]

Documentaries range from the individual report and its appeals to ethical and moral conscience, to the collective, naturalized report of an alien situation. Documentaries "take you around the world," reveal the story behind the actions, "show" and "capture" individuals' experience, show what really happened in order to "fight lies with truth," expose "the public to perspectives excluded from the corporate-owned media," and "bring our readers the stories that the corporate press never prints."[88] Their claims to authority are based on the authenticity of witnessing and on the expert skill of their writers as guides to knowledge, with representation taken as a transparent means of revealing if not revelation. Claims of authority rely on the institutional position of the maker, without questioning the form of what is made. For example, the activist news organization SourceCode asserts that, while "there is no longer any official media able to present an adequate version of reality," that means "if we want real information, we'll have to do it ourselves."[89] More immediate than documentaries are varieties of reportage, which often are repackagings of materials originated by other organizations. Others originate their own items largely by soliciting freelance work that meets professional technical and ethical guidelines.[90] While working within a different time scale, they rely nevertheless upon the same rationalist-empiricist claims to authority and capabilities of communications and representation.[91]

Commentaries are an individual report and moral appeal, with the source of authority the author's personal experience and history. They are an invitation to empathize and concur, with the means of communication seen as a direct expression of individual experience. Professional commentaries claim authority not simply by revealing what has been hidden, but also through the status of the writer as well as the keenness of argument. For example, the website TomPaine.com (whose namesake is, of course, the eighteenth-century radical-republican pamphleteer of *Common Sense* fame) sees itself as going "beyond the news to deliver well-informed, provocative, and credible progressive perspectives."[92] Common Dreams NewsCenter provides readers with an "eclectic mix of politics, issues, and breaking news" with an emphasis on commentary.[93] Nonprofessional commentaries claim the authority of experience and personal history, and thus rely on an engaged moral claim rather than one disengaged and forensic. Antiwar activist Cindy Sheehan's widely reprinted commentaries are one example, with her cultural authority based not on expert credentials, but on her experience as the mother of a United States soldier killed in Iraq.[94] Nonprofessional commentaries often shade into memoir due to their retrospective mode, with a current example in the United States being StoryCorps, an effort to create a collection of 250,000 individual stories that, by listening to them, "celebrates our shared humanity and collective identity."[95] The distinctiveness claimed for weblogs and listservs is more due to the technology than the form, which is a variation of serial commentary by nonprofessionals, and thus which relies upon the same claims to authority as commentary.

To address the social divisions of such practices, a number of attempts seek to rethink and remake the public isolation and disengagement from social-reform communications projects. The "grassroots political documentary" as developed by the Brave New Foundation coordinates 1,200 volunteer field producers, researchers, and organizers for the production and distribution of its various documentaries.[96] As an example, it drafted eight self-described "middle-aged citizens from different backgrounds," asking them to watch three months of Fox News and to offer their own sense of key recurring techniques that the newscast used. At the same time, the production company recorded the newscast around the clock, then used the volunteer researchers' observations to develop the key points of analysis embodied in the documentary released as *Outfoxed: Rupert Murdoch's War on Journalism*.[97] House parties, in which hosts agree to show a film or stage a meeting, seek to enable collective viewing and engagement.[98] Other uses of house parties as a means of sales and distribution as well as collective viewing include the environmental group The Sierra Club and its series of documentaries "Sierra Club Chronicles," and the producers of *The Ground Truth*.[99] Film Aid International is a recent iteration of the participatory documentary movement which gained new currency in the 1960s. It brings mobile motion-picture screens to refugee camps throughout the world, not only screening documentary and public-service films, but also enabling refugees to "tell their own story" in its Participatory Video Project.[100] An attempt to engage participation as the act of experiencing instead of just watching and listening can be seen in the project Sweatshopper. Its maker produced a series of downloadable podcasts critical of Wal-Mart. Instead of critical commentary, they are described as "a soundseeing tour designed to be experienced in your local Wal-Mart." The maker directs people who have downloaded her tour and who have a portable MP3 audio player to "drive to the store and press play after you have parked, but are still in your car. Follow the instructions as you walk through Wal-Mart, hearing, seeing and doing things you would not normally do!"[101] In this way, listeners hear her narration and thoughts about what she saw and touched while making the recording, with the listener seeing and touching the same things.[102]

Social Reform and the Problem of Democratic Communication

Despite these and many other innovations, to the extent that activist communications is seen as requiring professional direction and expertise, as requiring as a result significant resources and a large-scale organization, with all deriving from a commitment to democratic communications as a practice of broadcasting, such projects labor under the premises of a distinctly undemocratic conception and practice paradoxically in the service of an expanding, democratic one. Notable reform efforts have been and continue to be enabled, sustained by often immense degrees of personal commitment and belief. Yet, to the degree that social-reform communications is practiced in this way, not only does it continue to professionalize the practice and produce the social relations of expertise, it encourages an instrumental view of the purposes of a social movement and the devaluation of (unmanaged) popular participation.

A crucial response to these and other drawbacks of incorporated, centralized forms of social-reform communication is located in what can be called community media projects. While they refuse large-scale incorporation, professional elites, and the resulting immense need for support, the traditions and pressures that enable and shape community media derive nevertheless from social reform and commercialism, which work spatially to isolate and thus balkanize such activism. It is to this set of traditions, pressures, and implications, which together comprise another form of the struggle against market formations, to which we now turn.

Notes

1. Trevor Hugh Davis, "Coda?," *Crisis Pictures* <http://crisispictures.org/AFG/354/> (8 Mar. 2006).

2. Davis, "Coda?."

3. Davis, "Coda?."

4. Jessica Clark and Tracy Van Slyke, "Welcome to the Media Revolution; How Today's Media Makers Are Shaping Tomorrow's News," *In These Times*, 28 June 2006 <http://www.inthesetimes.com/site/main/article/2687/> (1 Oct. 2006).

5. For an overview, see Richard Viguerie and David Franke, *America's Right Turn: How Conservatives Used New and Alternative Media to Take Power* (Chicago, Ill.: Bonus Books, 2004).

6. Clark and Van Slyke, "Welcome to the Media Revolution."

7. Clark and Van Slyke, "Welcome to the Media Revolution."

8. Clark and Van Slyke, "Welcome to the Media Revolution."

9. Jason Miller, "Liberating the 'Liberal Media'," *ZNet*, 4 May 2005 <http://www.zmag.org/content/print_article.cfm?itemID=7780&sectionID=21> (29 Jun. 2005).

10. Indeed, progressive organizations such as the New Progressive Coalition see the radical Right as an exemplar of what progressive organizers should do. See Judith Siers-Poisson, "Funding the Left," *Center for Media and Democracy*, 28 July 2006 <http://www.prwatch.org/node/5034> (13 Oct. 2006).

11. Anthony Kaufman, "For a Few Dollars More; As Technologies Evolve and Priorities Shift, Media Nonprofits Struggle to Stay Afloat," *Village Voice*, 11 April 2006 <http://www.villagevoice.com/generic/show_print.php?id=72822> (31 Aug. 2006).

12. A collection of United States media criticism is Robert W. McChesney and Ben Scott, eds., *Our Unfree Press: 100 Years of Radical Media Criticism* (New York: Free Press, 2004).

13. "Seeing Is Believing; Using New Technologies to Change the World," *Seeing Is Believing* <http://www.seeingisbelieving.ca/> (12 September 2006); "Witness: See it. Film it. Change it," *Witness*, <http://www.witness.org/> (20 Mar. 2006).

14. Joseph C. Kiger, *Philanthropic Foundations in the Twentieth Century* (Westport, Conn.: Greenwood Press, 2000), 1.

15. Simon Cottle, "Media Organization and Production: Mapping the Field," in *Media Organization and Production*, ed. Simon Cottle (London: Sage, 2003), 3–20; Daniel Schiller, *Theorizing Communication; A History* (New York: Oxford University Press,

1996); Edward Said, *Covering Islam; How the Media and the Experts Determine How We See the Rest of the World*, rev. ed. (New York: Vintage, 1997); Edward S. Herman and Noam Chomsky, *Manufacturing Consent; The Political Economy of the Mass Media* (New York: Pantheon, 1988); Michael Schuson, *Discovering the News; A Social History of American Newspapers* (New York: Basic Books, 1978); William Solomon, "The Site of Newsroom Labor; The Division of Editorial Practices," in *Newsworkers; Toward a History of the Rank and File*, ed. Hanno Hardt and Bonnie Brennen (Minneapolis, Minn.: University of Minnesota Press, 1995), 110–34.

16. The mutual constitution of such oppositions was recognized some time ago by thinkers such as Marx. See Karl Marx, *A Contribution to the Critique of Political Economy*, transl. S.W. Ryazanskaya (New York: International Publishers 1970), 196.

17. An earnest attempt to develop a conception of social challenge other than the vanguard seizure of state power (thus replicating the very kind of society that is challenged) is John Holloway, *Change the World Without Taking Power* (London: Pluto Press, 2002).

18. Useful discussions of long-standing traditions of charitable practice include Demetrios J. Constantelos, *Byzantine Philanthropy and Social Welfare*, second revised ed. (New Rochelle, N.Y.: Aristide D. Caratzas, 1991); Demetrios J. Constantelos, *Poverty, Society and Philanthropy in the Late Mediaeval Greek World* (New Rochelle, N.Y.: Aristide D. Caratzas, 1992); Robert H. Bremner, *Giving; Charity and Philanthropy in History* (New Brunswick, N.J.: Transaction Publishers, 1994); J.B. Schneewind, ed., *Giving; Western Ideas of Philanthropy* (Bloomington, Ind.: Indiana University Press, 1996); Warren F. Ilchman, Stanley N. Katz and Edward L. Queen II, eds., *Philanthropy in the World's Traditions* (Bloomington, Ind.: Indiana University Press, 1998); and Mark R. Cohen, *Poverty and Charity in the Jewish Community of Medieval Egypt* (Princeton, N.J.: Princeton University Press, 2005).

19. Anthony Brundage, "Private Charity and the 1834 Poor Law," in *With Us Always; A History of Private Charity and Public Welfare*, ed. Donald T. Critchlow and Charles H. Parker (Lanham, Md.: Rowman & Littlefield, 1998), 99. See also Thomas Max Safley, *The Reformation of Charity: The Secular and Religious in Early Modern Poor Relief* (Boston, Mass.: Brill Academic Publishers, 2003); and Paul A. Fideler, *Social Welfare in Pre-Industrial England: The Old Poor Law Tradition* (New York: Palgrave Macmillan, 2006).

20. David C. Hammack, ed., *Making the Nonprofit Sector in the United States; A Reader* (Bloomington, Ind.: Indiana University Press, 1998), 9.

21. Kathleen D. McCarthy, *American Creed; Philanthropy and the Rise of Civil Society 1700–1865* (Chicago, Ill.: University of Chicago Press, 2003), 3.

22. McCarthy, *American Creed*, 21–2. See also Barry Hankins, *The Second Great Awakening and the Transcendentalists* (Westport, Conn.: Greewood Press, 2004).

23. McCarthy, *American Creed*, 49, 25.

24. David Paul Nord, *Faith in Reading; Religious Publishing and the Birth of Mass Media in America* (New York: Oxford University Press, 2004), 27–28.

25. McCarthy, *American Creed*, 2, 98–120; Hankins, *Second Great Awakening*, 85.

26. Nord, *Faith in Reading*, 98; Candy Gunther Brown, *The Word in the World; Evangelical Writing, Publishing, and Reading in America, 1789–1880* (Chapel Hill, N.C.: University of North Carolina Press, 2004), 46.

27. Margaret Finnegan, *Selling Suffrage; Consumer Culture and Votes for Women* (New York: Columbia University Press, 1999), 140.

28. Elliot Shore, *Talkin' Socialism; J.A. Wayland and the Role of the Press in American Radicalism, 1890–1912* (Lawrence: University of Kansas Press, 1988), 94–114.

29. Stephen Skowronek, *Building a New American State; The Expansion of National Administrative Capacities, 1877–1920* (New York: Cambridge University Press, 1982); Bruce Laurie, *Artisans into Workers; Labor in Nineteenth-Century America* (New York: Noonday Press, 1989); Josiah Bartlett Lambert, *"If the Workers Took a Notion; The Right to Strike and American Political Development* (Ithaca, N.Y.: Cornell University Press, 2005).

30. Robert Britt Horwitz, *The Irony of Regulatory Reform; The Deregulation of American Telecommunications* (New York: Oxford University Press, 1989), 49.

31. Horwitz, *Irony*, 49.

32. Alan Trachtenberg, *The Incorporation of America; Culture and Society in the Gilded Age* (New York: Hill and Wang, 1982).

33. Trachtenberg, *Incorporation*, 3–6.

34. Donald T. Critchlow and Charles H. Parker, "Introduction," *With Us Always; A History of Private Charity and Public Welfare*, ed. Donald T. Critchlow and Charles H. Parker (Lanham, Md.: Rowman & Littlefield, 1998), 5.

35. A critical account of a current example of this development is Samantha King, *Pink Ribbons, Inc.; Breast Cancer and the Politics of Philanthropy* (Minneapolis, Minn.: University of Minnesota Press, 2006).

36. Richard McCormick, *The Party Period and Public Policy: American Politics from the Age of Jackson to the Progressive Era* (New York: Oxford University Press, 1986), 313; Don S. Kirschner, *The Paradox of Professionalism; Reform and Public Service in Urban America, 1900–1940* (Westport, Conn.: Greewood Press, 1986), 27–28.

37. The literature on Progressivism is vast. Recent, notable works relevant to this argument include Maureen A. Flanagan, *America Reformed: Progressives and Progressivisms, 1890s–1920s* (New York: Oxford University Press, 2007); Michael McGerr, *A Fierce Discontent: the Rise and Fall of the Progressive Movement in America, 1870–1920* (New York: Oxford University Press, 2005); James Livingston, *Pragmatism, Feminism, and Democracy: Rethinking the Politics of American History* (New York: Routledge, 2001); and Katherine Adams, *Progressive Politics and the Training of America's Persuaders* (Mahwah, N.J.: Lawrence Erlbaum Associates, 1999).

38. Roy Lubove, *The Professional Altruist; The Emergence of Social Work as a Career 1880–1930* (Cambridge, Mass.: Harvard University Press, 1965), 12.

39. McCarthy, *American Creed*, 3.

40. A good overview is "Christian Socialism," in Domenica M. Barbuto, *American Settlement Houses and Progressive Social Reform; An Encyclopedia of the American Settlement Movement* (Phoenix, Ariz.: Oryx Press, 1999), 45–46.

41. Useful accounts include Charles Howard Hopkins, *The Rise of the Social Gospel in American Protestantism 1865–1915* (New Haven, Conn.: Yale University Press, 1940); Judith Ann Trolander, *Settlement Houses and the Great Depression* (Detroit, Mich.: Wayne State University Press, 1975); and Paul T. Phillips, *A Kingdom on Earth: Anglo-American Social Christianity, 1880–1940* (University Park, Pa.: Pennsylvania State University Press, 1996).

42. Barbuto, *American Settlement Houses*, 46.

43. Barbuto, *American Settlement Houses*, 46.

44. Quoted material is from Elizabeth N. Agnew, *From Charity to Social Work; Mary E. Richmond and the Creation of an American Profession* (Urbana, Ill.: University of Illinois Press, 2004), 9. In addition to Barbuto, *American Settlement Houses*, 45–46,

192–93; and Hopkins, *Rise of the Social Gospel*, see Robert T. Handy, ed., *The Social Gospel in America, 1870–1920* (New York: Oxford University Press, 1966); Cecil E. Greek, *The Religious Roots of American Sociology* (New York: Garland, 1992); and Jean Quandt, "Religion and Social Thought; The Secularization of Postmillennialism," *American Quarterly* 25, no. 4 (October 1973): 390–409.

45. Judith Sealander, "Curing Evils at Their Source: The Arrival of Scientific Giving," in *Charity, Philanthropy, and Civility in American History*, ed. Lawrence J. Friedman and Mark D. McGarvie (New York: Cambridge University Press, 2003), 217–39.

46. Warren Susman, *Culture as History; The Transformation of American Society in the Twentieth Century*, reprint ed. (Washington, D.C.: Smithsonian Institution Press, 2003), 91. A similar point is made in Stanley Wenocur and Michael Reisch, *From Charity to Enterprise; The Development of American Social Work in a Market Economy* (Urbana, Ill.: University of Illinois Press, 1989), 30–51.

47. Susman, *Culture as History*, 91; and Wenocur and Reisch, *From Charity to Enterprise*, 30–51.

The development of Taylorism as a means of organizing hierarchically assembly line production is important in this context. See for example Samuel Haber, *Efficiency and Uplift; Scientific Management in the Progressive Era, 1890–1920* (Chicago, Ill.: University of Chicago Press, 1964); Daniel Nelson, *Frederick W. Taylor and the Rise of Scientific Management* (Madison, Wisc.: University of Wisconsin Press, 1980); Hugh G.J. Aitken, *Scientific Management in Action: Taylorism at Watertown Arsenal, 1908–1915* (Princeton, N.J.: Princeton University Press, 1985); and Stephen P. Waring, *Taylorism Transformed: Scientific Management Theory Since 1945* (Chapel Hill, N.C.: University of North Carolina Press, 1991). More theoretical accounts include Harry Braverman, *Labor and Monopoly Capital; the Degradation of Work in the Twentieth Century* (New York: Monthly Review Press, 1975); and Stephen Wood, *The Degradation of Work?: Skill, Deskilling, and the Labour Process* (London: Hutchinson, 1982).

48. Kirschner, *Paradox of Professionalism*, 53, 181.

49. Edward Saveth, "Patrician Philanthropy in America: The Late Nineteenth and Early Twentieth Centuries," *Social Service Review* 54, no. 1 (March 1980): 78. See also Marvin E. Gettleman, "Philanthropy as Social Control in Late Nineteenth-Century America: Some Hypotheses and Data on the Rise of Social Work," *Societas* 5, no. 1 (Winter 1975): 49–59.

50. Kirschner, *Paradox of Professionalism*, 179–80.

51. Kirschner, *Paradox of Professionalism*, 183. A disagreement over the compatibility of professional elitism with democracy formed the basis of debates between the positions taken by early twentieth-century reformers represented by Walter Lippmann and by John Dewey. Compare Walter Lippmann, *Public Opinion* (New York: Harcourt, Brace and Company, 1922) to John Dewey, *The Public and Its Problems* (New York: Henry Holt and Company, 1927).

52. Kiger, *Philanthropic Foundations*, 41–42. See also Ellen Condliffe Lagemann, *The Politics of Knowledge; the Carnegie Corporation, Philanthropy, and Public Policy* (Middletown, Conn.: Wesleyan University Press, 1989).

53. Barry D. Karl and Stanley N. Katz, "Foundations and Ruling Class Elites," *Daedalus* 116, no. 1 (1987), 5. See also Judith Sealander, *Private Wealth and Public Life; Foundation Philanthropy and the Reshaping of American Social Policy from the Progressive Era to the New Deal* (Baltimore, Md.: Johns Hopkins University Press, 1997).

54. Bremner, *Giving*, 159.

55. Wenocur and Reisch, *From Charity to Enterprise*, 56. See also David C. Hammack and Stanton Wheeler, *Social Science in the Making: Essays on the Russell Sage*

Foundation, 1907–1972 (New York: Russell Sage Foundation, 1994); and Ruth Crocker, "From Gift to Foundation: The Philanthropic Lives of Mrs. Russell Sage," in *Charity, Philanthropy, and Civility in American History*, ed. Lawrence J. Friedman and Mark D. McGarvie (New York: Cambridge University Press, 2003), 199–215.

56. Such was the case even with foundations with explicitly radical goals. See Gloria Garrett Samson, *The American Fund for Public Service; Charles Garland and Radical Philanthropy, 1922–1941* (Westport, Conn.: Greenwood Press, 1996).

57. Wenocur and Reisch, *From Charity to Enterprise*, 73.

58. Relevant works include Nigel Smith, *Perfection Proclaimed; Language and Literature in English Radical Religion, 1640–1660* (Oxford: Clarendon Press, 1989); Michael Mendle, ed., *The Putney Debates of 1647; The Army, the Levellers and the English State* (Cambridge: Cambridge University Press, 2001); Kenneth Cmiel, *Democratic Eloquence: the Fight Over Popular Speech in Nineteenth-Century America* (New York: W. Morrow, 1990).

59. Relationships between capitalism and conceptions of rationality in the public sphere are explored in John Michael Roberts, *The Aesthetics of Free Speech; Rethinking the Public Sphere* (New York: Palgrave Macmillan, 2003). Useful accounts include Jeffery A. Smith, *Printers and Press Freedom: the Ideology of Early American Journalism* (New York: Oxford University Press, 1988); William E. Ames, *A History of the National Intelligencer* (Chapel Hill, N.C.: University of North Carolina Press, 1972); Gerald Baldasty, *The Press and Politics in the Age of Jackson*, Journalism Monographs no. 89 (Columbia, S.C.: Association for Education in Journalism and Mass Communication, 1984); and Kevin Barnhurst and John Nerone, *The Form of News; A History* (New York: Guilford, 2001). A detailed discussion of abolitionism in this context is Robert Fanuzzi, *Abolition's Public Sphere* (Minneapolis, Minn.: University of Minnesota Press, 2003).

60. John Durham Peters, "Witnessing," *Media, Culture and Society* 23, no. 6 (November 2001): 707–23; and Nick Couldry, *The Place of Media Power: Pilgrims and Witnesses of the Media Age* (New York: Routledge, 2000). A succinct account of theoretical issues related to naturalism, realism, and positivist theories of representation is Raymond Williams, *Marxism and Literature* (Oxford: Oxford University Press, 1977), 95–100.

61. John Durham Peters, "'The Only Proper Scale of Representation': The Politics of Statistics and Stories," *Political Communication* 18, no. 4 (October–December 2001): 433–49.

62. Janet Norwood, *Organizing to Count; Change in the Federal Statistical System* (Washington, D.C.: The Urban Institute, 1995), 8. See also James H. Cassedy, *Demography in Early America; Beginnings of the Statistical Mind, 1600–1800* (Cambridge, Mass.: Harvard University Press, 1969); and A. Ross Eckler, *The Bureau of the Census* (New York: Praeger Publishers, 1972).

63. Peters, "Only Proper Scale"; and John Tagg, *The Burden of Representation: Essays on Photographies and Histories* (Amherst, Mass.: University of Massachusetts Press, 1988).

64. Ronald C. White, Jr. and C. Howard Hopkins, *The Social Gospel; Religion and Reform in Changing America* (Philadelphia: Temple University Press, 1976), 136–37.

65. Michael Schudson, "The Objectivity Norm in American Journalism," *Journalism: Theory, Practice, Criticism* 2, no. 2 (August 2001): 149–70; Julie Robin Solomon, *Objectivity in the Making; Francis Bacon and the Politics of Inquiry* (Baltimore, Md.: Johns Hopkins University Press, 1998).

66. *Oxford English Dictionary*, 2nd ed., s.v. "documentary."

67. Barbara J. Shapiro, *A Culture of Fact; England, 1550–1720* (Ithaca, N.Y.: Cornell University Press, 2000).

68. Tagg, *Burden of Representation*, 12.

69. The literature of documentary is vast. Among the accounts most relevant here include Erik Barnouw, *Documentary: A History of the Non-Fiction Film* (New York: Oxford University Press, 1974); Lewis Jacobs, ed., *The Documentary Tradition* (New York: W.W. Norton, 1979); Stuart Hood, "John Grierson and the Documentary Film Movement," in *British Cinema History*, ed. James Curran and Vincent Porter (Totowa, N.J.: Barnes and Noble Books, 1983), 99–112; Thomas Waugh, ed., *"Show Us Life"; Toward a History and Aesthetics of the Committee Documentary* (Metuchen, N.J.: Scarecrow Press, 1984); Richard Barsam, *Nonfiction Film: A Critical History* (Bloomington, Ind.: Indiana University Press, 1992); Paula Rabinowitz, *They Must Be Represented: The Politics of Documentary* (London: Verso, 1994); and Brian Winston, *Claiming the Real: The Griersonian Documentary and its Legitimations* (London: British Film Institute, 1995).

A term increasingly current that seeks to distinguish the politically intentional documentary is "social documentary." See Julianne Burton, ed., *The Social Documentary in Latin America* (Pittsburgh: University of Pittsburgh Press, 1990), and Pat Aufderheide, "In the Battle for Reality: Social Documentaries in the U.S.," Center for Social Media, School of Communication, American University, Washington, D.C., December 2003 <http://www.centerforsocialmedia.org/> (29 Jan. 2004).

70. Jacob Riis, *How the Other Half Lives; Studies Among the Tenements of New-York*, reprint ed. (New York: Hill and Wang, 1957), 215. Originally published in 1890.

71. "Matters We Ought to Know," *The New York Times*, 4 January 1891, 19.

72. Williams locates a key point in the emergence of naturalism in the confluence of science and representation represented by the novels of Zola, which also placed great emphasis on the determination of human action by the environment. See Raymond Williams, *Keywords; A Vocabulary of Culture and Society* (New York: Oxford University Press, 1976), 181–84.

An early example of what would be called today a life-history or ethnographic approach as taken by Chicago School sociologists of the day is W.I. Thomas and Florian Znaniecki, *The Polish Peasant in Europe and America: Monograph of an Immigrant Group*, 5 vols. (Boston, Mass.: Gorham Press, 1918–1920).

73. In addition to Tagg, *Burden of Representation*, 153–83, and Rabinowitz, *They Must Be Represented*, 75–104, see James L. McCamy, *Government Publicity: Its Practice in Federal Administration* (Chicago, Ill.: University of Chicago Press, 1939); Carl Fleischhauer and Beverly W. Brannan, eds., *Documenting America, 1935–1943* (Berkeley, Calif.: University of California Press, 1988); and Nicholas Natanson, *The Black Image in the New Deal: the Politics of FSA Photography* (Knoxville, Tenn.: University of Tennessee Press, 1992). Broader, recent reflections on the political implications of documentary include Patricia Zimmerman, *States of Emergency: Documentaries, Wars, Democracies* (Minneapolis, Minn.: University of Minnesota Press, 2000); and Chad Raphael, *Investigated Reporting; Muckrakers, Regulators, and the Struggle Over Television Documentary* (Urbana, Ill.: University of Illinois Press, 2005).

74. Results here coincide with findings of a study commissioned by an activist collective interested in gaining increased funding for media work. See Nan Rubin and Sharon Maeda, "Funding Media for Social Change," *MediaWorks Initiative*, November 2003 <http://www.centerforsocialmedia.org/documents/FinalMWIreport.pdf> (11 Nov. 2006).

75. "Distribution of Foundation Grants by Subject Categories, circa 2005," *The Foundation Center* <http://foundationcenter.org/findfunders/statistics/> (4 Sept. 2007). Copyright © 2005 by the Foundation Center.

76. "Distribution of Foundation Grants by Subject Categories, circa 2004." *The Foundation Center* <http://foundationcenter.org/findfunders/statistics/> (4 Sept. 2007). Copyright © 2004 by the Foundation Center.

77. "Distribution of Foundation Grants by Subject Categories, circa 2005."

78. Calculations based on "Top 50 Recipients of Foundation Grants for Media and Communications, circa 2005," *The Foundation Center* <http://foundationcenter.org/findfunders/statistics/> (4 Sept. 2007). Copyright © 2005 by the Foundation Center.

79. "Top 50 Recipients of Foundation Grants for Media and Communications, circa 2005."

80. Mark Dowie, *American Foundations; An Investigative History* (Cambridge, Mass.: MIT Press, 2001), 3. See also Waldemar A. Nielsen, *Inside American Philanthropy; The Dramas of Donorship* (Norman, Okla.: University of Oklahoma Press, 1996), 208–20.

81. Richard Steckel and Robin Simons, *Doing Best by Doing Good; How to Use Public Purpose Partnerships to Boost Corporate Profits and Benefit Your Community* (New York: Dutton, 1992), 165, 173.

82. Quotation from "Witness Performance Dashboard Mid-Year Report (July-December, FY06)," *Witness*, 11.

83. Sheetal Matani and Gerri Spilka, "Strengths, Challenges and Collaboration: Advocacy Groups Organizing Together on Media Ownership and Beyond," report of the Media Diversity Convening, 12-13 September 2003, 29 <http://www. mediaactioncenter.org/?q=Publications> (10 Oct. 2006).

84. Alyce Myatt, "Report on the Public Media Caucus," *Center for Digital Democracy*, 3 May 2005, 9 <http://www.democraticmedia.org/ddc/PublicMediaCaucus.php #_Toc105405351> (19 Oct. 2006).

85. "SSRC: Necessary Knowledge for a Democratic Public Sphere: Program Background Papers," *Social Science Research Council* <http://www.ssrc.org/programs/ media/background/> (19 Sept. 2006).

86. Among recent examples are Hunter Cutting and Makani Themba-Nixon, *Talking the Walk; A Communications Guide for Racial Justice* (Oakland, Calif.: AK Press, 2006).

87. "Why Media Matters," and "Media Empowerment Organizing Manual; A Guide to Understanding Media Power and Organizing for Media Justice in Your Community," *Office of Communication, United Church of Christ* <http:// www.ucc.org/ocinc/> (19 Oct. 2006).

88. "Konscious: Video/Art/Action," *Konscious* <http://konscious.com/> (19 Oct. 2006); "The First Amendment Project," *The Sundance Channel* <http://www.sundancechannel.com/firstamendment/> (19 Oct. 2006); "The Ground Truth: About: About the Production," *The Ground Truth* <http://thegroundtruth.org/> (19 Oct. 2006); "WMD: The Film," *WMD* <http://www.wmdthefilm.com/mambo/index.php> (31 Mar. 2005); "Free Speech TV: About Us," *Free Speech TV* <http://www.freespeech.org/html/aboutus.shtml> (19 Oct. 2006); "Welcome to Counterpunch," *Counterpunch* <http://www.counterpunch.org/aboutus.html> (19 Oct. 2006).

89. "SourceCode | Reboot the System | Submission," *SourceCode* <http:// sourcecode.freespeech.org/submission/> (19 Oct. 2006).

90. "SourceCode: About," *SourceCode* <http://sourcecode.freespeech.org/ about> (19 Oct. 2006); "Free Speech Radio News/Pacifica Reporters Against Censorship," *Free Speech Radio News* <http://www.fsrn.org/> (19 Oct. 2006).

91. Examples current as of the time of writing include the Environmental News Network, the Information Clearing House, TruthOut, AlterNet, and Free Speech Network.

92. "TomPaine.com—About," *TomPaine.com* <http://tompaine.com/about/> (19 Oct. 2006).

93. "About Common Dreams," *Common Dreams* <http://www. commondreams.org/about.htm> (19 Oct. 2006).

94. A number of her commentaries are collected in Cindy Sheehan, *Not One More Mother's Child* (Kihei, Hawaii: Koa Books, 2005).

95. "StoryCorps: About," *StoryCorps* <http://www.storycorps.net/about/> (19 Oct. 2006).

96. "Meet Brave New Films," *Brave New Films* <http://www.bravenewfilms.org/who.php> (31 Aug. 2006); "About the Brave New Foundation," *Brave New Foundation* <http://www.bravenewfoundation.org/about.php> (19 Oct. 2006).

97. "News Hounds: News Hounds Manifesto," *News Hounds* <http://www. newshounds.us/2004/07/02/news_hounds_manifesto.php> (19 Oct. 2006); Robert Boynton, "How to Make a Guerrilla Documentary," *The New York Times Magazine*, 11 July 2004, 20–3; *Outfoxed: Rupert Murdoch's War on Journalism* (Brave New Films, directed by Robert Greenwald, 2004).

98. "How to Host a Screening," *Brave New Theaters* <http://bravenewtheaters .com/guide/host> (19 Oct. 2006).

99. "DVD Offer > Sierra Club Chronicles > Sierra Club," *Sierra Club* <http://www.sierraclub.org/tv/dvdoffer/> (19 Oct. 2006); "Ground Truth Events | Create Event," *The Ground Truth* <http://events.thegroundtruth.net/page/event/create> (19 Oct. 2006). The form has general presence in a range of uses and institutions, from progressive, to conservative, to commercial.

100. "Film Aid International's FAQs," *Film Aid* <http://filmaid.org/who/ faq.shtml> (19 Oct. 2006); Deirdre Boyle, *Subject to Change: Guerrilla Television Revisited* (New York: Oxford University Press, 1997); George C. Stoney, Barbara Abrash, Lynne Jackson, and Cara Mertes, *A Festschrift in Honor of George Stoney* (Baltimore, Md.: Johns Hopkins University Press, 2001).

101. "Sweatshopper.org," *Sweatshopper* <http://www.sweatshopper.org/podcast .htm> (24 Oct. 2006).

102. "Sweatshopper.org," *Sweatshopper* <http://www.sweatshopper.org/> (24 Oct. 2006).

Chapter 4

Community Media Projects and Their Containment Through the Mass-Culture Critique

In late 2006, Knowledge Networks, the company that for the past 32 years has produced the People's Choice Award by tallying public votes for favorite celebrities and entertainers, urged people to join its new project, the People's Choice Community. By providing a range of personal information about preferences and buying habits, members of this community will "shift the power to shape entertainment from a few people in Hollywood to a community of millions of fans" who will be able to "choose a cast member for a reality show," "pick a model for a major ad campaign," and "decide who should win a Hollywood makeover." In a promotional video, a fan identified as Laura from Boston, Mass. asserts that, by virtue of her belonging to the community, a cast member of a reality show will now be "the person I want to see, not some casting director."[1] As "the official voice of entertainment fans everywhere," the People's Choice Community lets "you talk" while "Hollywood listens!"[2]

Instead of the role of community in communications being simply a collective choice among given options, community media projects for the Alliance for Community Media are a means of "supporting our nation's communities and families and promoting effective communication through community use of media."[3] For this group, communities come in many different shapes and sizes. In addition to "local religious, community, and charitable groups," users of PEG (public, educational, and governmental) media centers on whose behalf the Alliance works includes "public schools, religious institutions, colleges and universities, government officials, the disabled, and second-language communities as well as national institutions such as N.A.S.A., the U.S. Department of Education, and the U.S. Army."[4] Although they consist of many different kinds, scales, and means of support, the primary goal of all is to use media to "advance their [own particular] causes."[5]

For Brad Will, community media projects exist by virtue of one's own feet and hands. His writing displays an anarchist-inspired disdain for attempts to change society by playing by the rules and meekly asking entrenched authorities (who benefit from keeping things as they are) to go against their own interests. In contrast to tame legislative letter-writing campaigns and officially allowed protest marches, "pure direct action," wrote Will in 2005, is when "you are not [simply] talking politics and yelling in the streets; you are doing it, making it real, and sharing it with the community."[6] For Will, direct action meant being involved in such community media projects as microradio and Independent Media Centers, but also in squatters movements, defense of old-growth forests against logging, movements to establish and protect urban community gardens reclaimed from deserted open lots, and protests in Seattle in 1999 and elsewhere as part of the global-justice movement.[7] It also meant traveling to Oaxaca, Mexico in 2006 to write and shoot video about a teachers' strike that had grown during the summer into a broader struggle against the state's administration.[8] By documenting sometimes violent repression by armed paramilitary squads, Will sought to help what he felt was a "genuine people's revolt" in ways similar to what he had done in Brazil. Video footage he had taken there during state armed repression in a squatters' encampment called Sohno Real (translated as "real dream") was later used to establish the need for an official investigation.[9] The social movement in Oaxaca was the last engagement for the 35-year-old Will, who was shot and killed by paramilitaries as he worked, his own video camera recording the attack, with Will himself behind the viewfinder.[10]

All three projects claim to be communities involved in and aided by communications, and descriptive studies often simply make this point. But the difficulties of analysis begin where generalities end and differences of practice begin.[11] The People's Choice Community is a pool of subjects joined to a national market-research project, with the project implying that consumers and their preferences determine all that is available in the capitalist marketplace. For the Alliance for Community Media, groups involved in community media projects range from neighborhood charities to N.A.S.A. and the U.S. Army, with the purpose of community media projects to promote their own narrow interests in the marketplace of ideas in an expression of a pluralist if not Tocquevillian associational polity.[12] For Brad Will, communities are people at sites struggling against authority, with writing, shooting video, and distributing it a means both of personal engagement and public confrontation.

To move beyond description to analysis, one must better understand the social formations in which these community media projects are based and from which they emerge. Doing so requires, first, a recognition that communities themselves are not naturally occurring, but instead are human projects. As Mulhern put it in response to contemporary, nostalgic laments about the loss of community, "if 'communities' are notoriously hard to find, it is because they are everywhere—not places but practices of collective identification whose variable order largely defines the culture of any social formation."[13] When seen in the terms of the insight that communities are not found as much as they are made, definitions of community media projects, such as in one example that defines community radio as "radio run by and for the community," become tautological and essentialized propositions, whose assumptions of the naturalness, stability, and discreteness of communities cannot be taken for granted.[14]

In addition to recognizing the multiplicity and historicity of communities, one must also recognize their specific social implications. As Raymond Williams observed, the fact that community is "one term which has never been used in a negative sense" (on what basis could one object to more community?) is reason for hope but also for suspicion. If the uniformly positive use of community reflected a uniformly positive and peaceful reality, "then," as Williams notes, "we'd be living in a world very different from this one."[15] However, in this world, just as communities can be a means of mutually sustaining collaboration and solidarity, they can also be a means of division, separation and, ultimately, of isolation.

Working from these premises, this chapter argues that communities as a tangible and recognized mode of human association arose as a response to capitalism, and that community media projects arose as a parallel response to market formations of communications. This claim relies in part on Anderson's argument that capitalism made possible the emergence of the modern nation as an imagined community, but I extend the insight about the formative relationship between capitalism and all kinds of communities as a creative response to capitalism's dislocations (in all senses of the word). I also take from Anderson the claim that communities in general "are to be distinguished, not by their falsity/genuineness, but by the style in which they are imagined," extending this insight as well to a matter of the style and the mode in which communities were practiced within their formative conditions, and to the corresponding conception of communications practice that enables it.[16] In his discussion of Dickens's novelistic rendering of Victorian-era London, Williams underscores the necessity of this insight. "It does not matter [. . . whether we claim] the experience of the city is the fictional method; or the fiction method is the experience of the city." Questions about primacy and directional causality sidestep the central insight about the social whole (without implying consistency or homogeneity) of material experience.[17]

Antebellum Community and Communications

The precondition of community media projects is the presence of communities themselves, an observation that places as much emphasis on analysis of communities as on the media projects in which they engage. While human settlements of varying population sizes have been present throughout human existence, many theorists emphasize the need to broaden demographic definitions of community to those more deeply social and cultural. Doing so emphasizes the importance of recognizing not simply different sizes of communities, but different kinds of community as ways of living, such as urban society or rural life, which must be seen as distinctive cultural and material inventions dependent on conditions brought about by capitalist expansion.[18] Such a distinction can be seen in the etymology of "community" in English, which emerged in the fourteenth century, its modern incarnation in social theory and popular commentary as an opposition to the amorphous and anonymous crowd of cities coinciding with the changes associated with industrialization, urbanization, and commercialization in the wake of the spread of capitalism.[19] Thus, to apply to communities a point that Williams has made about cities, "a connection of name and in part of function" indeed exists between communities of various times, but there is "nothing like [an] identity" between them.[20]

The need to historicize community extends to the equally compelling need to historicize social struggles engaged in by communities. It is through the form of community that varieties of social critique reimagine the possibilities of human association. Prior to the emergence of capitalism, struggles to assert or protect one's self and others of an explicitly identified "community" against an impersonal, centralizing "society" had little relevance. Political and cultural struggles of the precapitalist era in Anglo-American traditions were framed more typically within the distinction between the official (or royal/clerical) and the popular.[21]

By contrast, communities and community media projects emerged self-reflexively as a social-spatial means of embodying a critique of capitalist society. What remade the deterritorialized struggle was the steady consolidation of capitalism and the corresponding rationalization of the state, in which the struggle between the official and the popular was increasingly and variably modified and practiced in more overtly spatial terms as a struggle between a centralizing national administration and local social movements that resisted the presumptive superceding of previously local decisionmaking. When transported and transplanted to the Anglo New World and manifest in the newly formed United States, community resistance responded to the early capitalist contradiction between Federalist centralization and Anti-Federalist "oligarchical, particularistic, and parochially privileged approach"—neither of which had any place for fully democratic participation.[22] Community responses to Federalism and Anti-Federalism and the false opposition they embodied were based in varying mixes of radical republicanism, radical Protestantism, and natural-rights arguments, all of which asserted the necessity and morality of intentional, general, and direct human control over human affairs.[23]

As one challenge to Enlightenment liberalism as the ideological framework for laissez-faire capitalism, radical republicanism sought to reconcile direct democracy of the classic Platonic kind to the needs of a geographically much larger polity in which face-to-face deliberation was impracticable as a mode of political decisionmaking.[24] The form of community in which this effort was embodied was the cooperative commonwealth, which was seen to consist of self-sufficient producers who would collaborate in shaping conditions that would protect their individual freedom and livelihood for the good of all. Lambert elaborates the political significance of economic self-sufficiency, in that "free labor was not just a legal and economic condition but a requirement for republican self-government." The ability to conduct one's work as one saw fit was more than "the capacity to enter into labor contracts at will; it demonstrated that journeymen artisans were self-sufficient, productive individuals, capable of the self-directed activity necessary for self-rule."[25] By contrast to wage work, seen as "a system of wage slavery" and a "false liberty that trapped working Americans," the cooperative commonwealth would transform "the working men and women into their own employers" as well as eliminate "from the workplace those who lived off the productive labor of the working classes."[26] In ways directly corresponding to Jefferson's championing of yeoman farmers as the backbone of democracy, early labor leaders and reformers championed craftsmen and artisans as "'the yeomanry of the city.'"[27] Such a vision in the United States was institutionalized in a limited way by the 1820s, when all states except for Rhode Island did away with property qualifications and broadened suffrage to "all wage-earning white men."[28]

The kind of community imagined as the cooperative commonwealth diverged in important ways from liberal capitalism, but it also converged with it, too.[29] Thompson suggests that the brand of radical republicanism put forth by Paine and others asserted that "every man must have equal rights as a citizen," but that "in economic society he must naturally remain employer or employed," thus reproducing capitalist social relations and, in doing so, reigning in the form's deeper radical potential.[30] In a similar point, Lambert notes that the goal of radical republicanism was not to do away with property but to transform "working people into their own employers" and, in doing so, make "every employed person a capitalist." The state was to remain for a time, carrying out a limited role in overseeing "public policies that would foster the creation of self-governing producers' and consumers' cooperative enterprises, with the ultimate aim of establishing the cooperative commonwealth," which would make the state itself dissolve.[31] In this way, radical republicanism valorizes ostensibly self-sufficient individuals, with a readymade fit with reformist if not conservative, essentialist political thought.

While the cooperative commonwealth was a secular self-determined community that opposed the emerging capitalist society, Puritanism provided the impetus for an overtly religious challenge to it. It was within the radical Protestant tradition that a distinctive kind of community was conceived and practiced as a critique of the emerging capitalist social order. The spatial embodiment of its critique was a "city upon a hill," voiced in the proverbial 1630 sermon of

John Winthrop at the start of the Massachusetts Bay Colony settlement in New England. "We must consider," he said, "that we shall be as a city upon a hill. The eyes of all people are upon us."[32] Bestor links this form to the longstanding Christian (at least) conception that "believers constitute a separate and consecrated body set over against the sinful world," itself seen in the teachings of St. Benedict and his view of the monastery as "'a little State, which could serve as a model for the new Christian society'."[33] As such, a city upon a hill was meant to transform society not by forced imposition but by voluntaristic example. While, as Bestor notes, religious communitarians "were demanding reforms as far-reaching, as drastic, and as rapid as those that appeared in any revolutionary program," they "rejected completely the method of revolution" in favor of serving as a clear, rational example of a better way that should be voluntarily adopted.[34]

By virtue of their common intention of challenging liberal capitalism, the line between radical republicanism and Puritanism blurred, requiring terms such as "millennial republicanism" and "puritan radicalism" to emphasize hybridizations of the religious and the worldly in practice. As Claeys notes, such hybrids united religious enthusiasm with a "fervent devotion to the cause of political virtue," and thus took such mixed forms as a "propensity to interpret the American and French revolutions in terms of Providential development" and its implications for the millennial fulfillment of Christ's second coming.[35]

It is one thing to imagine a new kind of community, but quite another to realize it. What limited the substantial realization of the cooperative commonwealth and of the city upon a hill was their intended role as exemplar, which meant in practice isolating themselves in order to demonstrate a "pure" form, instead of engaging more directly with existing society. As a result, both the cooperative commonwealth and the city upon a hill were more a hope and orienting goal than a practiced means of organization. The cooperative commonwealth as a fully realized formation remained a hope, and the "city upon a hill" as lived day-to-day in New England looked and worked much like a regular town.[36]

A more aggressive spatial attempt in the antebellum United States to realize a counterpoint to the emerging capitalist society were utopian communities, themselves varying amalgams of natural-rights philosophy, radical republicanism and Puritanism.[37] The extent of communitarian, proto-socialist experimentation as a critique of liberal capitalism is alluded to by Emerson who, in an 1840 letter to Carlyle, commented that Americans "are all a little wild here with numberless projects of social reform. Not a reading man but has a draft of a new community in his waistcoat pocket."[38] Such a claim had its basis. Experiments in utopian communities ranged from Thoreau's one-person project at Walden to much larger efforts that involved all told by one estimate more than 100,000 people.[39]

Such communities also sought to become exemplars and lead the way to new kinds of human association. Although sometimes radically associationalist within their bounds, they sought to set themselves apart from the rest of society as an exemplar, instead of spreading and confronting existing society in concrete ways.[40] Hayden suggests that what these utopian communities had in common was the belief that "social change could be best stimulated through the organization and construction of a single ideal community"—in other words, a social experiment that, in its attempt to concretely demonstrate the viability of a new kind of social organization, in general intention was rooted in the assumptions of rationalist empiricism.[41] Fogarty connects further the role of exemplar to rationalist empiricism in his characterization that "communal and cooperative settlements could serve as social laboratories and that by experimenting on a small scale the reformers might teach the larger world some lessons."[42] Such a view fit well with voluntaristic, gradualist aims. Spann recalls the belief that by persuading people through example "to withdraw from the corrupting institutions of existing society and to adopt new ways," utopian-colony reformers "could effect a peaceful revolution in human affairs that would redeem the world."[43] Of course, the assumptions of individual autonomy and voluntarism on which depended the critical value of utopian communities—that people convinced of their value could join them or even start one of their own—belied the social and economic realities of life in the nineteenth-century United States, which included slavery, the restriction of suffrage to white men, and state-sanctioned territorial conquest and extermination of native populations under the guise of Manifest Destiny.

Critical Communities and Communications

While communications projects were hardly a primary means or a focal point of utopian communities, and specific cities upon a hill were a project in concert with but disengaged from evangelical publishing, it was within a diffuse, implicit radical republicanism that a community communications project emerged. Utopianism was clearly evident in reactions to the telegraph, deriving from popular wonderment regarding electricity and, more generally, admiration of engineering and mechanical accomplishments of the day.[44] Such reactions were utopian due in part to their sense that rational human invention had created a means of realizing an instant utopian community of size and scale far beyond the experimental colonies of Owenites, Shakers, Fourierists, and others. While promoting his invention to Congress in 1838, Samuel Morse argued that a national telegraph system would "diffuse, with the speed of thought, a knowledge of all that is occurring throughout the land; making in fact, one *neighborhood* of the whole country [emphasis original]." In the words of an 1846 newspaper account of the telegraph, the invention "leaves, in our country, no elsewhere—it is all *here* [emphasis original]."[45]

Such breathless and vociferous hopes for the telegraph drown out consideration of another communication system roughly contemporary with the emergence of the telegraph but that put into practice much more thoroughly the utopian ideas prevalent at the time. In its implicit embrace of the principles of what amounted to a producer cooperative, in its publicly funded and decentralized structure, and in its amenability to small-scale and artisanal producerism, the newspaper exchange was perhaps a more mundane but in key ways a critical project of forging a community media project as a counterpoint to capitalist forms.

In the antebellum United States, the national communications system was indeed consolidating, not as a centralized, urban node broadcasting to the rural hinterlands, but as an informal, artisanal consortium of local media projects.[46] Newspaper editors of the day were single-proprietor printers and postmasters with little time, energy, or resources to report original stories. As a result, in order to produce their papers, they routinely received newspapers from editors in other towns, reprinting items from these other papers.[47] The newspaper exchange has prompted less-than-admiring judgments from recent scholars, some of whom denigrate its derivative composition as "'scissors and paste pot' journalism."[48] In his studies of urban information systems of the nineteenth century, geographer Allan Pred refers to the newspaper exchange as "a well-entrenched system of mutual journalistic plagiarism."[49]

But the lack of respect for property rights was precisely the critical and practical value of the newspaper exchange, which was in effect a nonprofit news cooperative. It operated without central management and formal contract and instead by virtue of custom and convention, and without claims that news stories were private property for use only in exchange for payment. Indeed, as Kielbowicz notes, "the essence of the exchange system was reciprocity" instead of commercial gain.[50] In these ways, the newspaper exchange had much in common with radical-republican conceptions of small, independent proprietors as independent worker-citizens that, along with a vigorous public debate it enabled about issues of the day, sought to ensure the civic-republican guarantee of freedom.[51]

Although the practice may have been informal, its basis was quite intentional, enabled as it was by federal political decisions and resulting regulatory structures within the context of stridently partisan politics. To help bring together the disparate parts of the new nation, not only were steps taken to develop the transportation and postal systems, public subsidies in the form of preferential mailing rates and political patronage promoted the printing and circulation of magazines and newspapers.[52] Federal postal regulations also made possible the cheap delivery of magazines in comparison to private letters and the free delivery of newspapers.[53] To ensure some degree of parity between large urban newspapers and smaller rural ones, congressional action in 1845 established a local, postal-free zone that allowed newspapers to circulate for free only if delivered within 30 miles of its office of publication, the free-postage range by 1851 altering to consist of the county of publication.[54] In addition to systemic regulation, the ability of printer-postmasters to freely exchange newspapers was also promoted by permitting them to send correspondence related to their postal duties through the mails for free, a privilege that—given that the work of printer, newspaper editor, and postmaster was often done by the same person—was extended in practice to newspapers.[55]

In addition to preferential rates and privileges, what also enabled the newspaper exchange was political patronage in the form of federal government printing contracts for the printing and distribution of congressional laws and regulations.[56] Patronage was also distributed at the state and the local level, with newspapers receiving patronage from their affiliated political party even if it was not dominant and thus able to control the allocation of government printing contracts.[57] Due to what amounted in effect to public support, local newspapers were widely diffused throughout the country, viable even in recently or sparsely settled areas.[58]

Finally, such decisions and practices were possible due to news falling outside legal protection as private property. While the U.S. Constitution enshrines the protection of literary and intellectual property "to promote the Progress of Science and useful Arts," and while the Copyright Act of 1790 protects the work of authors and publishers, news accounts were not covered due to the belief that the information they contained was—unlike the creative product of a single author—a public property detectable by anyone.[59] It was not until 1918 that a property right related to news was established in the United States by virtue of a Supreme Court decision that held that unauthorized use of news reports was misappropriation.[60] The majority decision noted that, while "news of current events may be regarded as common property" and thus outside claims of copyright, the energy and costs incurred in collecting and distributing specific accounts of the news could indeed be protected by law.[61]

The net result of this combination of preferential rates, postal privileges, political patronage in addition to local advertising, and the legal status of news as outside copyright protection is by any measure impressive. By as early as 1820, the United States had the most post offices and newspapers per capita of any country.[62] By the 1840s the newspaper exchange had reached sizable proportions, as an 1843 count suggests. During October of that year, 518,676 newspapers (16.5 percent of all newspapers mailed) were exchanges between publishers, with each publisher receiving an average of 364 newspapers. Extended to the entire year, each publisher received an average of more than 4,300 newspapers, of course with major urban papers and those in Washington D.C. receiving more exchanges than those in rural towns. For example, the *National Intelligencer*, based in Washington, D.C., exchanged with close to 600 newspapers, who did so to gain access to and reprint the *Intelligencer*'s coverage of the federal government.[63] The newspaper exchange enabled the practice of a hybrid information exchange neither uniformly local nor national, but one composed of individual local efforts that together made possible a national system that embodied in many ways a radical-republican community media project.

Communities Remade

With the emergence of telegraphic networks and the increasing capitalization and concentration of newsgathering made possible by the expanding economy and the newly formed advertising industry, the newspaper exchange as the practice of a radical-republican critique of liberal capitalism was steadily displaced by a communications system increasingly expanded, consolidated, and commercialized.[64] By this time, as Barnhurst and Nerone summarize, "the press became an engine of wealth, controlled by increasingly powerful publishers," epitomized by the end of the century by Joseph Pulitzer and William Randolph Hearst, but in a myriad of less grandiose ways by smaller newspaper publishers throughout the country.[65] The importance of standardized content from news services as a way to share costs and thereby trim expenses grew with, for example, the Scripps chain of newspapers in the early twentieth century filling about two-thirds of its nonadvertising space with material from the chain's own news-features syndicate.[66]

These changes in the organization of communications were part of a larger set of developments with key implications for communities. With the post-Civil War consolidation and acceleration of capitalism came a spatial remaking both of means of the assertion of capitalism and the spatial means by which it was challenged. Enhancing trade by geographic extension has been a feature of the longstanding practices of imperialism and colonization, but its extent by this time was quite new and qualitatively different. Sanders notes that "the rapid industrialization of the post-Civil War period brought profound changes in the structure of manufacturing-belt industries."[67] Such changes included a transition between regional competition and small- to medium-scale industry to "a nationally unified monopoly capitalist order dominated by 'giant corporations' competing on a national, continental, and even world scale," with the scholarly consensus being that such changes in structure and organization amounted to a "second industrial revolution."[68] In addition to consolidation and extension, capitalism in the United States was also made more mobile, not only in the sense of more freely moving money or controlling manufacture from greater distances through improved systems of finance, communications, transportation, and management, but in deeper ways such as relocating factories from cities to newly created industrial suburbs and, in some cases, company towns (capitalism's (in)version of the utopian community).[69]

This expansion and reorganization was embarked upon in order to enhance profitability by securing lower-cost or more compliant labor.[70] By the latter half of the nineteenth century, doing so was an increasingly central concern. Efforts to consolidate and extend capitalist control were met by resistance from an increasingly large and militant industrial working class and small-scale farmers, who in varying ways responded to the further loss of personal control and autonomy, and to increasingly severe cycles of economic boom and bust.[71] The capitalist response to worker, reformist, and utopian challenges was to make capital more mobile, less subject to place-based resistance, and thus more controllable and profitable.

Like utopian communities before them, social movements that sought to resist the accelerating, respatialized assertion of capitalism in the Gilded Age assumed characteristic forms of community. These efforts took a variety of forms. One example already discussed in Chapter 3 is the social gospel, an associational reformism consisting of a merged Christianity, utopianism, and "soft socialism." Its spatial manifestation is worth noting. Contrary to the greater expansiveness and mobility of capitalism, the patchwork of social-gospel charity sought to meet the immediate needs of specific people in specific places. However, religious-social reformism of this kind sought to address the consequences of industrial capitalism rather than directly confront their causes.[72] In a similarly fragmented fashion also much the opposite of the respatialization of capitalism, the urban settlement-house movement (deriving from the utopian-community tradition of the early part of the century), like the social gospel, constituted a place-centered and primarily exemplary rather than engaged or confrontational approach to social problems.[73]

Fragmented, place-centered reformism of whatever stripe was largely ineffectual given the spatial extent of capitalist assertion and the resulting human costs.[74] For many people subjected to these changes, such conditions required not reformism, but direct confrontation, the most used tactic of which was the labor strike. Variable awareness of international examples such as English Chartism earlier in the century, the rise of syndicalism in France, and of the brief emergence of the Paris Commune served as sources of inspiration.[75] The years between 1881 and 1905, during which the consolidation of monopoly capitalism became widely recognized, saw some 37,000 labor strikes and riots "involving seven million workers on record," with 700,000 workers striking sometime in 1886 alone.[76]

Yet, trade unionism and tactics in the form of the strike had been calibrated to an earlier, regional-based capitalism within a comparatively weak national state—a much different situation. In the face of more frequent use of armed military troops and federal authority to quell strikes, and contrary to the more expansive aims of radical republicanism and early labor organizations such as the Knights of Labor, many labor leaders by the mid-1890s gave up the goal of challenging capitalism, seeking instead "a narrower, more instrumental right to strike for pure-and-simple union goals, such as better wages and working conditions, union recognition, and the closed shop."[77] Trade-union organization reflected this specific, more limited goal by being place-based and compartmentalized. Separate, distinct trade unions were organized around a distinctly defined skilled craft and were comprised of aptly named "locals" that were specific to a single workplace within that industry or trade. As Trachtenberg summarizes, by the end of the 1800s and the ascendancy of the American Federation of Labor as the pre-eminent labor organization, A.F. of L. head Samuel Gompers voiced his willingness to "accept the wage system in exchange for a secure place within the social order."[78]

The rise of industrial labor action as a critical response to the respatialization of capitalism was matched by the rise of agrarian populism.[79] Agrarian political organizations emerged with the Patrons of Husbandry (the Grange) in 1867, which was open to women as well as men, focusing initially on enabling cooperative buying of farming supplies, and later on extending in such ways as "arranging the shipment of wheat to foreign ports, operating grist mills, offering life and fire insurance, and even manufacturing their own farm machinery and organizing a few banks."[80] Sanders also notes the quite different spatial organization of agricultural radicalism compared to that of industrial labor radicalism, with resulting different and in some ways innovative forms of spatial struggle. Farmers' subjection to "powerful flows of interstate and foreign commerce" meant that "local economic collective action" typical of industrial workers such as the strike was of little use to farmers. As Sanders continues, "the farmer's enemy was not an employer but a *system*—a system of credit, supply, transportation, and marketing." Thus, agrarian radicals recognized that "to reorder such a system required political action at the highest level" instead of an on-site, geographically local challenge to a specific bank, lender, or supplier.[81] As such, agrarian radicals focused their attention on remaking delocalized systems of finance, distribution, and marketing and, in doing so, to remake their own position and role in such systems.[82]

Despite these and other differences in situation, organization, and tactics, and despite some notable legislative achievements, agrarian populism also was no match for a respatialized and increasingly deterritorialized capitalism, bound as it was by the much more place-bound nature of farming. For a time in the 1920s, the possibility emerged of a farmer-labor political party to pool strengths, but it soon flagged as well in the midst of such pressures as lack of perceived common interests, the fragmentation of grassroots organizations into political interest groups, and various federal legislative measures that increasingly subjected farmers and labor to federal control.[83]

While reformism, trade unionism, and agrarian populism were all significantly limited by being organized by discrete place, other more radical responses to the consolidation and extension of capitalism addressed more directly its mobility by also being more consolidated, expansive, and mobile. Instead of seeking to simply eke out the best wages and conditions possible within the current system, anarchism and syndicalism conceived of workers themselves owning the means of production, which would then be managed for the benefit of society.[84] To work toward this, labor radicals left behind the spatial fragmentation and place-boundedness of reformism, trade unionism, and agrarian populism. Instead of fragmenting the labor movement into separate trade unions, anarchism and syndicalism aimed to consolidate and expand their organization in much the same way as Gilded-Age monopoly capitalists had already done in steel, petroleum, railroading, and others. Such a vision had already been a part of American labor organizing in the form of the Knights of Labor.[85]

The clearest example of an anarcho-syndicalist response to an increasingly consolidated, expanded, and mobile capitalism was the Industrial Workers of the World.[86] Constituted by a novel and shifting set of European and indigenous anarchist and syndicalist influences, the I.W.W. carried the premises of anarchism and syndicalism—and their fluidity and mobility—further in a number of ways.[87] These included organizing as a federation of "mixed locals," each one of which could have members ranging across many shops and industries, while also including seasonal and unemployed workers.[88] A more flexible spatialization of organization also included so-called "jungles," which were improvised camps close to junctions of railroad lines on the outskirts of towns and cities.[89]

In addition to the syndicalist tactics of strikes and sabotage, the I.W.W. practiced and pioneered more flexible challenges to the very institutions of capitalism, and these initially improvised efforts were perhaps its most enduring contribution to the practice of community as a means of spatial struggle. Unlike the placed communities of utopian reformers, and the place-bound politics of reformers, trade unionists, and agrarian populists, the I.W.W. helped pioneer a flexible, deterritorialized resistance to capitalism that was much less about local workers confronting local employers, and much more about acting as a variable, general community of affinity at specific sites to be determined by choice and conditions. This fluidity manifest itself in work slowdowns and purposively damaging goods that were made.[90] By being de-institutionalized, such tactics and specific actions could spread person-to-person through word of mouth, becoming much less an official action of a union or party, and much more a fluid social movement and thus much more difficult for foremen, managers, and the state to detect, control, and punish.[91] Challenges to the institutions of capitalism included the free-speech fights, in which I.W.W. members from many states converged on a single town that had passed ordinances prohibiting street speeches in order to stop a handful of I.W.W. members from speaking out.[92] Upon arriving, they would all speak out, be arrested without resistance, fill the local jails and any other space, require food daily, demand separate trials, and ultimately tax the small town's ability to the extent that the offending ordinance would be rescinded.[93] At least twenty such challenges took place in many areas of the country between 1909 and 1913.[94]

While such actions hardly qualify as a significant blow to capitalism, and the I.W.W. was soon broken by concentrated police, military, and legislative action in the wake of a general crackdown on radicals at the start of World War One, the response of the I.W.W. and other radical labor organizations to the respatialization of capitalism marks an innovation in the form of community as a means of challenge and struggle.[95] They shared with utopian communities the goal of concretely remaking society. But, instead of withdrawn isolation from society, such respatialized communities directly confronted it. Instead of a place-bound locality, such respatialized communities were practiced as a discernable, yet dispersed and mobile solidarity that aspired to novel kinds of hybrid association consistent with, for example, Kropotkin's conception of "communal individuality."[96]

Respatialized Community Media Projects

Attempts to confront a respatialized capitalism had their correspondences in communications, particularly in the appearance of mass-circulation labor publications, which put to use the means and form of national religious publishing and commercial popular presses, with the exemplar being *The Appeal to Reason*, the largest-circulation socialist newspaper ever to appear in the United States.[97] In the shadow of the huge financial success of mass-circulation newspapers such as the *New York Journal* and the *New York World*, Shore notes that the reigning "models for the radical press to follow while seeking to develop a large audience sometimes came from the successful mainstream press." The *Appeal* editor J.A. Wayland did just this after trying reader contributions, party support, organizing as a cooperative, and donations from wealthy benefactors, but finding them all wanting.[98] Confronted with these choices, Wayland "came to the conclusion that a [financially] solvent paper could only be produced through 'sound' business practices." The goal of producing a "solvent paper" was not an egotistic one necessarily; business success was assumed to confirm the validity of the message and the viability of the social movement.[99] As Wayland put it in a January 1903 issue, "'The day has gone by for small mediums [*sic*] to tackle great undertakings, and we must prepare to propagate Socialism in just the proportions that Capitalism operates.'"[100]

The biggest step toward achieving this goal was to professionalize newsgathering by replacing the lower-cost option of commenting on capitalist news with the hiring of its own reporters and going full-steam into "the news business," running only "exclusive stories brought in by the *Appeal*'s own staff." The reorganized *Appeal* commissioned and ran serially Upton Sinclair's novel *The Jungle* and published original stories on political corruption, the Mexican Revolution, strikes and child-labor abuses, and the murder trial of "Big" Bill Haywood and other members of the Western Federation of Miners, the last of which it turned into what Shore calls a "cause celebré."[101] By 1902, circulation reached and passed 200,000, peaking prior to World War One at more than 750,000.[102]

Despite initial success, Wayland's efforts suffered due to the retrenchment of socialist politics in the wake of assassinations blamed on anarchists, the collapse of European socialism on the eve of World War One, and the domestic repression of socialist publications during the war primarily through the loss of their postal privileges.[103] Additional limitations arose as a result of adopting not only popular presses' means but form. While Wayland and other radical publishers at the time hoped to, in Ruff's words, "erase distinctions between press and reader"—with its erasure the only hope of building a social movement composed of and responsive to its participants and egalitarian in design—as long as vanguard forms of communications and the corresponding capital-intensive, professionalized form of organization were deemed necessary, hopes for achieving such goals remained limited.[104]

Despite the reliance on corporatized forms of media represented by Wayland and others, experimentation of more participatory kinds existed alongside it.[105] An example is the efforts of the Western Federation of Miners and, later, the I.W.W. under the leadership of Bill Haywood. Haywood felt his role in the union was not to direct the movement, but to encourage what he called "long-distance handshaking" between its members by sending them frequent bulletins of strikes, asking for information about member unions, and soliciting pictures of the strikers, union halls, and demonstrations to be reprinted in the magazine.[106] This openness continued with the I.W.W., which by one account published 66 different periodical titles between 1905 and 1919.[107] In addition to tracts and polemics penned by movement leaders, the pages of many contained contributions from the rank and file.[108] However, such experimentation was, like radicalism itself, contained as well as subsumed by the expanding commercial order, which soon would be presented with an entirely new situation in the form of broadcasting.

Remaking the Community Critique of Capitalism

It is with a respatialized capitalism and its national and increasingly international assertion in all realms of human experience that communications became directly relevant to the community-based critique of capitalism. For example, communications played a central role in the communitarian reformism of the early twentieth century.[109] An important scholarly arena for its development was the emerging academic discipline of sociology and its allied, applied partner, professional social work, which centered its inquiry on the emerging modern world order and the problems defined by the poles of community and society.[110] By the late nineteenth century, German scholarship was refracted through

American experience in the work of the influential Chicago School of sociology, which sought to explore means by which a new spatial synthesis of community and society—the "Great Community"—could be forged from the small-town tradition in the midst of the rapidly expanding and urbanizing United States.[111] As Quandt summarizes, early Chicago-School sociologists felt that the United States was undergoing "the breakdown of the small, close-knit group under the impact of urbanization and industrialization," with the "social organization based on family, neighborhood, and small-town solidarity" replaced increasingly by "the more impersonal and tenuous ties of the marketplace." [112] To address this problem, members of the Chicago School sought to explore how modern methods of communication could create a Great Community that, in size and scale, extended far beyond the small town but that nevertheless was composed of deep-seated, honest relationships heretofore known only in the small town.[113]

With the communitarian aims of the Chicago School largely unrealized, the trade-unionist, reformist mainstreaming of labor, and the decline of agrarian radicalism, the critique of capitalism enabled by forms of community changed apace. The key spatial struggle came with the emergence of radio broadcasting and the conflict between, on the one hand, a nationally consolidating commercial system and, on the other hand, advocating for decentralization, diversity, local control, and accountability.[114] Corresponding with this struggle was the emergence of an acceptable proxy for the community challenge to capitalism, which became the critique of mass culture, a critique enabled by the rise of localism as a set of principles with which to guide the development of communications in a rapidly changing society.

One key facet of localism as a critique was its objection to the commercial consolidation of the media industries, a development that, by the emergence of broadcasting networks in the 1920s, had become fully apparent. As Stavitsky observes, "though U.S. policy gives its official imprimatur to local broadcast service, in practice radio and television stations in the United States have long delegated the bulk of their programming responsibilities to networks and syndicators."[115] Such pressures and the networking and syndication tactics for responding to them had been recognized in the newspaper industry for some time, beginning in the mid-nineteenth century and the rise of telegraphic news services.[116] They had strengthened to the point where, as early twentieth-century newspaper-chain owner E.W. Scripps put it in 1903, "the ultimate end and ideal of the plan is to make-up and edit in one office all the matter fit for a national paper that can be published anywhere in the country under any name, without any support and co-operation of a local staff."[117] Such pressures of consolidation and commercialization in the newspaper industry spurred the formation of the early reformist project of support of the local newspaper, which under pressures of consolidation since the 1920s often sets itself in opposition to nonlocal owners who wish to homogenize content.[118]

Such pressures manifest themselves in similar ways in the radio industry. The ascendance of the economic pressure for centralization began with the 1927 Radio Act. Although seeming to enshrine local control by establishing "a system of broadcast regulation based upon the licensing of individual stations," the act did little to blunt the emerging commercial networks, which were already established by the late 1920s, attracted by "the economies of scale and high production values inherent in networking" that were possible through "the paradigm of [local] affiliation with national services."[119] Despite educators' opposition to wholesale commercialization, and despite a number of regulatory reaffirmations of the importance of localism throughout the 1930s and 1940s, "the battle [for localism in the geographical sense]," notes Stavitsky, "was already largely lost" by this time.[120] Fones-Wolf cites the fact that, even by 1931 and three years before the 1934 Communications Act, "the larger network-affiliated stations [already] had 76 percent of the power in the United States and the bulk of the advertising revenue."[121] The degree to which the system had closed to dissident views and uses was apparent as well by the 1930s in the general inability of socialists, pacifists, communists, advocates of birth control, and others to gain access to the airwaves, other than a handful of labor-owned stations.[122]

The struggle of localism against the deterritorialized commercial media system was and is more than a populist stand against monopolies, however. Localism is not only concerned about lack of diversity and accountability but also about mass culture as a symptom of deterritorialized commercialization run amok. The roots of the mass-culture critique as "reactions to mass culture that interpret it as either a symptom or a cause of social decay" have been traced to the eighteenth century and the rise of popular as well as popularizing institutions of communications in the context of an expanding capitalism and slow extension of a level of democratic accountability.[123] As one of a group of diverse intellectuals and reformers that ranged from Edward Shils to Robert Merton and the Frankfurt School, Dwight MacDonald voiced brashly in 1944 a sentiment generally shared among elites and policy makers aligned with the principle of localism that mass culture "is fabricated by technicians hired by businessmen; its audiences are passive consumers, their participation limited to the choice between buying and not buying."[124] As Brantlinger summarizes, MacDonald was concerned "with the deleterious effects of industrialism and commercialism on culture" as well as on the resulting prospects for a functioning popular democracy.[125]

While the regulatory tactics for preventing monopolies (geographical diffuseness and limits on multiple ownership) fit within the range of reformism and thus were able to be institutionalized to some degree, tactics for stemming the tide of mass culture were a great deal more controversial because they directly confronted capitalism. They called into question the assumption that the capitalist marketplace was the obedient servant to the individual needs and preferences of sovereign consumers. In the 1940s and 1950s, such a critique was increasingly raised and paid attention to with the publication of social criticism by David Riesman, David Potter, William Whyte, C. Wright Mills, and Vance Packard, among others, that exposed the anomie of middle-class life in the "American Dream" as well as that sought to attribute its cause to commercial elites and their aim of subordinating people to a life of commercial servitude.[126]

Due to this direct confrontation, chances of meaningfully mounting a challenge to capitalism via the mass-culture critique remained and remain quite slim. Not only does reformism limit the acceptable tactics to those that can be accommodated with little effort (such as consumer boycotts and the legislative letter-writing campaigns), reformism seriously compromises the ability of localism and the mass-culture critique to be a basis of substantial change, despite its continued resonance with activists and legislators.[127]

One set of reasons for this has to do with the relation of reformism to capitalism (or, for that matter, to any system). Within the reformist framework in the United States, the problem of commercialization is not seen as capitalism per se as much as uncontrolled capitalism. To counter the trend toward monopoly, regulatory intervention is needed not to produce an entirely new kind of system but to allow market competition to work as it is supposed to. In this sense, reformism shares with capitalism a faith in markets' ability to give people "what they want" as long as the market is open and undistorted by monopoly.[128] Reformism and its acceptance of capitalism can also accommodate the mass-culture critique to political conservativism. When attention to the roots in capitalism is stripped away, the mass-culture critique is easily adopted by cultural conservatives who lay the blame for the ills brought about by mass culture on intellectual laziness and liberal if not postmodern relativism.[129]

A second set of reasons why the mass-culture critique serves as a basis for reformism at best has to do with the spatial complexity of localism itself. Despite the clearly geographical sense of community in the Communications Act of 1934, which directs the F.C.C. to distribute "licenses, frequencies, hours of operation, and power among the several States and communities as to provide a fair, efficient, and equitable distribution of radio service to each of the same," the consolidation of a national system has been possible—and entirely consistent with the ideals of localism—due to the redefinition of "community" itself.[130] Instead of referring to residents of a discrete geographic locality, "community" has been increasingly seen in a deterritorialized sense. Stavitsky refers to the redefined community as "a social [as distinct from geographical] conception [. . .] defined by shared interests, tastes, and values."[131]

This redefined community corresponds to the imperatives of marketing, reflecting the broader problem of trying to reconcile localism in the geographic sense with a capitalist communications system.[132] Its coincidence with the imperatives of marketing becomes a basis for claiming paradoxically that highly centralized organizations are the best way to achieve localism in the sense of addressing diasporic communities, particularly with the multiplication of cable-

TV channels, satellite television, and radio, and the fragmentation of the "mass" audience into highly concentrated and narrowly defined niche markets. Despite its ostensible distinction from commercial broadcasting, arguments in favor of a national public-service broadcasting system also rely on this very same diasporic definition of community and champion the same definition of localism, which subjects it increasingly to commercial pressures of audience maximization and the embrace of "audience research as an essential management tool."[133]

As a result, by negating concerns about a national system (indeed, diasporic communities defined by interest, taste, and value range across the country), localism as a call for geographically local and distinct projects can no longer challenge "commercial," as both can be seen as highly centralized, professionalized national systems. For these reasons, a community media project that is commercial is no longer a contradiction in terms. A community scope and commercialism are not necessarily distinct nor opposites, a truism to which any local businessperson, newspaper publisher or broadcast-station owner can attest. Such a remaking of opposites is the underpinning of recent efforts to remake the regulatory environment. In the drive for deregulation since the 1980s, calls to protect localism have been dismissed as elitist attempts at protectionism and as anti-consumer.[134] Although the 2002 effort by the F.C.C. to relax media cross-ownership rules was blunted by a federal appeals court decision, in 2006 it not only reopened the effort to relax cross-ownership rules, but also asserted the primacy of market forces at the expense of local control by limiting the ability of local municipalities to freely negotiate conditions under which cable franchises could be granted.[135]

Community Challenges in the Age of Neoliberal Deregulation

An example of the insufficiency of reformist practices of community can be seen in the case of low-power FM radio (LPFM), a category of telecommunications service brought into being by the F.C.C. in 2000 in a contentious political struggle.[136] One class of LPFM radio stations broadcasts within a radius of approximately 3.5 miles, and a second in a radius of approximately 1-2 miles—miniscule by full-power radio standards, but intentionally so because of the emphasis on local as opposed to regional or national programming. Contrary to for-profit models marked by commercialization and consolidation, LPFM as a community project is exclusively nonprofit, with each station licensed only to a nonprofit educational or public-safety organization and meant to be responsive as well as accountable to a geographically local community instead of to a broadly defined, amorphous, deterritorialized public or market.[137] As of early 2007, 784 license applications have been approved.[138]

This spatial distinction between deterritorialized commercial and public media systems and the geographically localized LPFM service is not incidental nor implicit. Rather, it lies at the core of the rationale for the service and is seen most clearly in the requirements that must be fulfilled in order to receive a license. The intention of making a specific LPFM station accountable to people in a specific geographic locale is codified through the application process in such ways as requiring that the applicant organization be headquartered within ten miles of the transmission tower, have 75 percent of board members living within the same distance from the tower or, for public-safety uses, have legal jurisdiction in the service area of the proposed station. As a means of deciding between competing applications, organizations looked upon more favorably are those that have been in existence for the most recent two years as a nonprofit educational institution or organization, fulfilled the within-ten-miles stipulation noted earlier, and that pledge to produce at least eight hours of "locally originating" programming per day, with "local origination" meaning that production of the programming took place within ten miles of the station's transmission tower.[139]

While the geographic definition of "community" here is clear, the complexities of localism within existing traditions of broadcasting in a capitalist society have driven LPFM into a largely reformist direction. Although no programming schedules are generally available, as part of the license and construction permit application process applicants were invited to include exhibits and attachments with the application.[140] In these exhibits and attachments, many applicants explained not only how their application fit the guidelines for the service. They also elaborated on their conception of what a community radio station should be and what it should do. When interpreted in the context not only of regulatory imperatives but also of broader traditions and understandings of community media projects, an analysis of the comments from the initial set of 690-odd applications filed in 2000 provide a way of investigating how LPFM and, by extension, community media projects and communities themselves are currently defined and practiced.

Not surprisingly, the comments, exhibits, and attachments to license applications emphasize local and educational programming, thus coinciding clearly with the priorities of the system as set forth by the FCC. In their zeal to clearly demonstrate a fit with these requirements, a number of applicants offered hugely optimistic plans for their station's programming. For example, the application of the Star Tree Foundation lists a matrix of five different program types, five different content areas, and ten different formats, with specific programs being varying combinations of each of these.[141] For volunteer staffs composed of a small group of people who likely also hold outside jobs, programming the mandated thirty-six hours per week (and five hours per day on at least six days of any single week) is an immense burden, especially when considering that one estimate of the time needed to produce a seven-minute professional-style news piece by a person off the street is eight to ten hours.[142] The only means of fulfilling this requirement would be to use pre-recorded lectures, sermons, or music, as many applicants indeed proposed. Such proposals ranged from using more generic kinds of content such as an "extensive library of our locally produced programs that we have used in our television shows" to the rebroadcast of locally originating events such as high-school athletics or local government meetings.[143]

What's clear from an examination of these comments is the overwhelming conception of the geographic community as self-sufficient, bounded, and isolated. In such a conception, community needs and problems that the LPFM stations address emerge and exist wholly within this self-sufficient, bounded, and isolated community. Such a conception of community continues in the tradition of reformism by focusing in a purely local way on ameliorating the effects of conditions without inquiring into or addressing directly the very conditions that bring about the needs and problems in the first place.

Reformism in this sense appears in a wide variety of ways while also drawing upon long-standing traditions of communications. While the hybridizing of forms makes nonsensical an attempt to evaluate their comparative popularity by counting the frequency of different types, an holistic interpretation suggests the overwhelming predominance of reformist conceptions of station uses and, by implication, of reformist conceptions of communities, the majority of which rely upon rationalist-empiricism as a culture of truth in their use of combinations of reasoned argument and documentation.

For example, evangelism figures prominently among LPFM applicants, which draws upon the historical, formative relationship described in Chapter 2 between evangelism and broadcasting. Where some evangelical applicants assert the distinctiveness of their programming from commercial-station programming, many other evangelical applicants propose commercial-style programming, such as weather reports, current-issues talk shows, and music, along with more overtly Christian programming.[144] The more general point is the surprising ease with which commercial forms of programming apparently fit the expressed intentions not only of the station but of the evangelical organization. The recognition of this combination not only provides an example of the hybridity of forms in these comments. It also suggests that there is often no perceived conflict between evangelical intentions and those of an information or entertainment provider or educator—that the educational-sounding intentions of "religious instruction and moral teaching" are the means by which Howell Mountain Christian Radio, for example, seeks to fulfill its evangelical goal of fomenting religious conversion.[145]

A number of stations proposed themselves as in essence public utilities of information from and about the community, whether as the transparent voice (and willing servant) of the people, a common carrier, public utility, or even town booster. Such proposals resonate with early twentieth-century Chicago School theorizing of the necessity for full, accurate information as a common means by which a public can come into being and do its work.[146] For example, Rondaradio, Inc. seeks to be a neutral local communications utility that provides access for "the area's local governmental and non-profit organizations."[147] The Severe Storms Intercept Team justified its need for an LPFM station by emphasizing that communities outside the Oklahoma City metro area are just as in need of severe-weather information as is the metro area.[148] Other applicants who intend to become a transparent public communications utility express the desire to "enable a wide range of organizations and individuals to make their contribution to the educational, cultural and social life of the community," to "operate and manage the new LPFM service for use by a variety of governmental and community organizations," and to "establish a low power FM station for the transmission of audio from city and county governmental meetings."[149]

A more progressive form of public utility is to correct past inaccuracies or provide a more complete account that has relevance to the community—in short, to recover the repressed, hidden, or forgotten. An example is the Harry Bridges Institute, Inc. (the only labor-centered applicant in the selection of applications filed in 2000 examined here), which seeks not only to "educate students, union members, and the community of the contributions of organized labor and worker struggle in the United States" but also to "capture on video the testimonials of retired union members, primarily dockworkers, along the West Coast of the United States and make those testimonials available to students, union members, and the community through the video history project." An LPFM station would provide it with a means of reaching more people.[150]

Other applications proposed a more direct practice of social reformism. The intention implicit in social-reform uses is to provide useful, authoritative information but, in doing so, to empower residents to help themselves and thus become a means of achieving social, economic, environmental, and moral improvement. Proposed programs often focused on providing parents and children with appropriate informational and educational resources—a kind of combination of self-help, how-to books, and advice as well as children-specific stories and instruction. Beyond parents and children, other programming emphases included immigrant rights and environmental reform.[151] For example, the City Heights Community Development Corporation sees its community as a "virtual Ellis Island of California." As the application continues, "East Africans, Africans, African-Americans join with Indochinese and Latinos and other nationalities in this literal melting pot. A major challenge for this community is to help assimilate these 'new Americans' and to educate them and equip them to participate in their new home."[152] The community station (dubbed Radio City Heights) is to aid in this education and assimilation process. The need for such a station (as well as the overlaps with the form of information provider) are clear in the comment of a "young Somali gentleman [who] observed: 'we listen to BBC to hear news of our homeland (in Somalian language) but we don't know what is happening in our own neighborhood'."[153]

Reformist efforts that demonstrate a more overt combination with the faith-based nature of the social gospel are applicants who seek to use the LPFM station as a means of teaching life skills alongside religious doctrine. An example is the application of the First United Pentecostal Church of Redlands (Calif.), which plans programming on "practical life skills, such as how to obtain and keep a good job, how to manage small businesses and how to contribute to the social needs of low-income people in the listening area, [. . .] family relationships guidance, wholesome children's programming, family financial planning, [. . .] public service publicity for organizations serving the community, local weather information and announcements concerning emergency situations in the broadcast area." As if that extensive list of programming left anything out, the applicant concluded that "this unique format will be intermixed with appropriate light entertainment and inspirational music."[154]

Yet other reformist uses include using the station to help train people for careers in broadcasting, a direction with roots not only in apprenticeships but also as an extension of traditional, centralized public-works programs, most famously the variety of programs instituted during the New Deal of the 1930s, as well as the formation of vocational-technical institutions beginning with engineering and agriculture.[155] Staffing the station with students not only fits within the rubric of social reform, in that job skills can become a means by which a resident can improve herself and her situation, it is a readymade means of providing the volume of programming required by FCC regulations.

Finally, a number of stations seek to practice reformism by countering the deleterious effects of mass culture. Some applicants advocate their ability to provide insights into the world and its peoples. For example, the Midwestern Theater Troupe "believes that the tale is more powerful than the sermon, and that a simple story, simply told holds pure wisdom."[156] Music joins dramatic productions as a means of cultural uplift in the case of the National Jazz Hall of Fame Foundation, which seeks to "develop a national museum dedicated to preserve the memory of musicians who excelled in the art form of big band and/or jazz music." Its LPFM station will be used to "broadcast various artists' music to the general public" as well as "allow students to learn the broadcast industry" under supervision of radio professionals, thus blending cultural improvement with vocational training.[157]

Cultural uplift is sometimes hybridized with other forms of reformism. For example, the Unity World Production's proposal combines cultural programming with social reform and vocational training. Through "the presentation and production of high quality African-American cultural arts events and providing the resources for the artists to share their gifts and talents with the world," the organization seeks to create "an educational atmosphere that gives all students the opportunity and freedom to grow artistically and personally, that they may pursue college or a professional course in communication of the arts."[158] The application of the House of Gold Foundation hybridizes cultural uplift with evangelism and social reform by providing "faith-based education and activities primarily directed towards youth and families, specifically the cultural spread of the Gospel by such means as radio, theatre, art, and written forms."[159]

Despite the overwhelmingly dominant presence of reformism and corresponding definitions of community, a handful of dissenting statements implies a very different conception of community. Instead of as self-sufficient, bounded, and isolated, such applications imply a sense of community as relational rather than essential—that a community is embedded in broader spatial relationships and practices which have produced the very conditions to which the LPFM station is a response, and that these processes "outside" of the community must be addressed as well as their implications on the "inside."

The predominant form of embodying a relational sense of community was through media activism and its basis in the mass-culture critique. Granting the long-standing presence of media criticism in the U.S., media activism in its most recent incarnation is a response to the recent, accelerated pace of consolidation of commercial media companies. It is also an assertion of the need for an alternative source and mode of control outside conglomerates and rationales of commercial popularity.[160]

Within the commentary that drew upon this form, there was little attempt to recruit new adherents to this social cause, if for no other reason than overtly political solutions to problems of commercialization and consolidation could be interpreted by the FCC as advocating on behalf of a specific political purpose and thus a reason to disqualify the application. By contrast, the need to resist mass culture is simply a rationale for the need for the station, such as the aim of countering the "narrowness of mass media programming" and to give "voice to the thoughts, dreams, opinions, and community events not otherwise seen or heard on commercial and public radio, television, and other popular forms of media."[161] What is surprising here, too, is the apparent ease with which commercial-style programming apparently fits within such a form, although with some variations. While applications such as that from the Tulsa Free Radio Project seek to provide a quite conventional formula of "informational and editorial programming, public service messages, and in-depth news coverage," the application of Newspoetry serves as an example of distinctly unconventional programming.[162]

> Newspoetry's mission is to [. . .] promote the daily practice of writing poetry in response to the news in order to show that news is not simply unmediated information, but is a part of life that affects people deeply and is connected to other aspects of life from which it often claims to be objectively distanced. The work of "newspoets" furthermore constitutes a documentary record of our changing society which is richer in its variety of forms and attitudes than conventional forms of reportage and social science. Through our work, we hope to educate, enlighten, and move people who have become disenfranchised from

> the society they live in by the homogeneous, centralized media culture which dominates it.[163]

Other rationales and plans that relied on media activism minced few words in their denunciation of the commercial media landscape. The application of the Liberty Creek Homeowners Association stated that "the new station would be programmed to serve, inform, and to educate[,] which radio was originally intended [to do, unlike] a twenty-in-a-row, zip-me-off, rip-me-off, sterilized, sanitized, non-serving monster owned by ten companies."[164]

These were among the few applications with rationales that placed the station's geographic community within a web of spatial relationships (regional, national, international) and developments (homogenization, industry consolidation). Of the seven applications in 2000 based in such a rationale, all were cancelled or dismissed.[165]

Respatialized Community in the Context of Globalization

Stavitsky concludes his account of localism in U.S. broadcast regulations with the dismissive claim that "the comforting fable of a locally based broadcasting system that speaks to local concerns with local voices is chimerical."[166] This may indeed be true within the framework of broadcasting, but that does not exhaust the possible forms of community media projects. His solution to replace this chimera with a centralized, national system addressing niche audiences carved up into narrow, decontextualized sets of tastes, values, and preferences ignores the ways in which people in a geographical locale require a means of struggling against spatial pressures that affect their lives in profound ways. In more general terms, reformist practices of community—people in a locale, a diasporic group, or a combination—are ultimately an insufficient basis for community media projects as a means of significantly challenging market systems of communications, for they all presume not whether but in what way a particular group of people is to be addressed by a small group of other people.

Only with a very different sense of community can a tradition of community struggle against capitalism find purchase today in communications practice, and that is the sense of localism as an assertion of direct and popular access and control. In this sense, mobile anarchism and its radical rejection of institutionalized hierarchy and its replacement by a fluid voluntary association remains a resonant source for Independent Media Centers, which, despite their complexities and compromises, are the root of the novelty of Indymedia and what separates it from the reformism of community media such as LPFM.

Indymedia poses a clear challenge to reformist community media projects in its rejection of reformist conceptions of community. Although often seen as a global organization, the Indymedia collective is, in its own sense of itself, not a global organization at all, but a confederation of community media organizations.[167] While some Indymedia activists describe Indymedia as "a worldwide network," they also describe it as "a community based organization using media to facilitate political and cultural self-representation."[168] Such senses suggest in turn that the global and the local are no longer seen as separate, simply a matter of different scales (macro and micro), or simply imbricated, but instead are recognized as constituting each other.[169]

Other efforts also contribute to fluid approaches to programming and station operation. Westerkamp describes an attempt to broaden the conception of radio broadcasting from being a purveyor of content to a means of "environmental listening by broadcasting the soundscapes that listeners experienced in their daily lives."[170] A vibrant feminist practice in radio stretches possibilities in new directions. The broadcasting schedule for the illegal microradio station Women's Scéal Radio in Galway, Ireland was determined when "energy was stimulated by a chance meeting in the street—'Isn't it about time we did something?'" This freedom was intentional. "We were, at this stage, more interested in encouraging women onto the radio than in building up a listenership, so we did not plan programmes, and any woman who wanted to could drop in for a cup of tea. If she then wished to be part of the radio, she could be."[171]

In the critique of professionalism and the rationalization of communications, these and other examples not only engage in a significantly radical form of community media project. They are also some of many efforts that engage in a critique of rationalist empiricism as part of a struggle against market formations of communications. This critical project—which is roughly coterminous with the respatialization of capitalism and its challenges—is part of the more general formation of modernism.

Notes

1. "People's Choice Awards—Belong," *People's Choice Awards—Belong* <http://www.pcavote.com/belong/> (31 Dec. 2006).

2. "People's Choice Awards," *People's Choice Awards—Belong* <http://www. pcavote.com/> (30 Dec. 2006).

3. Throughout this chapter, I use the term "community media project" instead of the more typical "community media" in order to stress human practice over technology.

4. "A.C.M.: About the Alliance," *Alliance for Community Media* <http://www. alliancecm.org/index.php?page_id=3> (27 Nov. 2006). The acronym "N.A.S.A." stands for the National Aeronautics and Space Administration.

5. "A.C.M.: About the Alliance."

6. Brad Will, "Fragments of a Shattered Hope; Brazil Makes War on Sohno Real Squatters," *Brad Will in Boog City*, 25 April 2005 <http://bradwillboogcity .blogspot.com/> (28 Oct. 2006).

7. John Tarleton, "Brad Will's Remarkable Life," *The Indypendent*, November 2006, 9; Sarah Ferguson, "The Inconvenient Death of Brad Will," *Village Voice* <http://villagevoice.com/news/0646,ferguson,75024,2.html> (27 Nov. 2006).

8. Major-newspaper coverage of the situation at the time includes Sam Enriquez, "Who's in Charge in Oaxaca?," *Los Angeles Times*, 28 August 2006, A7; Marc Lacey, "Mexican Federal Police and Army Move to Retake Oaxaca," *The New York Times*, 30 October 2006, A3; and David Usborne, "What Went Wrong in Oaxaca?," *The Independent*, 31 October 2006, 26.

9. Will, "Fragments"; Brad Will, " 'One More Martyr in a Dirty War—One More Time to Cry and Hurt," *The Indypendent* November 2006, 8.

10. Among the many locations, see "Last Video of Brad Will," *Friends of Brad Will » Multimedia* <http://www.friendsofbradwill.org/category/multimedia/> (12 Nov. 2006).

11. William Barlow, "Community Radio in the United States: The Struggle for a Democratic Medium," *Media, Culture & Society* 10, no. 1 (January 1988): 81–105; Charles Fairchild, *Community Radio and Public Culture: Being an Examination of Media Access and Equity in the Nations of North America* (Cresskill, N.J.: Hampton Press, 2001); Federal Communications Commission, *Low-Power FM Radio: An Applicant's Guide* (Washington, D.C.: GPO, 2000), 2; Kevin Howley, *Community Media; People, Places, and Communication Technologies* (Cambridge: Cambridge University Press, 2005).

12. Alexis de Tocqueville, *Democracy in America* (New York: A. Knopf, 1994).

13. Francis Mulhern, "Towards 2000, or News From You-Know-Where," *New Left Review* no. 148 (November/December 1984): 24.

14. Cynthia Karena, "Get Heard: Get Involved in Community Radio," *Australian Screen Education* no. 37, Summer 2004, 44–51.

15. Raymond Williams, *Resources of Hope; Culture, Democracy, Socialism* (London: Verso, 1989), 112–13.

16. Benedict Anderson, *Imagined Communities; Reflections on the Origin and Spread of Nationalism*, rev. ed. (London: Verso, 1991), 6.

17. Raymond Williams, *The Country and the City* (London: Chatto & Windus, 1973), 154.

18. Henri Lefebvre, *The Urban Revolution*, transl. Robert Bononno (Minneapolis, Minn.: University of Minnesota Press, 2003), 1–22.

Among those who have explored the modernized and urbanized social relations in the sense meant here are Walter Benjamin, "On Some Motifs in Baudelaire," in *Illuminations; Essays and Reflections*, transl. Harry Zohn (New York: Schocken, 1969), 155–200; Raymond Williams, *Television: Technology and Cultural Form* (New York: Schocken, 1974), 26; and Raymond Williams, *The Year 2000* (New York: Pantheon Books, 1983), 187–89.

19. Raymond Williams, *Keywords; A Vocabulary of Culture and Society* (New York: Oxford University Press, 1976), 65–66.

20. Williams, *Country and the City*, 1.

21. An oft-cited study of the complication and play between distinctions of high-official and "low"-popular in premodern Europe is Mikhail Bakhtin, *Rabelais and His World*, transl. Hélène Iswolsky (Bloomington, Ind.: Indiana University Press, 1984).

22. Carol M. Rose, "The Ancient Constitution vs. the Federalist Empire: Anti-Federalism from the Attack on 'Monarchism' to Modern Localism," *Northwestern University Law Review* 84, no. 1 (1989–1990): 74–105. Paraphrase is from Paul Cowling, "An Earthly Enigma: The Role of Localism in the Political, Cultural and Economic Di-

mensions of Media Ownership Regulation," *Hastings Communications and Entertainment Law Journal* 27, no. 2 (Winter 2005): 283.

23. Gregory Claeys, *Citizens and Saints; Politics and Anti-Politics in Early British Socialism* (Cambridge: Cambridge University Press, 1989), 23–62.

24. Claeys, *Citizens and Saints*, 28–29.

25. Josiah Bartlett Lambert, *"If the Workers Took a Notion"; The Right to Strike and American Political Development* (Ithaca, N.Y.: ILR Press, 2005), 23.

26. Lambert, *Workers*, 26, 31–34; Bruce Laurie, *Artisans into Workers; Labor in Nineteenth-Century America* (New York: Noonday Press, 1989).

27. Quoted in Kim Voss, *The Making of American Exceptionalism: The Knights of Labor and Class Formation in the Nineteenth Century* (Ithaca, N.Y.: Cornell University Press, 1993), 24. "Yeomanry of the city" quotation cited in Lambert, *Workers*, 23.

28. Lambert, *Workers*, 24.

29. E.P. Thompson, *The Making of the English Working Class* (New York: Vintage, 1966), esp. 87–101; Christopher Lasch, *The True and Only Heaven; Progress and Its Critics* (New York: W.W. Norton, 1991), 168–225; Robert H. Wiebe, *Self-Rule: a Cultural History of American Democracy* (Chicago, Ill.: University of Chicago Press, 1995); and Sean Wilentz, *Chants Democratic: New York and the Rise of the American Working Class, 1788–1850* (Oxford: Oxford University Press, 1984).

30. Thompson, *Making*, 96. See also Eric J. Hobsbawm, *The Age of Revolution, 1789–1848* (Cleveland, Ohio: World Publishing Company, 1962).

31. Lambert, *Workers*, 35.

32. Quoted in Robert S. Fogarty, *All Things New; American Communes and Utopian Movements, 1860–1914* (Chicago, Ill.: University of Chicago Press, 1990), 8.

33. Quoted in Arthur Bestor, *Backwoods Utopias; The Sectarian and Owenite Phases of Communitarian Socialism in America: 1663–1829* (Philadelphia, Penn.: University of Pennsylvania Press, 1950), 5.

34. Bestor, *Backwoods Utopias*, 9.

35. Claeys, *Citizens and Saints*, 30–31.

36. The unremarkable, everyday experience of Puritan life is addressed in a range of works. See Perry Miller and Thomas H. Johnson, ed., *The Puritans*, rev. ed. (New York: Harper and Row, 1963), 381–95; Francis Bremer, *The Puritan Experiment: New England Society from Bradford to Edwards*, rev. ed. (Hanover, N.H.: University Press of New England, 1995); and John Demos, *A Little Commonwealth; Family Life in Plymouth Colony* (New York: Oxford University Press, 2000).

37. Edward K. Spann, *Brotherly Tomorrows; Movements for a Cooperative Society in America, 1820–1920* (New York: Columbia University Press, 1989), 1–2; Robert C. Schehr, *Dynamic Utopia; Establishing Intentional Communities as a New Social Movement* (Westport, Conn.: Bergin & Garvey, 1997), 26–43. An overview of the centrality of Parisian efforts to articulate a basis for utopian communities is David Harvey, *Paris, Capital of Modernity* (New York: Routledge, 2003), 65–80.

38. Ralph L. Rusk, ed., *The Letters of Ralph Waldo Emerson*, vol. 2 (New York: Columbia University Press, 1939), 353; quoted in Dolores Hayden, *Seven American Utopias; The Architecture of Communitarian Socialism, 1790–1975* (Cambridge, Mass.: MIT Press, 1976), 9.

39. Henry David Thoreau, *Walden, or, Life in the Woods* (Boston, Mass.: Ticknor and Fields, 1854); Hayden, *Seven American Utopias*, 9.

Useful discussions of the range and examples of such communities in the nineteenth-century United States include Bestor, *Backwoods Utopias*; Spann, *Brotherly Tomorrows*; Fogarty, *All Things New*; Carl J. Guarneri, *The Utopian Alternative; Fourier-*

ism in Nineteenth-Century America (Ithaca, N.Y.: Cornell University Press, 1991); Mark Holloway, *Heavens on Earth; Utopian Communities in America, 1680–1880* (New York: Library Publishers, 1951); Donald Pitzer, ed., *America's Communal Utopias* (Chapel Hill: University of North Carolina Press, 1997); and Anna Lisa Peterson, *Seeds of the Kingdom; Utopian Communities in the Americas* (New York: Oxford University Press, 2005).

40. In addition to Guarneri, *The Utopian Alternative*, see Gregory Claeys, *Machinery, Money, and the Millennium: from Moral Economy to Socialism, 1815–60* (Princeton, N.J.: Princeton University Press, 1987); Carol A. Kolmerten, *Women in Utopia: the Ideology of Gender in the American Owenite Communities* (Bloomington, Ind.: Indiana University Press, 1990); Barbara Taylor, *Eve and the New Jerusalem: Socialism and Feminism in the Nineteenth Century* (Cambridge, Mass.: Harvard University Press, 1993); and Edward Royle, *Robert Owen and the Commencement of the Millennium: a Study of the Harmony Community* (New York: Manchester University Press, 1998).

41. Hayden, *Seven American Utopias*, 9.

42. Fogarty, *All Things New*, 10.

43. Spann, *Brotherly Tomorrows*, 8.

44. Among the countless examples, those most relevant to the case being made here include Daniel Czitrom, *Media and the American Mind; From Morse to McLuhan* (Chapel Hill, N.C.: University of North Carolina Press, 1982), 3–29; and James W. Carey and John J. Quirk, "The Mythos of the Electronic Revolution," in James W. Carey, *Communication as Culture; Essays on Media and Society* (Boston, Mass.: Unwin Hyman, 1989), 113–41.

45. Quoted in Czitrom, *Media and the American Mind*, 11–12.

46. Kevin Barnhurst and John Nerone, *The Form of News; A History* (New York: Guildford, 2001), 36–41.

47. Richard B. Kielbowicz, *News in the Mail; The Press, Post Office, and Public Information, 1700–1860s* (Westport, Conn.: Greenwood Press, 1989), 149.

48. Quoted in Kielbowicz, *News in the Mail*, 155.

49. Kielbowicz, *News in the Mail*, 141–161; Allan Pred, *Urban Growth and City Systems in the United States, 1840–1860* (Cambridge, Mass.: Harvard University Press, 1980), 143.

50. Kielbowicz, *News in the Mail*, 149.

51. Barnhurst and Nerone, *Form of News*, 31–67; Lambert, *Workers*, 20; Laurie, *Artisans Into Workers*. See also William E. Ames, "Federal Patronage and the Washington D.C. Press," *Journalism Quarterly* 49, no. 1 (Spring 1972): 22–30; Wilentz, *Chants Democratic*; and Lasch, *True and Only Heaven*, 296–328.

52. Kielbowicz, *News in the Mail*, 31.

53. Richard John, *Spreading the News: The American Postal System from Franklin to Morse* (Cambridge, Mass.: Harvard University Press, 1995).

54. Kielbowicz, *News in the Mail*, 62–63, 84–86.

55. Kielbowicz, *News in the Mail*, 41.

56. In addition to Ames, "Federal Patronage," see William E. Ames, *A History of the National Intelligencer* (Chapel Hill, N.C.: University of North Carolina Press, 1972); Culver H. Smith, *The Press, Politics, and Patronage; The American Government's Use of Newspapers 1789–1875* (Athens, Ga.: University of Georgia Press, 1977); and Gerald J. Baldasty, *The Commercialization of News in the Nineteenth Century* (Madison, Wisc.: University of Wisconsin Press, 1992), 11–35.

Such a need existed prior to the establishment of the U.S. Government Printing Office. See Laurence Schmeckebier, *The Government Printing Office, Its History, Activi-*

ties, and Organization (Baltimore, Md.: Johns Hopkins University Press, 1925); Robert Kerr, *History of the Government Printing Office (at Washington, D.C.); With a Brief Record of the Public Printing for a Century, 1789–1881* (New York: Burt Franklin, 1970); and Robert E. Kling, Jr., *The Government Printing Office* (New York: Praeger, 1970), 3–17.

57. Baldasty, *Commercialization*, 20–24.

58. Carolyn Stewart Dyer, "Political Patronage of the Wisconsin Press, 1849–1861: New Perspectives on the Economics of Patronage," *Journalism Monographs* 109 (February 1989).

59. Siva Vaidhyanathan, *Copyrights and Copywrongs; The Rise of Intellectual Property and How It Threatens Creativity* (New York: New York University Press, 2001), 43–47.

60. *International News Serv. v. Associated Press*, 248 U.S. 215 (1918); Richard A. Schwarzlose, *The Nation's Newsbrokers, Volume 1: The Formative Years, from Pretelegraph to 1865* (Evanston, Ill.: Northwestern University Press, 1989); Richard A. Schwarzlose, *The Nation's Newsbrokers, Volume 2: The Rush to Institution, from 1865 to 1920* (Evanston, Ill.: Northwestern University Press, 1990), 233; Paul W. Sullivan, "News Piracy: Unfair Competition and the Misappropriation Doctrine," *Journalism Monographs* no. 56 (May 1978), 4–8.

61. In addition to Sullivan, "News Piracy," see Victor Rosewater, *History of Cooperative News-Gathering in the United States* (New York: D. Appleton and Company, 1930), 284–92.

62. Kielbowicz, *News in the Mail*, 57. See also Philip M. Katz, *From Appomattox to Montmartre; Americans and the Paris Commune* (Cambridge, Mass.: Harvard University Press, 1998), 62.

63. Kielbowicz, *News in the Mail*, 149.

64. Barnhurst and Nerone, *Form of News*, 103.

65. Barnhurst and Nerone, *Form of News*, 75.

66. Edward E. Adams and Gerald J. Baldasty, "Syndicated Service Dependence and a Lack of Commitment to Localism: Scripps Newspapers and Market Subordination," *Journalism and Mass Communication Quarterly* 78, no. 3 (Autumn 2001): 521.

67. Elizabeth Sanders, *Roots of Reform; Farmers, Workers, and the American State, 1877–1917* (Chicago: University of Chicago Press, 1999), 16.

68. Jason W. Moore, "Remaking Work, Remaking Space: Spaces of Production and Accumulation in the Reconstruction of American Capitalism, 1865–1920," *Antipode* 34, no. 2 (March 2002): 177–78; Marcel van der Linden and Wayne Thorpe, "The Rise and Fall of Revolutionary Syndicalism," in *Revolutionary Syndicalism; An International Perspective*, ed. Marcel van der Linden and Wayne Thorpe (Aldershot: Scolar Press, 1990), 4.

69. Moore, "Remaking Work," 178–79; Susan Eleanor Hirsch, *After the Strike: a Century of Labor Struggle at Pullman* (Urbana, Ill.: University of Illinois Press, 2003); David Ray Papke, *The Pullman Case: the Clash of Labor and Capital in Industrial America* (Lawrence, Kan.: University Press of Kansas, 1999).

70. Moore, "Remaking Work," 178–79.

71. Moore, "Remaking Work," 178; Sanders, *Roots of Reform*.

72. Paul Buhle, *Marxism in the United States; Remapping the History of the American Left* (New York: Verso, 1987), 62.

73. Domenica M. Barbuto, *American Settlement Houses and Progressive Social Reform: an Encyclopedia of the American Settlement Movement* (Phoenix, Ariz.: Oryx Press, 1999); Elisabeth Lasch-Quinn, *Black Neighbors: Race and the Limits of Reform in*

the American Settlement House Movement, 1890–1945 (Chapel Hill, N.C.: University of North Carolina Press, 1993); Allen F. Davis, *Spearheads for Reform: The Social Settlements and the Progressive Movement, 1890–1914* (New Brunswick, N.J.: Rutgers University Press, 1984).

74. Alan Trachtenberg, *The Incorporation of America; Culture and Society in the Gilded Age* (New York: Hill and Wang, 1982), xx.

75. A discussion of the interrelations and commonalities across national lines is Sima Lieberman, *Labor Movements and Labor Thought; Spain, France, Germany, and the United States* (New York: Praeger, 1986). Useful here among the immense literature of Chartism are Gareth Steadman Jones, "Rethinking Chartism," in *Languages of Class; Studies in English Working Class History 1832–1982* (Cambridge: Cambridge University Press, 1983), 90–178; Dorothy Thompson, *The Chartists; Popular Politics in the Industrial Revolution* (New York: Pantheon, 1984); and Owen Ashton, Robert Fyson, and Stephen Roberts, eds., *The Chartist Legacy* (Woodbridge, Suffolk: Merlin Press, 1999). Awareness in the United States of the Paris Commune is discussed in Katz, *From Appomattox to Montmartre*.

76. Trachtenberg, *Incorporation*, 80, 89.

77. Lambert, *Workers*, 43. The problem of inclusiveness is broached in Philip Foner, *Women and the American Labor Movement: From the First Trade Unions to the Present* (New York : Free Press, 1982); Philip Foner, *Organized Labor and the Black Worker, 1619–1981*, 2nd ed. (New York: International Publishers, 1981); and Paul D. Moreno, *Black Americans and Organized Labor: a New History* (Baton Rouge, La.: Louisiana State University Press, 2006).

78. Trachtenberg, *Incorporation*, 95. For a useful summary of the A.F. of L.'s development and political program, see Sanders, *Roots of Reform*, 71–100. Other labor leaders, such as William Z. Foster, had reached similar conclusions, at least at the tactical level. See James R. Barrett, *William Z. Foster and the Tragedies of American Radicalism* (Urbana, Ill.: University of Illinois Press, 1999), 46.

79. In addition to Sanders, *Roots of Reform*, valuable discussions include Seymour M. Lipset, *Agrarian Socialism* (New York: Doubleday, 1968); Lawrence Goodwyn, *The Democratic Promise* (New York: Oxford University Press, 1976); James Green, *Grass Roots Socialism* (Baton Rouge, La.: Louisiana State University Press, 1983); Norman Pollack, *The Just Polity: Populism, Law, and Human Welfare* (Urbana, Ill.: University of Illinois Press, 1987); Robert C. McMath, *American Populism: a Social History, 1877–1898* (New York: Noonday Press, 1993); and Michael Kazin, *The Populist Persuasion: an American History*, revised ed. (Ithaca, N.Y.: Cornell University Press 1998).

80. Sanders, *Roots of Reform*, 105. Lack of openness to nonwhite farmers, however, distinguished it from earlier, more inclusive labor organizations such as the Knights of Labor. See Gerald H. Gaither, *Blacks and the Populist Movement: Ballots and Bigotry in the New South*, revised ed. (Tuscaloosa, Ala.: University of Alabama Press, 2005).

81. Sanders, *Roots of Reform*, 101.

82. In addition to Sanders, *Roots of Reform*, see Gretchen Ritter, *Goldbugs and Greenbacks: the Antimonopoly Tradition and the Politics of Finance in America* (New York: Cambridge University Press, 1997).

83. Sanders, *Roots of Reform*, 414–19.

84. Among the more useful accounts in a vast literature are Alan Ritter, *Anarchism: A Theoretical Analysis* (Cambridge: Cambridge University Press, 1980); Jeremy Jennings, *Syndicalism in France; A Study of Ideas* (New York: St. Martin's Press, 1990); Marcel van der Linden and Wayne Thorpe, *Revolutionary Syndicalism; An International Perspective* (Aldershot: Scolar Press, 1990); George Crowder, *Classical Anarchism; The*

Political Thought of Godwin, Proudhon, Bakunin, and Kropotkin (Oxford: Clarendon Press, 1991); Kenneth H. Tucker, Jr., *French Revolutionary Syndicalism and the Public Sphere* (Cambridge: Cambridge University Press, 1996); Brian Morris, *Kropotkin; The Politics of Community* (Amherst, N.Y.: Humanity Books, 2004); Rudolf Rocker, *Anarcho-Syndicalism*, 6th ed. (Edinburgh: AK Press, 2004); and Seán M. Sheehan, *Anarchism* (London: Reaktion Books, 2003).

85. Craig Phelan, *Grand Master Workman; Terence Powderly and the Knights of Labor* (Westport, Conn.: Greenwood Press, 2000), 3. A useful summary of the Knights and their political program is Sanders, *Roots of Reform*, 33–55.

86. The Congress of Industrial Organizations (C.I.O.) also sought a more inclusive industrial-unionist structure, but yet was hampered by its reformist goals by comparison to the I.W.W. In addition to other general overviews already noted, an able, recent overview of the emergence of the C.I.O. is Clayton Sinyai, *Schools of Democracy; A Political History of the American Labor Movement* (Ithaca, N.Y.: ILR Press, 2006), 110–63.

87. John Graham Brooks, *American Syndicalism; the I.W.W.* (New York: Macmillan, 1913); Paul Brissenden, *The I.W.W.; A Study of American Syndicalism* (New York: Columbia University Press, 1919), 28–37; Salvatore Salerno, *Red November, Black November; Culture and Community in the Industrial Workers of the World* (Albany, N.Y.: State University of New York Press, 1989), 93–118; Melvin Dubofsky, "The Rise and Fall of Revolutionary Syndicalism in the United States," in *Revolutionary Syndicalism; An International Perspective*, ed. Marcel van der Linden and Wayne Thorpe (Aldershot: Scolar Press, 1990), 203–20.

88. Salerno, *Red November*, 7–8; Robert E. Weir, *Beyond Labor's Veil; The Culture of the Knights of Labor* (University Park, Penn.: Pennsylvania State University Press, 1996), 20.

89. Salerno, *Red November*, 7–8.

90. Salerno, *Red November*, 133–35.

91. Salerno, *Red November*, 134–40.

92. Salerno, *Red November*, 126–27.

93. Salerno, *Red November*, 129.

94. Brissenden, *The I.W.W.*, 263; Melvyn Dubofsky, *We Shall Be All; A History of the Industrial Workers of the World* (Chicago, Ill.: Quadrangle Books, 1969), 173–97.

95. Lambert, *Workers*, 64-83; David Montgomery, *The Fall of the House of Labor: The Workplace, the State, and American Labor Activism, 1865–1925* (New York: Cambridge University Press, 1987).

96. Morris, *Kropotkin*, 74. Quoted phrase is from Ritter, *Anarchism*, 3.

97. Elliot Shore, "Selling Socialism: The *Appeal to Reason* and the Radical Press in Turn-of-the-Century America," *Media, Culture and Society* 7, no. 2 (April 1985): 148.

98. Shore, "Selling Socialism," 148.

99. Elliot Shore, *Talkin' Socialism: J.A. Wayland and the Role of the Press in American Radicalism, 1890–1912* (Lawrence, Ks.: University of Kansas Press, 1988), 2.

100. Shore, "Selling Socialism," 158.

101. Upton Sinclair, *The Jungle* (New York: Jungle Publishing Company, 1906); Shore, "Selling Socialism," 164–65.

102. Jon Bekken, "The Working Class Press at the Turn of the Century," in *Ruthless Criticism: New Perspectives in U.S. Communications History*, ed. William S. Solomon and Robert McChesney (Minneapolis, Minn.: University of Minnesota Press, 1993), 158.

103. Shore, "Selling Socialism," 163–5; Shore, *Talkin' Socialism*, 208–26; Jon Bekken, "'These Great and Dangerous Powers': Postal Censorship of the Press," *Journal of Communication* 15, no. 1 (Winter 1991): 55–71.

104. Allen Ruff, *"We Called Each Other Comrade": Charles H. Kerr & Company, Radical Publishers* (Urbana, Ill.: University of Illinois Press, 1997), xvii.

105. Bekken, "Working Class Press"; Jeff Ferrell, "'The Song the Capitalist Never Sings': The Brotherhood of Timber Workers and the Culture of Conflict," *Labor History* 32, no. 3 (Summer 1991): 422–31.

106. William Haywood, *Bill Haywood's Book; The Autobiography of William D. Haywood*, reprint edition (New York: International Publishers, 1966), 153–54.

107. Joseph R. Conlin, ed., *The American Radical Press: 1880–1960*, Volume 1 (Westport, Conn.: Greenwood, 1974), 131. Quoted in Donald E. Winters, *The Soul of the Wobblies; The I.W.W., Religion, and American Culture in the Progressive Era, 1905–1917* (Westport, Conn.: Greenwood, 1985), 61.

108. Winters, *Soul of the Wobblies*; Salerno, *Red November, Black November*.

109. Among the most useful accounts is James W. Carey, "The Chicago School and the History of Mass Communication Research," in *James Carey; A Critical Reader*, ed. Even Stryker Munson and Catherine A. Warren (Minneapolis, Minn.: University of Minnesota Press, 1997), 14–33.

110. Ferdinand Tönnies, *Community and Civil Society (Gemeinschaft und Gesellschaft)*, transl. Jose Harris and Margaret Hollis (New York: Cambridge University Press, 2001).

111. Useful overviews of the Chicago School include Martin Bulmer, *The Chicago School of Sociology* (Chicago: University of Chicago Press, 1984); Fred Matthews, *Quest for an American Sociology; Robert E. Park and the Chicago School* (Montreál, Quebéc: McGill-Queens University Press, 1977); and Raymond Wilson, *In Quest of Community: Social Philosophy in the United States, 1860–1920* (New York: John Wiley and Sons, 1968). For a discussion of the early exchange between German and American social theory about communications and society, see Hanno Hardt, *Social Theories of the Press; Early German and American Perspectives* (Beverly Hills, Calif.: Sage, 1979).

112. Jean Quandt, *From the Small Town to the Great Community; The Social Thought of Progressive Intellectuals* (New Brunswick, N.J.: Rutgers University Press, 1970), 17.

113. A useful overview of the place of communications in the social theory of the Chicago School in addition to Czitrom, *Media and the American Mind*, 91–121 is Hanno Hardt, *Critical Communication Studies; Communication, History & Theory in America* (New York: Routledge, 1992), 31–76.

114. Useful accounts of the place of localism in U.S. broadcasting regulation in addition to Cowling, "Earthly Enigma" include Alan G. Stavitsky, "The Changing Conception of Localism in U.S. Public Radio," *Journal of Broadcasting & Electronic Media* 38, no. 1 (Winter 1994): 19–33; and David Silverman and David N. Tobenkin, "The FCC's Main Studio Rule: Achieving Little for Localism at a Great Cost to Broadcasters," *Federal Communications Law Journal* 53, no. 3 (May 2001): 469–507.

115. Stavitsky, "Changing Conception of Localism," 19.

116. Schwarzlose, *Nation's Newsbrokers*, v. 1 and v. 2.

117. Quoted in Adams and Baldasty, "Syndicated Service Dependence," 529.

118. Key works relevant to this position include Will Irwin, *The American Newspaper* (Ames, Iowa: Iowa State University Press, 1969); Alfred McClung Lee, *The Daily Newspaper in America; The Evolution of a Social Instrument* (New York: Macmillan, 1937); Commission on Freedom of the Press, *A Free and Responsible Press; A General Report on Mass Communication* (Chicago, Ill.: University of Chicago Press, 1947); Newton N. Minow, *Equal Time; The Private Broadcaster and the Public Interest* (New York: Atheneum, 1964); Herbert Schiller, *Culture, Inc.* (New York: Oxford University

Press, 1989); Ben Bagdikian, *The Media Monopoly*, 6th ed. (Boston, Mass.: Beacon Press, 2000); Richard McCord, *The Chain Gang: One Newspaper Versus the Gannett Empire* (Columbia, MO: University of Missouri Press, 1996); Robert W. McChesney, *Rich Media, Poor Democracy; Communication Politics in Dubious Times* (Urbana, Ill.: University of Illinois Press, 1999); and Robert W. McChesney and Ben Scott, eds., *Our Unfree Press: 100 Years of Radical Media Criticism* (New York: New Press, 2004).

119. Stavitsky, "Changing Conception of Localism," 19–20.

120. Stavitsky, "Changing Conception of Localism," 21. The authoritative account is Robert W. McChesney, *Telecommunications, Mass Media, and Democracy; the Battle for Control of U.S. Broadcasting, 1928–1935* (New York: Oxford University Press, 1993). An account of one legislator's efforts to protect localism is Jennifer M. Proffitt and Michael Brown, "Regulating the Radio Monopoly: Ewin Davis and His Legislative Debates, 1923–1928," *Journal of Radio Studies* 11, no. 1 (June 2004): 100–15.

121. Elizabeth Fones-Wolf, "Defending Listeners' Rights: Labor and Media Reform in Postwar America," *Canadian Journal of Communication* 31, no. 3 (2006): 501.

122. Fones-Wolf, "Defending Listeners' Rights," 501. An expanded discussion is Elizabeth Fones-Wolf, *Waves of Opposition; Labor and the Struggle for Democratic Radio* (Urbana, Ill.: University of Illinois Press, 2006). See also Nathan Godfried, *WCFL: Chicago's Voice of Labor, 1926–1978* (Urbana, Ill.: University of Illinois Press, 1997); and Matthew Lasar, *Pacifica Radio: The Rise of an Alternative Network* (Philadelphia, Penn.: Temple University Press, 1999).

123. Herbert Gans, *Popular Culture and High Culture; An Analysis and Evaluation of Taste* (New York: Basic Books, 1974), 4; Patrick Brantlinger, *Bread & Circuses; Theories of Mass Culture as Social Decay* (Ithaca, N.Y.: Cornell University Press, 1983), 17.

Other useful overviews and discussions include Bernard Rosenberg and David Manning White, eds., *Mass Culture; The Popular Arts in America* (Glencoe, Ill.: Free Press, 1957); Raymond Williams, "The Press and Popular Culture: An Historical Perspective," in *Newspaper History from the Seventeenth Century to the Present Day*, ed. George Boyce, James Curran and Paula Wingate (London: Constable, 1978), 41–50; and Tony Bennett, "Theories of the Media, Theories of Society," in *Culture, Society and the Media*, ed. Michael Gurevitch, Tony Bennett, James Curran, and Janet Woollacott, reprint edition (London: Routledge, 1995), 30–55.

The centralization was not only urban, but western, a process that was also recognized by in the nineteenth century. See Robert W. Rydell and Rob Kroes, *Buffalo Bill in Bologna: the Americanization of the World, 1869–1922* (Chicago, Ill.: University of Chicago Press, 2005). The mass-culture debate prefigured in important ways the later cultural-imperialism thesis of the 1960s and 1970s. Among the many useful accounts and critiques is John Tomlinson, *Cultural Imperialism: A Critical Introduction* (Baltimore, Md.: Johns Hopkins University Press, 1991).

124. Quoted in White and Rosenberg, *Mass Culture*, 60. MacDonald in a number of ways embodies the ideological range of contributors to the mass-culture debate. See Stephen J. Whitfield, *A Critical American: The Politics of Dwight MacDonald* (Hamden, Conn.: Archon Books, 1984); Michael Wreszin, *A Rebel in Defense of Tradition: the Life and Politics of Dwight MacDonald* (New York: Basic Books, 1994); and Gregory D. Sumner, *Dwight MacDonald and the Politics Circle: the Challenge of Cosmopolitan Democracy* (Ithaca, N.Y.: Cornell University Press, 1996).

125. Brantlinger, *Bread & Circuses*, 201.

126. David Riesman, *The Lonely Crowd: A Study of the Changing American Character* (New Haven, Conn.: Yale University Press, 1950); David M. Potter, *People of*

Plenty: Economic Abundance and the American Character (Chicago, Ill.: University of Chicago Press, 1954); William H. Whyte, *The Organizational Man* (New York: Doubleday, 1957); C. Wright Mills, *The Power Elite* (New York: Oxford University Press, 1956); and Vance Packard, *The Hidden Persuaders* (New York: D. McKay, 1957). A useful discussion of social criticism at this time is Daniel Horowitz, *Vance Packard and American Social Criticism* (Chapel Hill, N.C.: University of North Carolina Press, 1994).

127. Senate, Committee on Commerce, Science, and Transportation, *Preservation of Localism, Program Diversity, and Competition in Television Broadcast Services Act of 2003*, 108th Cong., 1st sess. (Washington, D.C.: Government Printing Office, 2003).

128. For a contrasting claim—that the market plays no such role—see Eileen R. Meehan, *Why TV is not Our Fault: Television Programming, Viewers, and Who's Really in Control* (Lanham, Md.: Rowman & Littlefield, 2005).

129. Prominent examples include Allan Bloom, *The Closing of the American Mind* (New York: Simon and Schuster, 1987); and a series of books by E.D. Hirsch, with the initial one being E.D. Hirsch, Jr., Joseph Kett, James Trefil, *Cultural Literacy: What Every American Needs to Know* (Boston, Mass.: Houghton Mifflin, 1987).

130. Robert L. Hilliard and Michael C. Keith, *The Quieted Voice: The Rise and Demise of Localism in American Radio* (Carbondale, Ill.: Southern Illinois University Press, 2005); Robert B. Horwitz, *The Irony of Regulatory Reform: The Deregulation of American Telecommunications* (New York: Oxford University Press, 1989), 154–95.

131. FCC quotation from David M. Silverman and David N. Tobenkin, "The FCC's Main Studio Rule: Achieving Little for Localism at a Great Cost to Broadcasters," *Federal Communications Law Journal* 53, no. 3 (May 2001): 474; Stavitsky, "Changing Conception of Localism," 20.

132. Horwitz, *Irony of Regulatory Reform*, 154–95.

133. Stavitsky, "Changing Conception of Localism," 25. General works on public broadcasting and such pressures include Michael Tracey, *The Decline and Fall of Public Service Broadcasting* (New York: Oxford University Press, 1998); B.J. Bullert, *Public Television: Politics and the Battle over Documentary Film* (New Brunswick, N.J.: Rutgers University Press, 1997); David Barsamian, *The Decline and Fall of Public Broadcasting* (Cambridge, Mass.: South End Press, 2001); James Ledbetter, *Made Possible By— : The Death of Public Broadcasting in the United States* (New York: Verso, 1997).

134. A useful account of the lead up to and the implications of the 1996 Telecommunications Act is Patricia Aufderheide, *Communications Policy and the Public Interest; The Telecommunications Act of 1996* (New York: Guilford, 1999).

135. Federal Communications Commission, *Notice of Proposed Rulemaking; In the Matter of 2002 Biennial Regulatory Review—Review of the Commission's Broadcast Ownership Rules and other Rules Adopted Pursuant to Section 202 of the Telecommunications Act of 1996; Cross-ownership of Broadcast Stations and Newspapers; Rules and Policies Concerning Multiple Ownership of Radio Broadcast Stations in Local Markets; Definition of Radio Markets*, MB Docket Nos. 02-277, 01-235, 01-317, 00-244, 23 September 2002; Federal Communications Commission, *Report and Order and Notice of Proposed Rulemaking; In the Matter of 2002 Biennial Regulatory Review – Review of the Commission's Broadcast Ownership Rules and Other Rules Adopted Pursuant to Section 202 of the Telecommunications Act of 1996; Cross-Ownership of Broadcast Stations and Newspapers; Rules and Policies Concerning Multiple Ownership of Radio Broadcast Stations in Local Markets; Definition of Radio Markets*, MB Docket Nos. 02-277, 01-235, 01-317, 00-244, 2 July 2002; Frank Ahrens, "FCC Releases Its New Media Ownership Rules," *Washington Post*, 3 July 2003, E01.

A useful account of the decision in *Prometheus Radio Project, et al. v. Federal Communications Commission*, 373 F.3d 372 (2004) is Stephen Labaton, "Court Orders F.C.C. to Rethink New Rules on Growth of Media," *The New York Times*, 25 June 2004, A1.

See also Federal Communications Commission, *In the Matter of 2006 Quadrennial Regulatory Review—Review of the Commission's Broadcast Ownership Rules and Other Rules Adopted Pursuant to Section 202 of the Telecommunications Act of 1996; 2002 Biennial Regulatory Review—Review of the Commission's Broadcast Ownership Rules and Other Rules Adopted Pursuant to Section 202 of the Telecommunications Act of 1996; Cross-Ownership of Broadcast Stations and Newspapers; Rules and Policies Concerning Multiple Ownership of Radio Broadcast Stations in Local Markets; Definition of Radio Markets*, Further Notice of Proposed Rulemaking, MB Docket Nos. 06-121, 02-277, MM Docket Nos. 01-235, 01-317, 00-244, 24 July 2006.

Materials regarding the relaxing of local conditions on cable franchises include Federal Communications Commission, "FCC Adopts Rules to Ensure Reasonable Franchising Process for New Video Market Entrants," news release, 20 December 2006; John Dunbar, "FCC Chairman Says Cities are Blocking Cable Competition," 18 December 2006, retrieved from Lexis-Nexis Academic <http://web.lexis-nexis.com/universe/> (18 Dec. 2006); and Alan Sipress, "FCC Vote A Victory For Phone Companies; TV Service Rule Hurts Cable Firms," *Washington Post*, 21 December 2006, final edition, D1.

136. James F. Hamilton, "Rationalizing Dissent? Challenging Conditions of Low-Power FM Radio," *Critical Studies in Media Communication* 21, no. 1 (March 2004): 44–63.

137. Barlow, "Community Radio"; Fairchild, *Community Radio*; Howley, *Community Media*; Federal Communications Commission, *Report and Order: In the Matter of the Creation of Low-Power Radio Service*, MM Docket No. 99-25, 20 January 2000, 2.

138. A running tally of LPFM applications in maintained at "Low Power FM (LPFM) Radio Service," *Federal Communications Commission* <http://www.fcc.gov/mb/audio/lpfm/> (1 Feb. 2007).

139. FCC, *Low-Power FM Radio*, 3–9.

140. FCC, *Low-Power FM Radio*, xx.

141. Startree Foundation, Inc., Costa Mesa, Calif., FCC 318—Application for Construction Permit for a Low Power FM Broadcast Station, File no. BNPL-20000606AAF, 28 October 2002.

142. Federal Communications Commission, *Memorandum Order and Reconsideration: In the Matter of Creation of Low Power Radio Service*, MM Docket No. 99-25, 20 September 2000, ¶ 182; Marc Fisher, "Low-Power to the People," *American Journalism Review*, 25 October 2000, 47.

143. Prophecy in the News, Inc., Oklahoma City, Okla., FCC 318—Application for Construction Permit for a Low Power FM Broadcast Station, File no. BNPL-20000602ADC, 2 June 2000; Brunswick High Pirate Pride Booster Club, Brunswick, GA, FCC 318—Application for Construction Permit for a Low Power FM Broadcast Station, File no. BNPL-20000531AAF, 24 June 2003; City of Logan, Logan, Utah, FCC 318—Application for Construction Permit for a Low Power FM Broadcast Station, File no. BNPL-20000605AAS, 1 June 2000.

144. Maranatha Fellowship Of Saddleback Valley, Mission Viejo, Calif., FCC 318—Application for Construction Permit for a Low Power FM Broadcast Station, File no. BNPL-20000605ADZ, 5 June 2000; Prophecy in the News, Inc., Oklahoma City, Okla., FCC 318—Application for Construction Permit for a Low Power FM Broadcast Station, File no. BNPL-20000602ADC, 2 June 2000.

145. Howell Mountain Christian Radio, Angwin, Calif., FCC 318—Application for Construction Permit for a Low Power FM Broadcast Station, File no. BNPL-20000608ACN, 8 June 2000.

146. Useful overviews include Hardt, *Critical Communication Studies*, 31–76; James W. Carey, "A Cultural Approach to Communication," in *Communication as Culture; Essays on Media and Society* (Boston, Mass.: Unwin Hyman, 1989), 13–36; and J. Herbert Altschull, *From Milton to McLuhan; The Ideas Behind American Journalism* (White Plains, N.Y.: Longman, 1990), 219–60. Important original works include Walter Lippmann, *Public Opinion* (New York: Harcourt, Brace, 1922); and John Dewey, *The Public and Its Problems*. New York: Henry Holt, 1927).

147. Rondaradio, Inc., Round Lake, Ill., FCC 318—Application for Construction Permit for a Low Power FM Broadcast Station, File no. BNPL-20000828AAW, 21 October 2003.

148. Severe Storms Intercept Team, Stillwater, Okla., FCC 318—Application for Construction Permit for a Low Power FM Broadcast Station, File no. BNPL-20000605ADE, 5 June 2000.

149. LSNET Educational Foundation, Galax, Va., FCC 318—Application for Construction Permit for a Low Power FM Broadcast Station, File no. BNPL-20000828ABS, 28 August 2000; Severe Storms Intercept Team, Stillwater, Okla., FCC 318—Application for Construction Permit for a Low Power FM Broadcast Station, File no. BNPL-20000605ADE, 5 June 2000; Monroe County Public Library, Bloomington, Ind., FCC 318—Application for Construction Permit for a Low Power FM Broadcast Station, File no. BNPL-20000602ACS, 2 June 2000.

150. Harry Bridges Institute, San Pedro, Calif., FCC 318—Application for Construction Permit for a Low Power FM Broadcast Station, File no. BNPL-20000605ABQ, 5 June 2000.

151. National Multi-Ethnic Families Association, Oxnard, Calif., FCC 318—Application for Construction Permit for a Low Power FM Broadcast Station, File no. BNPL-20000530AAI, 24 May 2000; Stockton Good Samaritan Service Center, Inc., Stockton, Calif., FCC 318—Application for Construction Permit for a Low Power FM Broadcast Station, File no. BNPL-20000605AAC, 2 June 2000; Kyle Cetrulo, San Francisco, Calif., FCC 318—Application for Construction Permit for a Low Power FM Broadcast Station, File no. BNPL-20000531AAW, 31 May 2000.

152. City Heights Community Development Corporation, San Diego, Calif., FCC 318—Application for Construction Permit for a Low Power FM Broadcast Station, File no. BNPL-20000605ADB, 1 June 2000.

153. City Heights Community Development Corporation.

154. First United Pentecostal Church of Redlands, Redlands, Calif., FCC 318—Application for Construction Permit for a Low Power FM Broadcast Station, File no. BNPL-20000605AAK, 1 June 2000.

155. Michael Denning, *The Cultural Front; The Laboring of American Culture in the Twentieth Century* (New York: Verso, 1996), 51–114. Examples include Governing Board Grossmont College, El Cajon, Calif., FCC 318—Application for Construction Permit for a Low Power FM Broadcast Station, File no. BNPL-20000531AAI, 31 May 2000; and Shishmaref School, Shishmaref, Alaska, FCC 318—Application for Construction Permit for a Low Power FM Broadcast Station, File no. BNPL-20000606AAL, 6 June 2000.

156. Midwestern Theater Troupe, Inc., Tulsa, Okla., FCC 318—Application for Construction Permit for a Low Power FM Broadcast Station, File no. BNPL-20000607AAB, 7 June 2000.

157. National Jazz Hall of Fame Foundation, Palm Springs, Calif., FCC 318—Application for Construction Permit for a Low Power FM Broadcast Station, File no. BNPL-20000531ABG, 31 May 2000.

158. Unity World Production, Incorporated, Fort Wayne, Ind., FCC 318—Application for Construction Permit for a Low Power FM Broadcast Station, File no. BNPL-20000608ACU, 5 June 2000.

159. House of Gold Foundation, East Providence, R.I., FCC 318—Application for Construction Permit for a Low Power FM Broadcast Station, File no. BNPL-20000608ABN, 8 June 2000.

160. McChesney and Scott, *Our Unfree Press*; Andrew Opel and Donnalyn Pompper, eds., *Representing Resistance: Media, Civil Disobedience, and the Global Justice Movement* (Westport, Conn.: Praeger, 2003); T.V. Reed, *The Art of Protest; Culture and Activism from the Civil Rights Movement to the Streets of Seattle* (Minneapolis, Minn.: University of Minnesota Press, 2005), 240–85.

161. Anchor City Radio, Anchor City, Alaska, FCC 318—Application for Construction Permit for a Low Power FM Broadcast Station, File no. BNPL-20000605AAA, 1 June 2000; Sacramento Community Cable Foundation DBA Access Sacramento, Sacramento, Calif., FCC 318—Application for Construction Permit for a Low Power FM Broadcast Station, File no. BNPL-20000607AAY, 7 June 2000.

162. Tulsa Free Radio Project, Tulsa, Okla., FCC 318—Application for Construction Permit for a Low Power FM Broadcast Station, File no. BNPL-20000530AAY, 30 May 2000.

163. Joseph M. Futrelle, Urbana, Ill., FCC 318—Application for Construction Permit for a Low Power FM Broadcast Station, File no. BNPL-20000828ACF, 28 August 2000.

164. Liberty Creek Homeowners Assoc., Indianapolis, Ind., FCC 318—Application for Construction Permit for a Low Power FM Broadcast Station, File no. BNPL-20000605AAG, 8 June 2000.

165. Applications denied include those submitted by Anchor City Radio, Sacramento Community Cable Foundation DBA Access Sacramento; Tulsa Free Radio Project; Joseph M. Futrelle; Liberty Creek Homeowners Assoc., Glenn Austin, Minneapolis, Minn., FCC 318—Application for Construction Permit for a Low Power FM Broadcast Station, File no. BNPL-20000901ABU, 1 September 2000; and Washtenaw Community College, Ann Arbor, Mich., FCC 318—Application for Construction Permit for a Low Power FM Broadcast Station, File no. BNPL-20000901ABZ, 31 August 2000 (permit cancelled).

166. Stavitsky, "Changing Conception of Localism," 26–28.

167. For key discussions, see Graham Meikle, *Future Active: Media Activism and the Internet* (London: Routledge, 2002); John D.H. Downing, "The Independent Media Center Movement and the Anarchist Socialist Tradition," in *Contesting Media Power; Alternative Media in a Networked World*, ed. Nick Couldry and James Curran (Lanham, Md.: Rowman and Littlefield, 2003), 243–57; Dorothy Kidd, "Indymedia.org: A New Communication Commons," in *Cyberactivism: Online Activism in Theory and Practice*, ed. Martha McCaughey and Michael D. Ayers (New York: Routledge, 2003), 47–69; Sara Platon and Mark Deuze, "Indymedia Journalism: A Radical Way of Making, Selecting and Sharing News?," Journalism 4, no. 3 (August 2003): 336–55; Chris Atton, *An Alternative Internet* (Edinburgh: Edinburgh University Press, 2004); Wim Van de Donk, Brian D. Loader, and Paul G. Nixon, and Dieter Rucht, eds., *Cyberprotest: New Media, Citizens and Social Movements* (New York: Routledge, 2004); Douglas Morris, "Globalization and Media Democracy: The Case of Indymedia," in *Shaping the Network Society,*

ed. Douglas Schuler and Peter Day (Cambridge, Mass.: MIT Press, 2004), 325–52; Kate Coyer, "If It Leads It Bleeds: The Participatory Newsmaking of the Independent Media Centre," in *Global Activism, Global Media*, ed. Wilma De Jong, Martin Shaw, and Neil Stammers (London: Pluto Press, 2005), 165–78; Victor W. Pickard, "Assessing the Radical Democracy of Indymedia: Discursive, Technical, and Institutional Constructions," *Critical Studies in Media Communication* 23, no. 1 (March 2006): 19–38; and Victor W. Pickard, "United Yet Autonomous: Indymedia and the Struggle to Sustain a Radical Democratic Network," *Media, Culture and Society* 28, no. 3 (May 2006): 315–36.

168. "What Is Indymedia?," *The Indypendent*, November 2006, 2.

169. Akhil Gupta and James Ferguson, "Beyond 'Culture': Space, Identity, and the Politics of Difference," *Cultural Anthropology* 7, no. 1 (February 1992): 6–23; Lisa Law, "Transnational Cyberpublics; New Political Spaces for Labour Migrants in Asia." *Ethnic and Racial Studies* 26, no. 2 (March 2003): 234–52.

170. Hildegard Westerkamp, "The Soundscape on Radio," in *Radio Rethink; Art, Sound, Transmission*, ed. Daina Augaitis and Don Lander (Banff: Walter Phillips Gallery, 1994), 88. See also Douglas Kahn and Gregory Whitehead, eds., *Wireless Imagination: Sound, Radio, and the Avant-Garde* (Cambridge, Mass.: MIT Press, 1992); and Dan Lander and Micah Lexier, *Sound By Artists* (Toronto: Art Metropole, 1990).

171. Margaretta D'Arcy, *Galway's Pirate Women; A Global Trawl* (Galway: Women's Pirate Press, 1996), 3, 7. Additional, recent studies and commentary include Margaretta D'Arcy, "Playing With the Airwaves," *TDR: The Drama Review* 34, no. 4 (Winter 1990): 179–82; Caroline Mitchell, ed., *Women and Radio: Airing Differences* (London: Routledge, 2000); Michelle A. McKinley and Lene O. Jensen, "In Our Own Voices: Reproductive Health Radio Programming in the Peruvian Amazon," *Critical Studies in Media Communication* 20, no. 2 (June 2003): 180–203; Susan Carter, "A Mic of Her Own: Stations, Collectives, and Women's Access to Radio," *Journal of Radio Studies* 11, no. 2 (December 2004): 169–83; Margaret E. Thompson, Katerina Anfossi Gómez, and María Suárez Toro, "Internet Audience Perceptions of Feminist International Radio Endeavour (FIRE)," *Feminist Media Studies* 5, no. 2 (July 2005): 215–36; and Linda Steiner, "The Feminist Cable Collective as Public Sphere Activity," *Journalism: Theory, Practice, Criticism* 6, no. 3 (August 2005): 313–34. For a valuable review essay that places this and similar work in a larger context, see Bonnie J. Dow and Celeste Condit, "The State of the Art in Feminist Scholarship in Communication," *Journal of Communication* 55, no. 3 (September 2005): 44.

Chapter 5

Modernism and the Aestheticization of Dissent

"For two decades now," noted the organizers of the group Step It Up in early 2007, "the Intergovernmental Panel on Climate Change has been the gold standard for climate research. So when the IPCC issued its new five-year update this morning, it was tough reading for the planet."[1] To help galvanize a response, the group organized for mid-April 2007 a National Day of Climate Action. Instead of seeking to disrupt carbon production and emission by blockading car manufacturers, coal-fired power plants, or freeways, the group advocated hundreds of rallies at "meaningful, iconic places" to "demand, in a unified voice, strong action on climate change." Examples included "canyoneering in the West, forming a human line depicting what high water levels might look like in Manhattan fifty years from now, [or] standing together in a town park, on the steps of churches and synagogues, or on a college campus." Suggestions for the gatherings included having "some music, a few speakers, and most importantly [. . .] to pose for a picture that will be sent 'round the world."[2] The organizers claimed that, if successful,

> it's going to be truly striking for Americans to turn on the TV and see their countrymen in iconic spots across the nation—some spectacular, like the peaks of icy mountains or underwater on coral reefs. But many more in places that look like home to them: on church steps and in city parks, along the tidelines of our coastal cities, in the fields where our food comes from. This is the America we want to defend against the massive change that would come with global warming—it's the America we dearly love.[3]

The goal of generating iconic publicity images is the flip side of the goal of creating underground, anonymous videos that seek not to make but to disassemble iconic images. An example is a critique of the policies and actions of U.S. President George W. Bush that takes the form of a video clip widely available on the Internet. It portrays the president delivering the 2003 State of the Union speech to a joint session of Congress. Despite clearly being video footage from C-SPAN, the disjunction between the veracity of the video and the outrageous assertions signals immediately that something is askew.[4] At the same time, however, given strident domestic and global objections to the U.S. invasion of Iraq, the vanishing of the key reason given for the invasion (that Saddam Hussein possessed weapons of mass destruction), the selective emphasis given to pre-war intelligence, the brutality directed toward Iraqi civilians by the insurgencies as well as by some U.S. troops and military advisers, the open-ended imprisonment without trial of hundreds of detainees at the U.S. prison camp at Guantanamo Bay, the U.S. military torture of prisoners at Abu Ghraib, C.I.A. kidnappings and subsequent transport of suspects for interrogation to countries that allow torture, and a number of other actions taken by the White House and its Republican-controlled Congress to obscure if not undermine decades-old initiatives with broad public support, this speech remix—in all its outrageousness—seemed to reveal the heretofore hidden, real White House agenda as well as congressional and class collusion.[5]

After welcoming various members of the audience, we hear and see in the video clip the president saying "Every year, by law and by custom, we meet here to threaten the world," with a cut to a scene showing a standing ovation from the legislators. The clip continues with the president asserting that "the American flag stands for corporate scandals, recession, stock-market declines, blackmail, terror, burning with hot irons, dripping acid on the skin, mutilation with electric drills, cutting out tongues, and rape" (cut to another standing ovation). Later in the 4:08 clip, after noting that "our first goal is to show utter contempt for the environment," the president reminds Congress that "I have sent you a comprehensive energy plan to devastate communities, kill wildlife, and burn away millions of acres of treasured forests" (applause). On the topic of the terrorist group Al-Qaeda, the president asserts that "one by one, the terrorists are learning we are building a culture to encourage international terrorism," and, later, that "Secretary of State Powell will plant information to incite fear about Iraq's links to terrorist groups" (applause). The clip concludes with the president forcefully emphasizing that "trusting in the sanity and restraint of the United States is not a strategy, and it is not an option" (wild cheering and a standing ovation).[6]

Despite their differences, each of these examples places great importance on representation as a tactic if not the substance of critique and social change. Where the National Day of Climate Action seeks to generate publicity images for use by commercial communications organizations, the remixed State of the Union speech seeks to generate critical awareness of the disjunction between White House assertions and White House actions. Both examples also build upon authoritative materials (the IPCC report for one, and video of the presidential speech for the other), but recognize as well that information on its own is not enough to galvanize social change—that resistance requires compelling representation.

These and many other examples struggle against market formations, but also conduct a critical engagement with rationalist empiricism in ways suggested by the two examples above. Philanthropically funded communications and community media projects struggle to provide a form and means of democratic communications not subject to the market but at the same time uncritically grounded in rationalist empiricism, the acceptance of which is beyond question and which constitutes the starting point of practice—and the primary point of critique. In brushing against the grain the critical value of rationalist empiricism, such protests and remixed video clips exemplify a critical reconstruction of the very premises of the production of truth.[7]

This chapter addresses these complexities by first outlining the basis of what can be called a critical-modernist culture of truth as a critique of the epistemology, form, and organization of rationalist empiricism, and applies this to a redefinition of the possibilities for democratic communications. After recovering the pre-modern roots of quintessentially current forms of critical-modernist dissident as a fusion of art and politics, the chapter concludes by describing how the rationalizing pressure of publicity seeks to channel such work into manageable forms.

Critical Modernism as a Culture of Truth

Just as the formation of rationalist empiricism and its facets as a culture of truth required an extensive discussion in Chapter 1, a similar effort is needed here to adequately characterize modernism, its variants, and how it challenges rationalist empiricism by becoming the basis for a quite different culture of truth.

Despite their variety, assessments and explanations of modernism have in common the notion of a break between an existing era and a new one.[8] As Kellner summarizes, modernity has been seen from Hegel on as "the 'new age' which emerged from feudalism and the Renaissance."[9] Bauman identifies the beginnings of modernity "in Western Europe with a series of profound sociostructural and intellectual transformations of the seventeenth century [. . . which] achieved its maturity: (1) as a cultural project—with the growth of the Enlightenment; (2) as a socially accomplished form of life with the growth of the individual (capitalist, and later also communist) society."[10] In his review, Sheppard notes key developments including the emergence of the modern public in the wake of the French Revolution; immense technological advance and industrial growth along with increasing waves of political resistance and revolution and thus the radical questioning of such developments as uniformly progressive; and the collapse of faith in absolute reason in the wake of increasingly dystopian developments linked to unchecked urbanization, industrialization, and the onset of World War One.[11]

Given such changes, the key challenge has been, in Harvey's words, "how to discover the [so-called] 'eternal and immutable' elements [of modernity] in the midst of such radical disruptions."[12] Efforts to meet this challenge often take the form of attempts at greater precision. For example, Berger suggests modernity's immense range of contexts and forms by noting such facets as

> an interlocking world system of imperialism; opposed to it, a socialist international; the founding of modern physics, physiology and sociology; the increasing use of electricity, the invention of radio and the cinema; the beginnings of mass production; the publishing of mass-circulation newspapers; the new structural possibilities offered by the availability of steel and aluminum; the rapid development of chemical industries and the production of synthetic materials; the appearance of the motor-car and the aero plane.[13]

To this list (published in 1969), one could add practices and developments unforeseen at the time such as the diffusion of digital computation, the global ecological crisis, world health issues of HIV/AIDS and SARS, and the equally expansive range of changes referred to as globalization, as well as resistance to it.[14] In an effort to further specify modernity, Doyle and Winkiel, Hegeman, and Brooker and Thacker all emphasize the chronologically and geographically uneven presence and diverse manifestations of modernism(s), a point that Harvey emphasizes by noting the "tensions [in modernism] between internationalism and nationalism, between globalism and parochialist ethnocentrism, between universal and class privileges," thus adding up to a "complex historical geography of modernism (a tale yet to be fully written and explained) [which] makes it doubly difficult to interpret exactly what modernism was all about."[15] Williams concurs by noting that assertions of the "undifferentiated twentieth-century world" as modern is "now at best anachronistic, [and] at worst archaic."[16]

In the same way that attempts to adequately understand the immense range and complexity of modernity can, in Berger's words, all too easily "lead to despair," attempts to understand the practice and implications of what can be called a modernist struggle against market formations of communications pose a similar challenge.[17] Even a highly selective mapping of key developments is fraught with difficulty because of the relationships of communications to so many of the areas noted by Bauman, Sheppard, Harvey, Berger, and others. However, while setting aside the unachievable goal of comprehensiveness, key points in the development and its implications for democratic communications can still be proposed, keeping in mind the looming presence of the immense range only suggested here.

The first key point concerns the highly contradictory character of modernity, due to its being deeply entwined with similarly contradictory processes of capitalism and the Enlightenment. This integration and contradiction helps explain the corresponding divergence in evaluations of modernity. In some cases, modernity embodies a utopian faith in people's abilities to improve the world, their lives, and themselves, with such abilities enhanced through technical innovation and development, accelerated and mechanized mass production, and the valorization of secular society. Such a faith cannot be pinned exclusively to conservative or radical politics; the dream of a cornucopia of goods, services, and wellbeing available to everyone is the hallmark not only of capitalist justifications for the system but also of anti-capitalist hopes for the future. However, in other cases, such developments are not a means of liberation, but of enslavement. This dystopian critique views scientific discovery and accelerated mass production as the source not of improvement and progress, but of deadening rationalization, dislocation and estrangement, and as an expression of inhuman instrumental reason that paves the way for apocalyptic disaster symbolized by the nuclear bomb.[18] Continuing his characterization, Sheppard remarks more specifically that modernism "has been viewed as a continuation of or a reaction against Romanticism; as a reaction, in its extreme avant-garde forms, against Aestheticism; as an inversion of the conventions of Realism; and as a reaction against Naturalism. Conversely, modernism has also been viewed as both the antecedent of postmodernism and as the phenomenon to which postmodernism forms the reactive contrast."[19] In his more pointed characterization, Kellner notes that, "while some theories of modernization present modernity as a purely positive force bound up with progress in science, industry, and technology, others see it as a wholly negative force bound up with dehumanization, destruction of the environment, technology out of control, and totalitarian political development."[20]

More nuanced attempts to evaluate modernity seek to overcome altogether the exclusive choice between the utopian and the dystopian. Many of these attempts are in dialog with Marx and Engels' influential, dialectic assessment of modernization, in which, again as Kellner summarizes, modernity is seen as both "'progressive'—creating new needs, connections, possibilities, and so forth—and destructive—creating commercial crises, the alienation of the working class, impoverishment and proletarianization of the 'lower strata of the middle class' and the peasantry leading to class war."[21] In such a view, the destructive does not simply come along with the progressive (as in claims about technology having both good and bad effects). Rather, the progressive is a destructive force that clears away the existing social order, just as the destructive is a progressive force necessary for bringing about a new synthesis—in this case, for enabling the proletarian revolution, the dethroning of capitalism, and the installment of communist socialism.

Yet, during the twentieth century, faith in the mechanical inevitability of a new, utopian synthesis arising from the dialectical struggle became much less persuasive. While the promise of modernity has been acknowledged in the miracles of modern engineering and science, in the emergence of nations and, later, multinational bodies such as the League of Nations, the United Nations, and the World Social Forum, at the same time it has been refuted by the darker implications of developments alluded to above. Modernity is the era of Francis Bacon and John Stuart Mill, but also of Herbert Spencer, Sigmund Freud, and Friedrich Nietzsche. Modernity in all its contradictions was born of the Enlightenment, but today remains even more haunted by it as well.

Adequately understanding the range of modernity is essential for understanding the similar range claimed for market formations of communications. The most affirmative assessments attach to broadcasting and the Internet a utopianism that is epitomized in such forms as the "global village."[22] Such utopian glorification of modern means and organization of communications underlies the defense of market formations of communications, which use economistic, utilitarian conceptions to argue that the current regime has modernized and thus democratized access and information. As the argument goes, these developments have enabled greater individual freedom, thus rendering moot calls for democratic communications, because the future is already here and the goal is immanent if not fully realized.[23]

In contrast to this affirmative position that uses the utopian promise of modernity as a justification for the present system of communications, two general kinds of critical responses also indebted to modernism dispute the degree to which such goals have been reached or even can be reached. The first kind of critical response accepts the value of the current extension of means and organization of communications, but argues that their utopian potential has been perverted by commercial incorporation. As a result, the progressive goal within this view is to use these tools but for completely different purposes—in short, to outdo the dominant by using currently available means and organization for dissident instead of dominant purposes.[24] As discussed in earlier chapters, such efforts range from calls for noncommercial funding and for enhanced program diversity to establishing separate but parallel systems. Efforts in the United States have ranged from J.A. Wayland's late-1800s project of the radical mass-circulation newspaper *The Appeal to Reason* to the current effort of Free Speech TV as a nationally available network aspiring to a scale equal to the commercial system, but with the intention of "exposing the public to perspectives excluded from the corporate-owned media."[25]

These and other examples of what could be called a reformist-modernist critique typify philanthropically funded projects and community media projects while they work largely through the culture of truth of rationalist empiricism, from which emerges projects' intention, organization, composition, distribution, and interpretation. Such a position seeks paradoxically to undermine if not replace the existing system, but to do so through the same rationalist-empiricist culture of truth which helped give it rise.

A second, more radical response urges a wholesale rejection of market means and organization and, further, of capitalism as a system and, ultimately, of all such systems. Not only is the utopian potential of market formations of communications rejected, the very premises of broadcasting, of centralized and professionalized organization, and their legitimation in rationalist empiricism are rejected as well. Philosophical and political bases of this radical critique range from Romanticism and anarchism and their valorization of humans' abilities either to liberate themselves individually or through uncoerced collective action, to nihilism and the rejection of any claim that humans can overcome their flawed nature either individually or collectively. Such positions are voiced for example in the modernism-postmodernism debate of the past thirty or so years.[26]

To accomplish its aim, the radical critique of modernism encompasses not only concerns about intention, funding, organization, and composition, but also to its philosophical justifications, thus directly challenging rationalist empiricism and its support of specific kinds of communications premises, forms, and practices. In the place of rationalist empiricism, what can be called a critical-modernist culture of truth, takes issue with the very forms of practice and, with them, rationalist-empirical premises about the very nature and possibilities of knowledge about the world.

One key challenge of critical modernism is epistemological, summarized by Hebdige as a Nietzschean negation of totalization, teleology, and utopia.[27] In it, the hope of locating dependable facts in the world is critiqued as nonsensical, given not only a protean, alien world characterized only by flux, but also the rejection of the positivity of knowledge as existing whole and already meaningful. Furthermore, humans are not seen as neutral gatherers of knowledge, but as driven and shaped by forces beyond their control. In the wake of the dethroning of the integrated Enlightenment individual by Darwin, Marx, Nietzsche, and Freud is a being that is subject to biological, historical, and psychic dynamics wholly beyond conscious, intentional control and management, thus erasing the validity of claims of disinterested truth. The resulting recognition and embrace of irrationality, intuition, spontaneity, and urge set radical modernism against rationalist empiricism and its conception of the consciously deliberative Enlightenment individual.

What enables as well as validates such negations is the increased recognition of the irrational core of capitalism itself. Such a recognition corresponded with the maturation of the quintessential capitalist cultural form of advertising, which, by the early twentieth century, responded to as well as helped accelerate the development of consumer culture and the impossible-to-achieve imperative of limitless expansion that, even when confronted as the world is today with the resulting ecological and climatological disaster, has yet to be widely accepted as such.

More generally, capitalism and its inequalities were seen as driven by social and cultural pressures instead of moral or biological needs. In addition to Marx, Thorstein Veblen was an early figure in the United States who recognized the essential irrationality of capitalism. In his flawed but prescient analysis of consumption by the leisure class, Veblen dispensed with moralistic and economistic critiques (claims that social inequalities were due simply to the greed and/or selfishness of rich people). By contrast, his explanation emphasized the social and the cultural dimensions of consumption, recognizing goods not as a means of fulfilling avaricious urges or meeting basic needs, but as an index for social status in the vortex of competition without end, thus placing it wholly outside providential or empirically rational explanation.[28]

Veblen's socializing of heretofore moral failings or natural processes was only one example of a much broader revolution in epistemology. Such a revolution extended to science, for example, in the formation of postpositivism assisted by Wittgenstein and his linguistic turn, and its incorporation through Popper and others into philosophies of science, as well as by Kuhn and his interpretation of scientific enterprise as discontinuous instead of cumulative.[29] A corresponding radical relationalism was underwritten and legitimated in theoretical physics by Einstein and Heisenberg; in sociology by the symbolic interactionism of Charles Cooley and George Herbert Mead at the Chicago School of sociology; and in the rise of cultural anthropology and recognition of different cultural systems as logically valid within themselves.[30]

In addition to rejecting the positivist epistemology of rationalist empiricism, critical modernism rejected its form as well. Instead of reconciling or synthesizing differences through rationalist logic and summary exposition, critical-modernist forms embody not only contradictions and differences but also the undesirability if not impossibility of comprehensive understanding, reconciliation, or synthesis. Lunn summarizes key compositional assumptions of modernism as aesthetic self-consciousness or self-reflexiveness; the use of simultaneity, juxtaposition, and montage; the evocation of paradox, ambiguity, and uncertainty; and, in the wake of the demolition of the unified rational Enlightenment actor, "the demise of the integrated individual subject or personality."[31] Such a form promotes not a single, authoritative answer but a troublingly plural engagement that serves as an open invitation to confrontation while refusing reconciliation, synthesis, or resolution.

While some iterations of critical modernism descend into the wholly subjective, others hybridize empiricism and subjectivism, forming efforts in which the central organizing principle for embodying knowledge is not the linear logic tree but the unidirectional and multilayered mosaic. As a result, comprehensive accounts underwritten by rationalist empiricism are discarded in favor of collections of local, often deeply subjective stories/knowledge, with the intention not of reconciling differences or generating a universal explanation, but paradoxically to document the impossibility of a comprehensive, complete account. In addition to avant-garde examples, such uses of mosaic forms extend in liberal-pluralist ways throughout even government and nonprofit practice. For example, by being a collection of countless individual stories and situations, the attempt to "know" the people of a nation in the form either of the Depression-era U.S. Works Progress Administration (W.P.A.) Federal Writers Project or the currently running philanthropically funded U.S. radio series StoryCorps (which promotes itself as a "W.P.A. for the twenty-first century") accepts in its very mosaic form the impossibility of any universal definition or synthesis, thus validating an irreducible pluralism.[32]

Finally, in addition to rejecting the epistemology and the form of rationalist empiricism, critical modernism also rejects collective, rationalized, and professionalized organization, seeing it as embodying illegitimate claims of superior knowledge and a will to power which is used to dominate others. Much of this point owes its insights to the work of Max Weber regarding bureaucratic rationality, which were developed further through the critical theory of the Frankfurt School.[33] As Kellner summarizes, the Enlightenment support of "criteria of utility, efficiency, and success" defined by the ability to dominate nature became the key criteria for the formation and operation of organizations and modes of human participation.[34] "The social implications of this mind-set," concludes Kellner, "concern attempts by social organization and administration to control individuals in ways that abstract from individuality and uniqueness through the imposition of formal rules and regulations."[35]

Thus, critical modernism sees little chance of utopian collective action, with the only possible action instead (and this is its connection to the Romantic vision of the tragic artist as isolated genius) never-ending negation in a whirlwind of isolation or, as the postmodern turn, a never-ending immersion in the recycled irony of hyperrealism. More generally, critical modernism indicates to Hebdige not only that "modernity was always a Janus-faced affair," but that, when pushed to its extreme, critical modernism "is modernity without the hopes and dreams which made modernity bearable."[36] The dilemma we are left with is, however, a quintessentially modern one. Although used to make a slightly different point about realism and modernism, Jameson puts the dilemma about the conflict between different conceptions of modernism (affirmative, reformist, and critical) clearly in his comment that "we may feel that each position is in some sense right and yet that neither is any longer wholly acceptable."[37]

Critiquing the Break

The range of forms of critical-modernist practice were and remain an open possibility, due to contradictory contextual determinations and the resulting historical variability. To enable an analysis of this range, what must first be done is to critique the account of modernism as a clean break, in order to make available an understanding of the many ways in which practices in the ostensibly "premodern" era form substantial resources for its practice in the "modern" and even the "postmodern" era.

Too great a reliance on the foundational premise of a clean break between the pre-modern and the modern downplays, obscures, or even denies what persists despite what has changed. Dettmar and Watt note this problem by remarking that "the positing of such a gulf [. . .] is not only inaccurate, but also exceptionally damaging insofar as it influences matters of [literary] canonization and reception."[38] Other kinds of cultural developments require equally great care in this regard. For example, Harvey seems at first to posit a clean break in his discussion of how Rousseau, Kant, Nietzsche, and other thinkers and artists helped articulate "a radical shift from a rational and instrumentalist to a more consciously aesthetic strategy for realizing Enlightenment aims" in which "the pursuit of aesthetic experience as an end in itself became, of course, the hallmark of the romantic movement." As a result, the Romantic movement "paves the way for active aesthetic interventions in cultural and political life" of the kind indicated by the examples of the National Day of Climate Action and the remixed State of the Union speech.[39]

However, such a point must be grasped precisely. Harvey and others do not claim that the "aesthetic" did not exist prior to modernism, or that human action before the break was only raw, animal impulse or utilitarian, rationalist assessment of cost versus benefit. Instead, the less sweeping but more profound reading of this claim is that, as a response to modernity's pressures toward the instrumentalization of human experience, the "aesthetic" was produced as a category of knowledge and practice that did not replace other realms, but that was distinguished from them and, as in the case of Romanticism and some forms of modernism, unduly elevated above them.[40]

Reading the break as a complication instead of a narrative of succession underscores the necessity of refusing the modernist separation and abstraction of "art" from "society," as Williams and many others assert. Regarding the aesthetic (or, more broadly, culture as a whole) as an autonomous realm idealizes its making (forgetting that flesh-and-blood human beings born into and working within specific social and historical conditions enact it) while fetishizing the purported talismanic power of charmed objects of "art" to exert their own intrinsic (aesthetic) "aura."[41] Equally debilitating theoretically and politically is regarding society as a realm separate from culture. Doing so idealizes society as a natural and automatic perpetual-motion machine that is not made, but that simply is, thus putting the current social order outside of human ability to change it. In both cases, culture and society are reduced to naturalized, idealized objects instead of recognized more fully as mutually constituting practices within what Williams calls an "indissoluble social material process."[42] Put more concretely, Williams asserts the need to see cultures as socially produced but, at the same time, to see societies as culturally produced, with the end result not the entity "society" or objects of "culture," but a whole process of their continual, historical realization.

Such an insight enables one to regard, for example, the social organization required to compose, perform, and interpret an artistic performance as much a cultural production as the performance itself, whether conventionally "artistic" or conventionally placed in a realm distinct from art.[43] For example, Williams sees the making of the social organization known as "the collective democratic institution" as itself "a remarkable creative achievement," with its makers themselves not as prefigured, essentialized actors but instead as making themselves through the activity of making their social and cultural order. As such, this social and cultural process of the production of social organization is seen as dialectical and historical, and immensely more contradictory, complex, and contingent than overly rationalistic assessments would have it, while at the same time every bit as creative an accomplishment as an artistic performance.[44]

Such a retheorization, commencing for Williams in the 1950s, has been elaborated more recently by scholars who draw on a variety of perspectives. Studies investigating the "politics of performance" are increasingly concerned also with the performance of politics, with their combination and hybridization seen as "a form of representation that constructs political discourse and [also] as a social activity that interacts with other sectors of society," and thus seeking to address "the 'broad spectrum' of performance" as a general human activity.[45] The presence and political significance of ritualized public action has also been noted by historians seeking to counter explanations of human action based on "mechanistic economism" or irrationalism.[46] For example, the recent reevaluation of irrational mobs as purposive crowds emphasizes the necessity of investigating the frameworks of understanding that enabled such action and within which it took place.[47]

A critique of the break clears the way for a wider-ranging inquiry. The process described as the "aestheticization" of dissidence cannot be regarded as the replacement of one kind of intervention (a rational, utilitarian, "real" one) by another (a self-reflexive, aesthetic, "representational" one), but rather as a reconfiguration of the whole social material process in which many modes of intervention were practiced in variable form. Of particular usefulness in such an inquiry is a focus on the ostensibly modern forms of the protest demonstration and of montage. By addressing them as historical practices unbounded by assertions of the modernist break, one can substantiate the claim that differences in forms of public dissidence are defined not by an epochal, linear chronology that locates the inchoate (riots) in the premodern, the utilitarian (strikes) in the early modern, and the aesthetic (image events) in the high modern, but as simultaneously available but highly variable forms of practice in specific conditions.[48]

Contrary to assertions about protest demonstrations' quintessentially modern nature is the recognition that public action has always been representational.[49] Along with affirmative, incorporative pre-modern public rituals such as coronations, commemorations, marches, pageants, pilgrimages, worship, and warfare, examples of forms of dissident action employed by medieval European peasants who sought to challenge the authority of lordship range from petitions and litigation to rent strikes, non-cooperation, revolt, and flight, all of which had a ritual as well as a utilitarian value—and, more precisely, had a utilitarian value because of the ritual value, and vice versa.[50] Many scholars have also commented on the ritual nature and carnivalesque value of popular festivals, which are seen to have constituted an important public inversion of authority.[51] In these and other cases, the utilitarian and ritualistic must be seen dialectically—such actions do something because they mean something, while at the same time they mean something because they do something.

In more particular ways, the protest demonstration must be seen in relation to theater. Prior to and as a response to its bourgeois enclosure in playhouses, theater had literally been in the street.[52] But, just as the street came to the theater first in an organizational sense through its commercial popularization and, later, in a performative sense through the emergence of naturalist and realist modes, theater came to the street in the form of more or less choreographed pageants that sought to convey a focused message.[53] This redefinition can be seen in the gradual shift in usage of "demonstration" from a medieval sense of a logical proof supporting a proposition, to the roughly early modern sense of an expression or assertion of an individual or collective view, and finally to the linkage in the mid-nineteenth century to mass-public politics.[54]

Premodern bases of montage are similarly apparent. Although commonly regarded as an invention of the early twentieth-century European avant-garde or based in early theories about the aesthetics of film, montage has much longer, transnational roots.[55] Generally speaking, the appropriation and refashioning of found materials into new signifying relationships describes the full range of human work on the natural world. More specifically, when considering montage as a general practice of composition, a diverse set of examples can be related. Among those referred to by various scholars include African tribal emblems, twelfth-century Japanese text-collages, and a mosaic-style textual practice of thirteenth-century Kabbala. More recent precursors include eighteenth- and nineteenth-century butterfly-wing collages; German folk-art weather charms and lace valentines; widely practiced European and American hobbyist experimentation with photography in the nineteenth century; personal scrapbooks of newspaper articles and advertisements; and other kinds of assembled works that, in the words of one author, were "practiced almost everywhere that paper might be found," using such diverse materials as cigar bands, lithographed scenes, advertisements, box and bottle labels, and postage stamps.[56]

Not only did protest demonstrations and montage have significant premodern bases, they were also reformed through the emergence, expansion, and transformation of capitalism and the capitalist remaking of societies and experience, which centered most dramatically in major cities. The extent to which modernism was an urban phenomenon can be seen in the ways in which its generalized dislocations and disruptions became the grist for émigré and exiled artists who converged in modern imperial metropolises such as Paris, Berlin, London, and New York, while also becoming a highly variable condition enabling the recognition of works of art so conceived.[57]

The extension of cities was however but one facet of increasing centralization, urbanization, and industrialization, as well as forms of capitalist consumerism that forged a relation between consumption and spectacle.[58] Such forms include covered shopping arcades in nineteenth-century European cities, which Buck-Morss describes as "the first international style of modern architecture, hence part of the lived experience of a worldwide, metropolitan generation," with the U.S. iteration being the department store as a temple of consumption.[59] One can easily imagine the material realization and embodiment of these discontinuities and dislocations through the increasingly extensive experience of the juxtaposition of wealth and despair as well-heeled urban shoppers brush past or step over beggars on the sidewalk outside downtown luxury stores.

Within these and other immense changes, of particular relevance to protest demonstrations and to montage as critical-modernist forms of democratic communications were two complex developments. First is the accelerating role and increasing centrality of communications organizations (later, industries) in the emerging consumer capitalism, first in the form of popular, commercial newspapers and the many related popular presses and publishers, then joined in the first decades of the twentieth century by motion pictures and by radio. These, of course, relied in varying ways on the ever-expanding role of advertising.[60] Such growth and ubiquity indicated while also contributed to the growing common public experience of the flood of commercial images that embodied a culture of consumerism.[61]

Second is a corresponding skepticism toward the mimetic claims of representation, due not so much to a loss of faith in reality, but to an increasing awareness of how much more "reality" outside of personal sensory awareness there was. For example, as noted in Chapter 2, the seemingly miraculous technological innovations of the telegraph and, later, the wireless made palpable a paradoxical "ethereal reality," as forces of nature undetectable by human senses came to be harnessed for human use. Awareness grew as well of the ways in which representations were not simply neutral reflections but could themselves have quite dramatic effects—a belief seen, for example, in an increasing concern about the extension of popular publishing and the moral effects of popular culture, and about publicity as a tactic used in politics (such as propaganda) and entertainment (the innovations of circus promoters Barnum & Bailey as particularly prominent), and their emergence and progressive deployment in the advertising industry.[62]

In the experience of such changes, familiar Enlightenment oppositions became increasingly untenable. While artistic practice from the Renaissance through nineteenth-century impressionism posited variations of a profound separation of humans from the world, the "moment of Cubism" in the early twentieth century posited, in Berger's estimation, a profound sense of the "qualitatively different relationship" between humans and the world.[63] The crux of this insight is, for Berger, the realization that the artist's "awareness of nature was part of nature," thus proposing an integrative fusion instead of a definitive separation.[64] Cubist art focused not on objects in their unitary, empirical totality, but on "interaction between objects," which extends not only to "different aspects of the same event, between empty space and filled space, between structure and movement," but also and most importantly "between the seer and the thing seen"—this last one relinquishing a stable external reality for one produced within and through the relationship between people and the world.[65] Such a proposition gained all the more resonance due to the degree "it coincided with the new scientific view of nature which rejected simple causality and the single permanent all-seeing viewpoint."[66]

Surrealism was another effort that took issue with not only mimetic claims or pretensions of images, but with the rationalist-empiricist basis of everyday life based on the fidelity and direct correspondence of sensory apprehension to the world. In the wake of the work of Freud, Breton argued for the relevance of the psychic: "If the depths of our mind contain with it strange forces capable of augmenting those on the surface, or of waging a victorious battle against them, there is every reason to seize them—first to seize them, then, if need be, to submit them to the control of our reason."[67] In a 1930 statement, Breton made the more strident claim that Surrealism "aims quite simply at the total recovery of our psychic force by a means which is nothing other than the dizzying descent into ourselves, the systematic illumination of hidden places and the progressive darkening of other places, the perpetual excursion into the midst of forbidden territory [. . .]."[68]

The sum embodiment and popular extent of the challenge posed by new means of communications, the rise of popular culture, of propaganda and advertising, Cubism, and Surrealism to the rationalist-empiricist proposition of a sensory-immanent, external, singular reality that could be fully represented is characterized by Leja, who suggests that "adjusting to modern life in New York circa 1900 meant learning to see skeptically."[69] As he continues, "to function successfully, even to survive, every inhabitant of the modern city, every target of competitive marketing, every participant in the new mass culture, every beneficiary of modern science and technology, every believer in spiritual realms had to process visual experiences with some measure of suspicion, caution, and guile."[70] Thus, in Leja's estimation, the full range of visual arts including "the fine arts and the larger visual culture of commercial amusements, photographic illustrations, and pictorial advertising" both fostered and responded to this skepticism, helping "make 'looking askance'—to use the period lingo—a generally shared habit."[71]

Act(ion)s of Dissidence

Under the pressures of such immense changes, protest demonstrations and montage underwent similarly immense revision and realignment, with perhaps the most important being not the separation of "art" from "real political action" (as overly literal accounts of the "modernist break" would have it) but the re-made fusion of the two. The Armory Show of modernist art and the Patterson Strike and pageant, both of which took place in New York in June 1913 and involved many of the same people, serves as a key transitional point.[72] The confluence of the two events was most clearly apparent on June 7, when "twelve hundred of the striking workers marched up Fifth Avenue to Madison Square Garden [. . .] and only a few blocks from the Armory" where the art show was taking place, where they then "put on a pageant in which they depicted their sufferings and announced their demands before a capacity audience of fifteen or sixteen thousand."[73] The related fusion of art and activism in workers' theater—itself drawing upon already existing traditions of theater for working-class audiences—became apparent in its didactic attention to ongoing situations and concerns in the context of politics of the Depression and of the postwar periods.[74] Traditions of a broader, informal practice of street theater re-emerged in the 1960s, with many modes since of community-based theater evident in many areas of the world.[75]

A similarly transitional point for montage was Berlin during the Weimar era. By the early twentieth century, and despite the persistence of the Romantic vision of the solo artist-as-genius, a critical-modernist conception of art as political practice developed an institutional critique by calling attention to the contextual determination of meaning and value, and thus the social construction of representation. In few other cities of the world did the crucible of capitalism, mass media, and urbanism prove as compelling as in Berlin at this time, where "almost no chronicle of the period failed to note or recall the sensations of life being speeded up and profoundly transformed." As Lavin continues, "montage was thought to echo the pace, the multiplicity, the disorientation, the thrill, and of course the fragmentation of modern everyday life."[76]

While montage has been put to use for many purposes, including propagandizing on behalf of Soviet consolidation of power in the 1920s and for commercial commodity advertising soon after in many Western countries, Raoul Hausmann and other Berlin Dadaists in the early twentieth century saw montage as an aesthetic-political practice that sought to undo authority and professionalized hierarchies.[77] As Ades notes, "*montage* in German means 'fitting' or 'assembly line', and *Monteur* 'mechanic', [or] 'engineer'." Such a semantic heritage underlies Hausmann's explanation of the term: "'This term translates our aversion at playing the artist, and, thinking of ourselves as engineers (hence our preference for workmen's overalls) we meant to construct, to assemble [*montieren*] our works'."[78]

By calling themselves engineers, Hausmann and others rejected the Romantic premise of art as created by an individual genius. By using found materials instead of their own, they also sought to turn the output of commodity capitalism against itself. This was of course an intentional, critical strategy. For them, montage was an important means of publicly critiquing not only the dominant, but also the role of media in the perpetuation of a generalized social, political, and economic order intertwined with the media industries.[79] By using widely known commercial artifacts as their basic material, but by adding, editing, or otherwise recontextualizing them, montage critiques not only dominant modes of production (visually demonstrating the possibility of a non-commercial mode of media production) but dominant social conventions and relations of capitalism and other cultures of oppression. If consumerism, the aura of authenticity and transparency of photographs, and, perhaps most importantly, the role of mass media as agents of social control were to be generally understood as social conventions that perpetuated specific interests instead of as natural and neutral facts, the inevitability and legitimacy of the existing social order would be seriously compromised, and possibilities for its challenge would have greater room to emerge.[80] The practice of montage was also a deeper claim about the importance of the unseen "real" that was embodied in fragments, which, if critically assembled, could allow one to more fully and deeply see.

A prominent example of the political use of photomontage can be seen in the work of John Heartfield and others in Berlin in the 1920s and 1930s, who used commercially available images and artifacts in such a way as to critique the social conventions upon which such images depended for their interpretation and acceptance.[81] For example, through montage, Heartfield revealed Hitler's cocked-wrist, platter-carrying variation of the Nazi salute as an open palm waiting for yet another payoff from collaborating industrialists.[82] Another consisted of a portrait of a working man seated in a chair, his body relaxed despite having his head wrapped tightly with newspapers, which, unbeknownst to him and contrary to the dominant convention of newspapers being a window on the world, actually hid the world with their obscuring accounts.[83] In yet another piece, titled "Mimicry," Heartfield portrayed Hitler's propaganda minister Joseph Goebbels dressed in a tailored suit and standing behind and above Hitler as a beautician stands above a customer, draping a Karl Marx look-alike beard onto his boss, with an adjacent excerpt from a newspaper story noting how Hitler used Bolshevik symbols to "win over for the regime any workers tending to opposition."[84] Montage was deemed to be such a powerful means of critique that it "was used increasingly by all political factions in Europe and Russia in the decades before the Second World War."[85] Nazis regarded Heartfield's critique so threatening that, to avoid capture, Heartfield soon had to operate in hiding in Hungary, enlisting the help of others to smuggle posters, reprints, and stickers back into Nazi Germany.

Yet, avant-garde Europe was not the exclusive centre of the practice of montage, as other examples illustrate. When combined with the spread of mass media and capitalism on the coattails of colonialism, the transnational folk-art heritage and accessible means of production characteristic of montage provided conditions for its use in non-West situations, such as the struggle of India to free itself from British colonial rule. For example, as pictured in colonial-Indian school books, the British Isles were equal in size or larger than India. As a result, suggests one account, "children going to school had no sense of the size of their own country." But an underground montage circulated during this time used popular knowledge with this depiction in order to critique its implicit judgments. The montage was an image that showed a map of India, but with all of Western Europe fitted inside it. As one account of this concludes, "For the first time, we realized what a power we were, how big we were [. . .]. The fight for independence somehow took on a new meaning [after seeing it]."[86]

The development of protest demonstrations and montage as forms of popular political action are single instances of a much broader process. Both expanded their form and practice by becoming a self-conscious performance with the emergence of "performance art" as an attempt to turn the object of art into the practice of culture and critique. Scholars have traced a trajectory from bohemian cafes of the nineteenth century to Dada and various other early-twentieth century avant-garde modernist artistic movements, the recognition in the 1950s of the performative composition of abstract-expressionist Jackson Pollock, the emergence in the 1960s of "happenings" and performance art, conceptual art of the 1970s, and beyond.[87] Newman characterizes these developing sources and conceptions of aesthetic-political action as including the "writings of the Situationist International (1957–1972) in France and the neo-Dada anti-art activities, including 'happenings,' of the artists associated with Fluxus (1961–). Both were concerned to use art as social action, through *détournment*, refunctioning and recontextualizing of elements, and direct intervention in the city [emphasis original]."[88]

This variable recognition of the modernist interrelations between theater, protest demonstration, and performance art were furthered by the availability of translations in English of key Western Marxist writings ("little known," according to Lunn, "until 1955 or 1960"), which in turn were evidence of the self-reflexive intellectualization of the practice due in large part to the general extension of college-level education.[89] Gitlin's recounting of the actions of the Diggers, Abbie Hoffman, Jerry Rubin, and others emphasizes the popularization of a complex practice that was part avant-garde art, part anarchist direct action, and part radical-revolutionary theory all wrapped up in a McLuhan-inspired "politics of *display* [emphasis original]."[90]

Yet, this display was not self-referential, but seen to directly engage the real—another instance of the insufficiency of conceptions of critical modernism as solely "aesthetic." A general, critical-modernist conception of public dissidence thus was not simply "aesthetic," but sought to fuse representation with political action in the forge of creative practice. Doing so did not mean the abandonment of the "real," but its critical recognition. For example, at the same time that the Surrealism of Breton championed the unconscious, this critical stance also needed to preserve a sense of the real, if for no other reason than to maintain a basis for the Surrealist critique. As Breton put it, "it is absolutely essential for us to act as though we were really 'part of the world,' in order thereafter to dare formulate certain reservations [about it]."[91]

Another instance that demonstrates how such iterations of critical modernism do not seek to replace the real with representation is how the critical premise and practice of montage inspired the ostensibly naturalist practice of documentary photography and its capture of contradictions present in everyday life. Instead of creating critical juxtapositions using scissors, a paste pot, and a pile of images, such photographers sought critical juxtapositions during the course of their work, isolating and emphasizing them through selective framing of images, and, by implication, calling attention to the contradictions of American society that were embedded in the fabric of everyday life. Such a usage underscores the broader importance of montage as not simply a specific technique, but a general approach underwritten by critical-modernist premises.

One such instance is the public-relations efforts of the federal Farm Security Administration (F.S.A.) during the catastrophe of Great Depression.[92] A number of photographs from the era exhibit how the critical premise of montage was embodied by an unflinching social realism. For example, F.S.A. photographs of makeshift shelters papered with newspaper and magazine ads to keep out rain and cold "served to underscore bitter distinctions between commercially promulgated ideals and the realities of people's lives."[93] Margaret Bourke-White's 1937 photograph of a line of African-American flood refugees seeking food and water also reveals the line's backdrop, which is an immense billboard erected by the National Association of Manufacturers that shows a scrubbed Anglo family complete with family dog packed tightly together in their car, grinning, and driving toward what one can only assume is greater prosperity. The banner headline reads "World's Highest Standard of Living; There's no way like the American Way."[94] The detection of contradictions in everyday life also informs a recent photograph taken of the front of the City Hall in San Francisco, with the stairs to the building blocked by a double line of police in riot gear and a banner hung on the building above the front doors that reads "SAN FRANCISCO: #1 World's Best Destination."[95] The premise also informs efforts that do not use images, in such ways as a humorist recently remarking how U.S. President George W. Bush flew on Air Force One (a pollution-belching Boeing 747) a mere 98 miles from Washington, D.C. to Wilmington, Delaware to give a speech on the dangers of global warming.[96]

Publicity and the Rationalization of Dissent

While these and other examples exemplify a critique of market formations of communications, rationalist empiricism, and capitalism, they also exist within specific historical contexts, which determine their resonance. The intensification of consumer capitalism is the facet of greatest importance to critical-modernist practices of democratic communication in the United States, and such a context has a range of implications.

On one hand, the imperative to expand markets has not only extended access to the products of commercial communications systems, but has also spurred the development and greater accessibility of means of communications (from small-scale offset printing and photocopying to various digital capabilities today of composition, sampling, reproduction, and circulation). In addition, the generative energy of a variety of social movements that focus on issues of identity, the environment, anti-war, labor, global justice, and many others and their various combinations continues to respond vigorously to the current neoliberal social and economic order.[97] A number of spectacular public events have taken place, with some of the most spectacular in recent years the set of actions at various locations of the World Trade Organization meetings and marches against U.S. invasion of Iraq in late March 2003 that took place throughout the world involving in aggregate millions of people.[98] These are all exceptionally significant in their embodiment of a general social willingness and ability.

On the other hand, while these and other related developments are positive developments, the very same access and means and, through them, the social movements themselves are increasingly subject to limits and pressures of rationalization. Understanding the various ways in which rationalization operates in relation to protest demonstrations, montage, and their hybrid combinations underscores the extent to which critical modernism, like rationalist empiricism, constitutes not a formula to follow but a site of struggle. While the effectiveness of specific examples can be debated, of greater importance here is the potential of the cultural forms through which innumerable examples work.

Such a consideration of the rationalization of critical modernism must center, first, on the emergence of publicity, which is the key form through which communications is professionalized and transformed into the representational complement of bureaucracy.[99] Publicity is not simply certain kinds of images, a particular industry, or a technique, but a particularly resonant cultural form enabling specific intentions, composition, and interpretation. It encourages an instrumental, positivist sense of representation consistent with broadcasting as discussed in Chapter 2—communications regarded as a tool with which to engineer a desired effect on the "masses."[100] Despite the error noted earlier of dismissing all critical-modernist activism as irretrievably aestheticized, publicity is a form that indeed embodies what critics of modernist activism call the aestheticization of dissent, if by that is meant the acceptance of the professional justification for the practice: that all that is ever needed is a well-done promotions campaign. The reduction can be traced to a conflation of the sense of "image" as a picture and as a valuable property. Recognizing the materialization of a commercial "brand image" as a property to build, maintain and protect, and thus for activists to attack, enables a key tactic, while getting pictures of an event on the evening television news or in the newspaper does little beyond reproducing the system of publicity.[101]

Conceptions of the self-sufficiency of publicity persist today in a range of overtly confrontational efforts, perhaps most clearly in the glorification of semiological warfare in the form of "culture-jamming."[102] DeLuca identifies an important moment of its development in the formation of the activist environmental group Greenpeace in the late 1960s, which drew inspiration from the notions of Marshall McLuhan, just as the advertising industry had done.[103] One of Greenpeace's early members made clear the connections between the form of publicity, the assumption of direct media effects, and the aestheticization of activism by claiming that "the more dramatic you can make it, the more controversial it is, [and] the more publicity you will get."[104] More recent claims remain mired in a McLuhanesque fantasy of publicity alone as a revolutionary tactic. A plan of first jamming the corporate American "'image factory until the day it comes to a sudden shattering halt'," then of building upon "'the ruins of the old'" a new culture "'with a noncommercial heart and soul'" is as hearty in earnestness as it is thin in an understanding of the dilemmas of publicity.[105] Even the very contention that, as long as an event is outrageous enough, news organizations will cover it is increasingly dubious. Such a claim relies on the assumption of equitable treatment of the promotional efforts of, on the one hand, "environmentalists and political activists promoting humanist values and the sanctity of place" and, on the other hand, of "celebrities, politicians and prominent businesspeople" and the corresponding "media managers, public relations consultants, spin doctors, stage-managed events and image manipulation [which] are [by contrast] an accepted reality—'part of the game'—of engagement with the media." As Hutchins and Lester conclude, contrary to the assumption of equitable treatment in the marketplace of ideas, quite "different rules of engagement" apply to the treatment of publicity that is critical of capitalism *as a system*.[106]

To adequately grasp its shortcomings, publicity must be recognized not as a neutral tactic nor a means of direct connection to "the public" or of "getting your word out," but as the consumer-capitalist transformation of traditions of democratic public action, which have from the start been subject to the pressures of rationalization, containment, and incorporation.[107] As Meyer argues, its rationalization commenced in the United States with the institutionalization of dissent, which meant in effect "bringing political conflict into the government in order to confine the boundaries of claims that activists might make and to invite partisans to struggle using conventional political means rather than taking up arms or opting out of the system."[108] A resulting, early rationale for quelling critical public gatherings was developed in reaction to labor strikes, based on the common-law doctrine of conspiracy and strikes' ostensible challenge to state sovereignty.[109] Courts struck down this rationale in the 1840s, refusing to regard collective co-ordination to exercise acknowledged rights as criminal conspiracy.[110] This change corresponded with the republican conception that citizens' coordinated, public confrontation of laws or official actions that impaired their status as "free, self-governing citizens" was not only legal, but necessary.[111]

However, with the consolidation and transformation of the American state "from a republican polity into a modern liberal state," the republican legitimacy of popular dissent waned.[112] As Lambert continues, not only does the modern state have "an autonomous interest [. . .] in asserting its claim to a monopoly of the legitimate use of force," it also has a coalescing interest in consolidating the emerging social order, which would be directly threatened by significant citizen autonomy.[113]

With this change in the nature of the state came an intensification of the rationalization of public assembly, which had a direct bearing on the efficacy of protest demonstrations as a means of significant dissidence. The implications of such a transformation were apparent by the late 1870s and the new tactic of using troops to suppress labor strikes.[114] Labor dissidence was increasingly seen as cut from the same cloth as political dissidence, with state power used to quell and suppress public dissident speech as well as assembly, particularly during times of war.[115] More generally, between the 1860s and 1940s judges throughout the country typically issued court orders preventing labor picketing "because of the 'potential' for violence" and enforced "statutes on the books that made even peaceful union picketing illegal in most situations."[116] By the 1960s, examples of the overt use of coercive state power extended to riot squads unleashed on civil-rights marches in the South and on antiwar rallies.[117] Such tactics were part of the "escalated-force" strategy used by police and troops within such a legal framework—the assumption that force by the irrational mob must be met with escalating force until it is quelled.[118]

However, beginning in the 1940s, a series of legal decisions began to put limits on repressive and arbitrary bans. By the 1970s, law-enforcement strategy had also moderated, with many adopting a "negotiated management" approach, which "was created in response to the civil rights and antiwar movements of the 1960s and 1970s" and the criticism by prominent national commissions of the escalated-force strategy.[119] Instead of stressing confrontation, control, and arrests, such a strategy advocated a high tolerance for community disruption, emphasized contact and negotiation, and promoted self-policing by demonstration organizers. Also reorienting the repressive tactics of the escalated-force strategy was concurrent litigation by the American Civil Liberties Union (A.C.L.U.) that broadened the right to demonstrate in "'traditional public forums' such as public streets, sidewalks, and parks."[120]

By placing greater emphasis on less confrontational means, this more collaborative form effectively rationalizes protest demonstrations, helping them fit smoothly into the dominant sometimes to the degree that the demonstration becomes an overt partnership between protesters and local authorities. For example, a march in September 2006 in favor of expanding a union for hotel workers adjacent to Los Angeles International Airport became what a journalist judged was "a highly choreographed episode of street theater, timed for news broadcasts and peaceful enough to persuade but not enrage the public," with the involvement of the Los Angeles Police Department "at nearly every stage." Such a protest demonstration exemplifies the aestheticization of dissent into publicity, with the key measure of success the successful staging of the event as planned. As the article puts it, "both sides are keenly aware of the public relations advantages of keeping the protest orderly and peaceful."[121]

Such rationalization and incorporation has provoked its own critical-modernist responses. One is to dispense altogether with organizing and instead individualize resistance, such as in the effort titled Sweatshopper discussed in prior chapters, which is a series of digitized monologues ("podcasts") on "the themes of globalization, labor activism, paranoia, and literature."[122] Among the offerings is a self-administered critical travelogue through Wal-Mart retail stores, to be listened to on personal digital players while walking through the store.[123]

Another response to rationalization is to deprofessionalize access and practice, such as is done on various collaborative efforts including Wikipedia and various video-sharing websites.[124] However, what continues to rationalize these efforts is their use of existing, professionalized forms, such as how the Wikipedia, in seeing itself as an encyclopedia, provokes debates about the degree of dependability of knowledge, or how offerings available on video-sharing sites not only mimic short dramatic films or faux-commercials but are evaluated with their own systems of rankings and ratings pioneered by commercial, broadcasting practice. For example, Current TV (which promotes itself as "about what's going on: stories from the real world, told by you") includes as part of its viewer-created content listings a promotion called "Watch and Vote" by which viewers can vote for their favorite listing as part of a weekly competition among video makers for a $1,000 prize.[125]

Additional attempts to counter rationalization include political performance art and street theater, which seeks in its small size and fluid practice to evade policing.[126] However, it remains subject to rationalization of more diffuse but equally powerful kinds, with the commercial incorporation of performance art and "happenings" as a publicity tactic diluting its radical potential. The hoped-for shock value has been increasingly incorporated as a form of entertainment—and often a not very daring example when measured up to the big-budget spectacles that are now a staple of guerrilla marketing campaigns in major cities.[127] Small-scale efforts such as the Surveillance Camera Players, who stage short, iconic plays critical of state video surveillance while being filmed by security cameras (with a member of the group filming the performance as seen on a security monitor) are assessed sympathetically through an evaluation of how satirically funny they are—the claim, for example, that "the [surveillance] cameras don't smile, but you might."[128] Less dour efforts such as the Yes Men, who spout inane official-sounding proposals while impersonating bureaucrats at policy meetings, and Reverend Billy and the Church of Stop Shopping, in which a white-suited, bouffant-haired faux-televangelist preaches against the sins of consumerism are both inspired theater, but are also easily reconciled not only with guerrilla-marketing events but with the carnival that is antiauthoritarian, individualist commercial culture.[129] To the extent that the ubiquity of publicity as an industry, an intention, and a cultural form reconciles these and other such demonstrative efforts as public displays of ironic, satirical cleverness, their critical purchase within consumer capitalism remains slight indeed.

In similar ways, montage is increasingly rationalized through satirical humor as witty "spoofs."[130] Such amenability to the dominant social order was clear early on. In 1929, an American advertising-industry commentator observed that, because photomontage "best reveals itself to the quickly moving eye," it therefore is "ideally suited to the spirit of this age in motion. We no longer sit and contemplate. We move, and at ever increasing speed."[131] The fit of montage was part of a more generally perceived fit by the late 1930s of "modernist forms [. . .] as alternatives to mainstream narrative and sentimentality" that had been the current formula in advertising.[132] Although the current practice of spoof ads by AdBusters and others were resonant for a time as challenges to the equity of brand images and to overly protective interpretations of copyright, the publicity industry has made increasing use of the tactic as well.[133] A business reporter noted that 2006 was "the year of anti-advertising, when marketers and their ad agencies went to great lengths to make sure their ads didn't look like typical Madison Avenue handiwork."[134] Even the commercial incorporation of the spoof ad appears to have run its course, notes a recent industry commentary, who suggests the extent to which it has already become passé for an industry restlessly searching for an innovation to set clients and campaigns above their competitors.[135]

In sum, claims of the aestheticization of dissent can be traced not to the embrace of specific tactics of representation, but to the pervasiveness of publicity, which is an outcome of long-standing efforts of rationalization. The very amenability and recognizability of dissident practices within consumer capitalism suggests not only that dissident publicity has been incorporated, nor only that the dominant has been successfully challenged and changed, but that the practice itself is part of a deeper formation in which capitalism generates and provides conditions not only for its perpetuation but its challenge.

Possibilities for Dissent

Recognizing the contextual determination of publicity on critical-modernist forms of democratic communications avoids textualist essentialism as well as the reductionism of the modernist break. It also avoids the tactical reduction of activism to realistic or achievable goals, which ignores how the benchmark of realistic and achievable is established and thus already accommodated by the dominant. Instead, a contextual understanding emphasizes the historical open-endedness of practice while also pointing to the struggle against market formations as a relational one, intimately tied to the emergence of new contexts and practices within them.

The way ahead is not only to think past the realistic and achievable but to do so by reconsidering present practices within new contexts as much as by generating practices that are new. Either way, as Epstein puts it, "protest politics must be utopian, in the sense that it must hold out a vision of a nonviolent and egalitarian society, and that it must build the new society within the shell of the old by creating a space within which these values can be realized as far as possible."[136]

And this is precisely where an exploration of the possibilities of democratic communications must now focus, given the contradictory, molten process that is capitalism, particularly in the face of globalization. For it is within emergent contradictions and conditions that not only the potential for re-energizing traditional dissident forms emerges, but that the possibility of producing new kinds of dissident practice out of what has been viewed as the dominant can be realized.

Notes

1. For the report and additional materials, see "Intergovernmental Panel on Climate Change," *Intergovernmental Panel on Climate Change* <http://www.ipcc. ch/> (7 Feb. 2007).

2. "Step It Up 2007: Index," *Step It Up* <http://www.stepitup2007.org// index.php> (7 Feb. 2007); "Step It Up 2007: Sign Up to Organize a Day of Action," *Step It Up* <http://www.stepitup2007.org//article.php?id=38> (7 Feb. 2007).

3. "Step It Up 2007: Organizer FAQs," *Step It Up* <http://stepitup2007.org/ article.php?id=76> (7 Feb. 2007).

4. The program provider C-SPAN describes itself as a "private, non-profit company, created in 1979 by the [U.S.] cable television industry as a public service. Our mission is to provide public access to the political process. C-SPAN receives no [U.S.] government funding; operations are funded by fees paid by cable and satellite affiliates who carry C-SPAN programming." See "C-SPAN—ABOUT US," *C-SPAN* <http://c-span.org/about/ index.asp?code=About> (31 Jan. 2007).

A similar remix of the 2002 State of the Union speech can be found at "/bsh," *Fuck It All* <http://www.fuckitall.com/bsh/> (21 Jan. 2007). The year of the appearance of the raw footage used in this remix was determined by comparing the clip to the transcript of the original broadcast. One useful location for such speech transcripts is "C-SPAN: STATE OF THE UNION," *C-SPAN* <http://www.c-span.org/executive/stateoftheunion.asp> (21 Jan. 2007).

5. Particularly comprehensive is U.S., House, Minority Staff, "The Constitution in Crisis; The Downing Street Minutes and Deception, Manipulation, Torture, Retribution, Coverups in the Iraq War, and Illegal Domestic Surveillance," Minority Report Prepared at the Direction of Rep. John Conyers, Jr., August 2006.

6. Locations of this clip at the time of writing include "State of the Union—George Bush Remix," *eBaum's World* <http://www.ebaumsworld.com/presaddress.html> (31 Jan. 2007).

7. Walter Benjamin, "Theses on the Philosophy of History," in *Illuminations; Essays and Reflections*, ed. Hannah Arendt, transl. Harry Zohn (New York: Schocken, 1968), 257.

8. George Trey, *Solidarity and Difference; The Politics of Enlightenment in the Aftermath of Modernity* (Albany, N.Y.: State University of New York Press, 1998), 3. Among many such claims are Andreas Huyssen, *After the Great Divide: Modernism, Mass Culture, Postmodernism* (Bloomington, Ind.: Indiana University Press, 1986). The vast range of concerns is embodied in Malcolm Waters, ed., *Modernity: Critical Concepts*, 4 vols. (New York: Routledge, 1999).

9. Douglas Kellner, *Critical Theory, Marxism, and Modernity* (Baltimore, Md.: Johns Hopkins University Press, 1989), 3.

10. Zygmunt Bauman, *Modernity and Ambivalence* (Ithaca, N.Y.: Cornell University Press, 1991), 4. Quoted in Richard Sheppard, *Modernism—Dada—Postmodernism* (Evanston, Ill.: Northwestern University Press, 2000), 8.

11. Sheppard, *Modernism—Dada—Postmodernism*, 8–9.

12. David Harvey, *The Condition of Postmodernity; An Enquiry into the Origins of Cultural Change* (Cambridge, Mass.: Basil Blackwell, 1989), 12.

13. John Berger, "The Moment of Cubism," in *The Sense of Sight; Writings by John Berger* (New York: Vintage International, 1993), 162.

14. Arjun Appadurai, *Modernity at Large: Cultural Dimensions of Globalization* (Minneapolis, Minn.: University of Minnesota Press, 1996); John Tomlinson, *Globalization and Culture* (Chicago, Ill.: University of Chicago Press, 1999); Arjun Appadurai, ed., *Globalization* (Durham, N.C.: Duke University Press, 2001); David Harvey, *Spaces of Global Capitalism; Towards a Theory of Uneven Geographical Development* (New York: Verso, 2006).

15. Laura Doyle and Laura Winkiel, eds., *Geomodernisms: Race, Modernism, Modernity* (Bloomington, Ind.: Indiana University Press, 2005); Susan Hegeman, *Patterns for America: Modernism and the Concept of Culture* (Princeton, N.J.: Princeton University Press, 1999); Peter Brooker and Andrew Thacker, eds., *Geographies of Modernism: Literatures, Cultures, Spaces* (New York: Routledge, 2005); and Harvey, *Postmodernism*, 24–25.

16. Raymond Williams, "Metropolitan Perceptions and the Emergence of Modernism," in *The Politics of Modernism*, ed. Tony Pinkney (London: Verso, 1996), 38.

17. Berger, "Moment of Cubism," 162.

18. Traditions of such views are extremely diverse. For a summary of key strands and critiques in the United States, see Christopher Lasch, *The True and Only Heaven; Progress and Its Critics* (New York: W.W. Norton, 1991).

19. Sheppard, *Modernism—Dada—Postmodernism*, 5. For another useful attempt at synoptic overview, see Eugene Lunn, *Marxism and Modernism; An Historical Study of Lukács, Brecht, Benjamin, and Adorno* (Berkeley, Calif.: University of California Press, 1982), 37–71.

20. Kellner, *Critical Theory*, 4.

21. Kellner, *Critical Theory*, 3.

22. Marshall McLuhan, *Understanding Media; the Extensions of Man* (New York: Signet Books, 1966); Marshall McLuhan and Quentin Fiore, *The Medium is the Massage* (New York: Random House, 1967). For a critique of the term and its premises, see Raymond Williams, *Television: Technology and Cultural Form* (New York: Schocken, 1974).

23. Denning has made the point that a democratization of buying has come at the price of a severe restriction in making. See Michael Denning, "The End of Mass Culture," *International Labor and Working Class History*, no. 37 (Spring 1990): 4–18.

24. Kellner, *Critical Theory*, 94, notes the presence of such a position among radicals, who perceived them as "progressive forces of production that could be used to criticize irrational relations of production and as forces leading to the creation of a more rational society."

25. At the same time it seeks to serve as an alternative to commercial systems, it relies for its distribution on Dish Network, a commercial provider of satellite entertainment. See "Free Speech TV: About Us," *Free Speech TV* <http://www.freespeech.org/html/aboutus.shtml> (21 Jan. 2007).

26. For example, Foster's formulation maps to three positions what is referred to above as a reformist critique and a radical critique, positing three levels of rejection of modernism in neo-conservative postmodernism, anti-modernist postmodernism, and critical postmodernism. See Hal Foster, *The Anti-Aesthetic: Essays on Postmodern Culture* (Port Townsend, Wash.: Bay Press, 1983), xii.

27. Dick Hebdige, *Hiding in the Light; On Images and Things* (New York: Routledge, 1988), 186–203.

28. Thorstein Veblen, *The Theory of the Leisure Class; An Economic Study in the Evolution of Institutions* (New York: Macmillan, 1899).

29. Raymond Morrow and David Brown, "Postempiricist Critiques of Positivism and Empiricism," in *Critical Theory and Methodology* (Thousand Oaks, Calif.: Sage, 1994), 62–81; Karl Popper, *The Open Society and Its Enemies*, revised edition (Princeton: N.J.: Princeton University Press, 1950); Karl Popper, *The Logic of Scientific Discovery* (New York: Basic Books, 1959); Jürgen Habermas, "Rationalism Divided Into Two," in *Positivism and Sociology*, revised edition, ed. Anthony Giddens (Aldershot: Gower, 1987), 195–223; Thomas S. Kuhn, *The Structure of Scientific Revolutions* (Chicago, Ill.: University of Chicago Press, 1962).

30. David Kadlec, *Mosaic Modernism; Anarchism, Pragmatism, Culture* (Baltimore, Md.: Johns Hopkins University Press, 2000); Stephen Kern, *The Culture of Time and Space 1880–1918* (Cambridge, Mass.: Harvard University Press, 1983); Linda Dalrymple Henderson, *The Fourth Dimension and Non-Euclidean Geometry in Modern Art* (Princeton, N.J.: Princeton University Press, 1983); and Hegeman, *Patterns for America*.

31. Lunn, *Marxism and Modernism*, 34–37.

32. Christine Bold, *The W.P.A. Guides: Mapping America* (Jackson, Miss.: University Press of Mississippi, 1999); Jerrold Hirsch, *Portrait of America: a Cultural History of the Federal Writers' Project* (Chapel Hill, N.C.: University of North Carolina Press, 2003); "StoryCorps®—Interview Your Loved Ones," *StoryCorps* <http://www. storycorps.net/> (19 Jan. 2007).

33. Wolfgang J. Mommsen, *The Age of Bureaucracy; Perspectives on the Political Sociology of Max Weber* (Oxford: Blackwell, 1974); Stanislav Andreski, ed. and transl., *Max Weber on Capitalism, Bureaucracy, and Religion: A Selection of Texts* (Boston, Mass.: Allen & Unwin, 1983); Robert J. Antonio and Ronald M. Glassman, eds., *A Weber-Marx Dialogue* (Lawrence, Ks.: University Press of Kansas, 1985); Max Horkheimer and Theodor Adorno, *Dialectic of Enlightenment*, transl. John Cumming (New York: Continuum, 1972).

34. Kellner, *Critical Theory*, 89–90.

35. Kellner, *Critical Theory*, 97.

36. Hebdige, *Hiding in the Light*, 195.

37. Fredric Jameson, "Reflections in Conclusion," in Ernst Bloch, Georg Lukács, Bertolt Brecht, Walter Benjamin, and Theodor Adorno, *Aesthetics and Politics*, ed. and transl. Ronald Taylor (London: New Left Books, 1977), 196.

38. Kevin J.H. Dettmar and Stephen Watt, "Introduction: Marketing Modernisms," in *Marketing Modernisms; Self-Promotion, Canonization, Rereading*, ed. Kevin J.H. Dettmar and Stephen Watt (Ann Arbor, Mich.: University of Michigan Press, 1996), 6.

39. Harvey, *Postmodern Condition*, 19.

40. Raymond Williams, *Marxism and Literature* (Oxford: Oxford University Press, 1977), 151.

41. Walter Benjamin refers to the auratic quality of high art in Walter Benjamin, "The Work of Art in the Age of Mechanical Reproduction," in *Illuminations*, ed. Hannah Arendt, transl. Harry Zohn (New York: Schocken, 1969), 217–51.

42. Raymond Williams, *Marxism and Literature* (Oxford: Oxford University Press, 1977), 152.

43. Williams, *Marxism and Literature*, 90–94.

44. Raymond Williams, *Culture and Society, 1780–1950* (New York: Columbia University Press, 1958), 327; E.P. Thompson, *The Making of the English Working Class* (New York: Vintage, 1966).

A useful, brief summary of sociological work regarding the rationalism of the process and the unity of the outcome is Donatello Della Porta and Mario Diani, *Social Movements; An Introduction*, second edition (Malden, Mass.: Blackwell, 2006), 11–14. In addition to Thompson, *Making*, examples of the usefulness of historical work in resisting oversystematization of social movements' cultural work includes Gareth Steadman Jones, "Rethinking Chartism," in *Languages of Class; Studies in English Working Class History 1832–1982* (Cambridge: Cambridge University Press, 1983), 90–178; and Michael Denning, *The Cultural Front; The Laboring of American Culture in the Twentieth Century* (New York: Verso, 1996).

45. David A. Schlossman, *Actors and Activists; Politics, Performance, and Exchange Among Social Worlds* (New York: Routledge, 2002), 5; Richard Schechner, *The Future of Ritual; Writings on Culture and Performance* (New York: Routledge, 1993), 21.

46. The phrase is from Andy Wood, *The Politics of Social Conflict: The Peak Country, 1520–1770* (New York: Cambridge University Press, 1999), 249.

47. Recent scholarship has redefined collective action quite substantially. In contrast to the tendency by the 1950s and 1960s to "classify under the same heading phenomena as diverse as crowds, movements, panics, manias, fashions, and so on," by the 1970s and in the wake of social upheavals of the 1960s that included substantial collective action in the form of protest demonstrations, sociological theorizing of collective action increasingly left behind the functionalist emphasis on social movements as "irrational, largely reactive phenomena" for a sense of them as based in some kind of general conception,

purposeful and organized. For a useful overview of such a sociological redefinition, see Della Porta and Diani, *Social Movements*, 1-32. Quotation is from page 11.

A second key area of similar redefinition and reassessment has been in the social history of early modern England as developed in works such as Thompson, *Making of the English Working Class*; George Rudé, *The Crowd in History; a Study of Popular Disturbances in France and England, 1730–1848* (New York: Wiley, 1964); E.P. Thompson, "The Moral Economy of the English Crowd in the Eighteenth Century," *Past and Present* 50 (1971): 76–136; Mark Harrison, *Crowds and History: Mass Phenomena in English Towns, 1790–1835* (Cambridge: Cambridge University Press, 1988); Nicholas Rogers, *Crowds, Culture, and Politics in Georgian Britain* (Oxford: Clarendon Press, 1998); Adrian Randall and Andrew Charlesworth, eds., *Moral Economy and Popular Protest: Crowds, Conflict and Authority* (New York: St. Martin's Press, 2000); and Ian Munro, *The Figure of the Crowd in Early Modern London: The City and Its Double* (New York: Palgrave Macmillan, 2005).

48. Such points build upon the emergence of significant literatures on dissidence and popular culture, particular in history and in sociology. Among the vast number of historical studies in addition to Thompson, *Making of the English Working Class*, representative works include Herbert G. Gutman, *Power and Culture; Essays on the American Working Class* (New York: Pantheon, 1987), and Salvatore Salerno, *Red November, Black November: Culture and Community in the Industrial Workers of the World* (Albany, N.Y.: State University of New York Press, 1989).

Useful overviews of the study of culture as part of the sociological study of social movements, in addition to Della Porta and Diani, *Social Movements*, 64–88, include Marco Giugni, Doug McAdam, and Charles Tilly, eds., *How Social Movements Matter* (Minneapolis, Minn.: University of Minnesota Press, 1999); and T.V. Reed, *The Art of Protest; Culture and Activism from the Civil Rights Movements to the Streets of Seattle* (Minneapolis, Minn.: University of Minnesota Press, 2005), 286–315.

49. For one such example, see John D. McCarthy, Clark McPhail, and Jackie Smith, "Images of Protest: Dimensions of Selection Bias in Media Coverage of Washington Demonstrations, 1982 and 1991," *American Sociological Review* 61, no. 3 (June 1996): 478.

50. Phillipp R. Schofield, *Peasant and Community in Medieval England, 1200–1500* (New York: Palgrave Macmillan, 2003), 163–65.

51. Mikhail Bakhtin, *Rabelais and His World*, transl. Hélène Iswolsky (Bloomington, Ind.: Indiana University Press, 1984).

52. Michael Bristol, "Shamelessness in Arden: Early Modern Theater and the Obsolescence of Popular Theatricality," in *Print, Manuscript & Performance; the Changing Relations of the Media in Early Modern England*, ed. Arthur F. Marotti and Michael D. Bristol (Columbus, Ohio: Ohio State University Press, 2000), 279–306; Brean S. Hammond, *Professional Imaginative Writing in England, 1670–1740; 'Hackney for Bread'* (Oxford: Clarendon Press, 1997), 21–24.

53. Bristol, "Shamelessness in Arden"; Hammond, *Professional Imaginative Writing*, 21–24.

54. *Oxford English Dictionary*, 2nd ed., s.v. "demonstration."

55. Christine Poggi, *In Defiance of Painting: Cubism, Futurism, and the Invention of Collage* (New Haven, Conn.: Yale University Press, 1992); Katherine Hoffman, "Collage in the Twentieth Century: An Overview," in *Collage: Critical Views*, ed. Katherine Hoffman (Ann Arbor, Mich.: UMI Research Press, 1989), 1–37.

56. Douglas Kahn, *John Heartfield: Art and Mass Media* (New York: Tanam Press, 1985), 106–13; Diane Waldman, *Collage, Assemblage, and the Found Object* (New

York: Harry N. Abrams, 1992), 7–15; Cynthia Wayne, *Dreams, Lies, and Exaggerations; Photomontage in America* (College Park, Md.: Art Gallery at the University of Maryland, 1991), 14; Eddie Wolfram, *History of Collage; An Anthology of Collage, Assemblage, and Event Structures* (New York: Macmillan, 1975), 7–14; Thomas Leonard, *News for All; America's Coming-of-Age with the Press* (New York: Oxford University Press, 1995), 125; Norman Laliberté and Alex Mogelon, *Collage, Montage, Assemblage; History and Contemporary Techniques* (New York: Van Nostrand Reinhold, 1971), 1; Claude M. Bromley, *Crazy Camera; Secrets of Photomontage* (New York: Focal Press, 1943); Michel Frizot, "The Ancestry of Photomontage," in *Photomontage; Experimental Photography Between the Wars*, transl. Shaun Whiteside (London: Thames and Hudson, 1991) [unpaginated]; and Harriet Janis and Rudi Blesh, *Collage: Personalities, Concepts, Techniques* (Philadelphia, Penn.: Chilton, 1967), 1–4.

57. Raymond Williams, "When Was Modernism?," in *The Politics of Modernism*, ed. Tony Pinkney (London: Verso, 1996), 53.

58. Gunther Barth, *City People; The Rise of Modern City Culture in Nineteenth-Century America* (New York: Oxford University Press, 1980).

59. In addition to Barth, *City People*, 110–47, see Susan Buck-Morss, *The Dialectics of Seeing; Walter Benjamin and the Arcades Project* (Cambridge, Mass.: MIT Press, 1989), 39; Johann Friedrich Geist, *Arcades, the History of a Building Type* (Cambridge, Mass.: MIT Press, 1983); Elaine S. Abelson, *When Ladies Go A-Thieving: Middle-Class Shoplifters in the Victorian Department Store* (New York: Oxford University Press, 1989); Simon J. Bronner, ed., *Consuming Visions: Accumulation and Display of Goods in America, 1880–1920* (New York: Norton,1989); William Leach, *Land of Desire: Merchants, Power, and the Rise of a New American Culture* (New York: Pantheon, 1993); Bill Lancaster, *The Department Store: A Social History* (New York: Leicester University Press, 1995); Geoffrey Crossick and Serge Jaumain, eds., *Cathedrals of Consumption: the European Department Store, 1850–1939* (Aldershot: Ashgate, 1999); and Susan Lomax, "The View From the Shop: Window Display, the Shopper and the Formulation of Theory," in *Cultures of Selling: Perspectives on Consumption and Society Since 1700*, ed. John Benson and Laura Ugolini (Aldershot: Ashgate, 2006), 265–92. Compare such accounts with the nostalgic but well-detailed account of the precapitalist local store in Gerald Carson, *The Old Country Store* (New York: Oxford University Press, 1954).

60. Roland Marchand, *Advertising the American Dream; Making Way for Modernity 1920–1940* (Berkeley, Calif.: University of California Press, 1985); Susan Strasser, *Satisfaction Guaranteed; The Making of the American Mass Market* (New York: Pantheon, 1989); Ralph M. Hower, *The History of an Advertising Agency; N.W. Ayer & Son at Work 1869–1949* (Cambridge, Mass.: Harvard University Press, 1949); Daniel Pope, *The Making of Modern Advertising* (New York: Basic Books, 1983); Michael Schudson, *Advertising, the Uneasy Persuasion: Its Dubious Impact on American Society* (New York: Basic Books, 1984); Raymond Williams, "Advertising: The Magic System," in *Problems in Materialism and Culture* (London: Verso, 1980), 170–95.

61. In addition to Bronner, *Consuming Visions*, see Stuart Ewen, *All Consuming Images; The Politics of Style in Contemporary Culture* (New York: Basic Books, 1988); Jackson Lears, *Fables of Abundance; A Cultural History of Advertising in America* (New York: Basic Books, 1994); and various essays in Martyn J. Lee, ed., *The Consumer Society Reader* (Oxford: Blackwell, 2000).

62. Jennifer Wicke, *Advertising Fictions: Literature, Advertisement & Social Reading* (New York: Columbia University Press, 1988).

63. Berger, "Moment of Cubism," 185.

64. On this point in relation to the counterculture of the 1960s, see Julie Stephens, *Antidisciplinary Protest; Sixties Radicalism and Postmodernism* (Cambridge: Cambridge University Press, 1998).

65. Berger, "Moment of Cubism," 178.

66. Berger, "Moment of Cubism," 180.

67. Andre Breton, *Manifestoes of Surrealism*, transl. Richard Seaver and Helen R. Lane (Ann Arbor, Mich.: University of Michigan Press, 1972), 10.

68. Breton, *Manifestoes of Surrealism*, 136–37.

69. Michael Leja, *Looking Askance; Skepticism and American Art from Eakins to Duchamp* (Berkeley, Calif.: University of California Press, 2004), 1. See also Ewen, *All Consuming Images*.

70. Leja, *Looking Askance*, 1.

71. Leja, *Looking Askance*, 1.

72. Martin Green, *New York 1913; The Armory Show and the Paterson Strike Pageant* (New York: Scribner's, 1988), 11.

73. Green, *New York 1913*, 4.

74. Bruce A. McConachie and Daniel Friedman, eds., *Theatre for Working-Class Audiences in the United States, 1830–1980* (Westport, Conn. Greenwood, 1985); Raphael Samuel, Ewan MacColl, and Stuart Cosgrove, *Theatres of the Left, 1880–1935: Workers' Theatre Movements in Britain and America* (Boston, Mass.: Routledge & Kegan Paul, 1985); Lorraine Brown, ed., *Liberty Deferred and Other Living Newspapers of the 1930s Federal Theatre Project* (Fairfax, Va.: George Mason University Press, 1989); Colette A. Hyman, *Staging Strikes: Workers' Theatre and the American Labor Movement* (Philadelphia, Penn.: Temple University Press, 1997); Kirk W. Fuoss, *Striking Performances/Performing Strikes* (Jackson, Miss.: University Press of Mississippi, 1997).

75. Useful expressions and studies of street theater of the 1960s include John Weisman, *Guerrilla Theater; Scenarios for Revolution* (Garden City, N.Y.: Anchor Press, 1973); Bradford D. Martin, *The Theater is in the Street: Politics and Public Performance in Sixties America* (Amherst, Mass.: University of Massachusetts Press, 2004); Claudia Orenstein, *Festive Revolutions: the Politics of Popular Theater and the San Francisco Mime Troupe* (Jackson, Miss.: University Press of Mississippi, 1998); and Harry J. Elam, Jr., *Taking It to the Streets: the Social Protest Theater of Luis Valdez and Amiri Baraka* (Ann Arbor, Mich.: University of Michigan Press, 1997).

For work on community-based activist theater, see Susan C. Haedicke and Tobin Nellhaus, eds., *Performing Democracy; International Perspectives on Urban Community-Based Performance* (Ann Arbor, Mich.: University of Michigan Press, 2001); and Jan Cohen-Cruz, ed., *Radical Street Performance; An International Anthology* (New York: Routledge, 1998).

76. Maud Lavin, *Cut with the Kitchen Knife; The Weimar Photomontages of Hannah Höch* (New Haven and London: Yale University Press, 1993), 47–49; and Peter Fritzsche, *Reading Berlin 1900* (Cambridge: Harvard University Press, 1996).

77. Matthew Teitelbaum, ed., *Montage and Modern Life, 1919–1942* (Cambridge, Mass.: MIT Press, 1992); Lears, *Fables of Abundance*, 299–344; and Thomas Frank, *The Conquest of Cool; Business Culture, Counterculture, and the Rise of Hip Consumerism* (Chicago, Ill.: University of Chicago Press, 1997).

78. Dawn Ades, *Photomontage* (New York: Pantheon, 1976), 7.

79. Often the two methods were combined, as in Heartfield's work. See David Evans and Sylvia Gohl, *Photomontage; A Political Weapon* (London: Gordon Fraser, 1986), 21.

80. Kahn, *John Heartfield*, 73–103.

81. Peter Selz, "John Heartfield's Photomontages," in John Heartfield, *Photomontages of the Nazi Period* (New York: Universe Books, 1977), 11.

82. Heartfield, *Photomontages*, 46, 120.

83. Heartfield, *Photomontages*, 70.

84. Heartfield, *Photomontages*, 123. A complete, annotated English-language catalogue of Heartfield's magazine work is David Evans, *John Heartfield AIZ; Arbeiter-Illustrierte Zeitung Volks Illustrierte 1930–38* (New York: Kent, 1992).

85. Ades, *Photomontage*, 8, 12.

86. "Remembering Independence," interview with Madhur Jaffrey, *Morning Edition – National Public Radio*, 14 August 1997 <http://www.npr.org/> (2 Feb. 2001).

87. Jayne Wark, *Radical Gestures; Feminism and Performance Art in North America* (Montreál, Quebéc: McGill-Queens University Press, 2006), 29; and Allan Kaprow, *Assemblages, Environments and Happenings* (New York: Harry N. Abrams, 1966).

88. Michael Newman, "Revising Modernism, Representing Postmodernism," in *Postmodernism: ICA Documents*, ed. Lisa Appignanesi (London: Free Association Books, 1989), 105–106, emphasis in original. See also Guy Debord, *Society of the Spectacle* (Detroit, Mich.: Black and Red, 1983); Tom McDonough, ed., *Guy Debord and the Situationist International* (Cambridge, Mass.: MIT Press, 2002); Mike Sell, *Avant-Garde Performance & the Limits of Criticism: Approaching the Living Theatre, Happenings/Fluxus, and the Black Arts Movement* (Ann Arbor, Mich.: University of Michigan Press, 2005); Geoffrey Hendricks, ed., *Critical Mass: Happenings, Fluxus, Performance, Intermedia, and Rutgers University, 1958–1972* (New Brunswick, N.J.: Rutgers University Press, 2003); Owen F. Smith, "Fluxus Praxis: An Exploration of Connections, Creativity, and Community," in *At a Distance; Precursors to Art and Activism on the Internet*, ed. Annmarie Chandler and Norie Newmark (Cambridge, Mass.: MIT Press, 2005), 116–138.

89. Lunn, *Marxism and Modernism*, 5.

90. Todd Gitlin, *The Sixties; Years of Hope, Days of Rage* (New York: Bantam, 1989), 233; Jerry Rubin, "24: Revolution is Theater-in-the-Streets," in *Do It!; Scenarios of the Revolution* (New York: Simon and Schuster, 1970), 132–43; and Richard Schechner, "The Street is the Stage," in *The Future of Ritual; Writings on Culture and Performance* (New York: Routledge, 1993), 45–93.

91. Breton, *Manifestoes of Surrealism*, 137–38.

92. Carl Fleischhauer and Beverly W. Brannan, eds., *Documenting America, 1935–1943* (Berkeley, Calif.: University of California Press, 1988).

93. Stuart Ewen, *PR!: A Social History of Spin* (New York: Basic Books, 1996), 283.

94. The photograph is reproduced and discussed in Sean Callahan, ed., *The Photographs of Margaret Bourke-White* (New York: New York Graphic Society, 1972), 136–37.

95. Reproduced in Richard Edmondson, *Rising Up; Class Warfare in America from the Streets to the Airwaves* (San Francisco, Calif.: Librad Press, 2000), second page following 184 (unnumbered).

96. "A Prairie Home Companion for January 27, 2007 from American Public Media," *A Prairie Home Companion* <http://prairiehome.publicradio.org/programs/2007/01/27/> (5 Feb. 2007).

97. George McKay, *Senseless Acts of Beauty; Cultures of Resistance Since the Sixties* (New York: Verso, 1996); Andrew Opel and Donnalyn Pompper, eds., *Representing Resistance: Media, Civil Disobedience, and the Global Justice Movement* (Westport, Conn.: Praeger, 2003).

98. In addition to popular-press accounts, see Tom Mertes, ed., *A Movement of Movements; Is Another World Really Possible?* (London: Verso, 2004).

99. In addition to Ewen, *PR!*, see Inger L. Stole, *Advertising on Trial: Consumer Activism and Corporate Public Relations in the 1930s* (Urbana, Ill.: University of Illinois Press, 2006).

100. Raymond Williams, *Culture and Society, 1780–1950* (New York: Columbia University Press, 1958), 297–99.

101. Tonya Couch Roy, "Coke Is It? (Dis)Locating Identity in Corporate Identity Discourse" (unpublished Ph.D. diss., University of Georgia, 2004). Less nuanced but still useful sources include Jarol B. Manheim, *The Death of a Thousand Cuts: Corporate Campaigns and the Attack on the Corporation* (Mahwah, N.J.: Lawrence Erlbaum, 2001); Celia Lury, *Brands: the Logos of the Global Economy* (New York: Routledge, 2004); Jarol B. Manheim, *Biz-War and the Out-of-Power Elite: the Progressive-Left Attack on the Corporation* (Mahwah, N.J.: Lawrence Erlbaum, 2004); and Liz Moor, *The Rise of Brands* (Oxford: Berg, 2007).

102. Umberto Eco, "Towards a Semiological Guerrilla Warfare," in *Travels in Hyper Reality: Essays*, ed. Helen Wolff and Kurt H. Wolff, transl. William Weaver (London: Harcourt Brace Jovanovich, 1986), 135–44; Mark Dery, *Culture Jamming: Hacking, Slashing and Sniping in the Empire of the Signs* (Westfield, N.J.: Open Pamphlet Series, 1993); and Christine Harold, *OurSpace; Resisting the Corporate Control of Culture* (Minneapolis, Minn.: University of Minnesota Press, 2007).

Such a mediacentric conception coincides with the semiotic focus of the work of Melucci regarding social movements and the technological determinism of Castells. See Alberto Melucci, *Challenging Codes: Collective Action in the Information Age* (New York: Cambridge University Press, 1996); Manuel Castells, *The Rise of the Network Society* (Cambridge, Mass.: Blackwell, 1996); Manuel Castells, *The Power of Identity* (Malden, Mass.: Blackwell, 1997); and Manuel Castells, *End of Millennium* (Malden, Mass.: Blackwell, 1998).

103. McLuhan, *Understanding Media*; McLuhan and Fiore, *Medium Is the Massage*.

104. Quoted in Kevin DeLuca, *Image Politics: The New Rhetoric of Environmental Activism* (New York: Guilford Press, 1999), 4–5.

105. Kalle Lasn, *Culture Jam: How to Reverse America's Suicidal Consumer Binge—and Why We Must* (New York: Quill/HarperCollins, 2000), xvi. Quoted in Michael Strangelove, *The Empire of Mind; Digital Piracy and the Anti-Capitalist Movement* (Toronto: University of Toronto Press, 2005), 119.

106. Brett Hutchins and Libby Lester, "Environmental Protest and Tap-Dancing with the Media in the Information Age," *Media, Culture and Society* 28, no. 3 (May 2006): 446. A comprehensive study of such situations is Todd Gitlin, *The Whole World Is Watching; Mass Media in the Making and Unmaking of the New Left* (Berkeley, Calif.: University of California Press, 1980).

107. Jürgen Habermas, *The Structural Transformation of the Public Sphere: An Inquiry into a Category of Bourgeois Society*, translated by Thomas Burger and Frederick Lawrence (Cambridge: Mass.: MIT Press, 1989); and Craig Calhoun, ed., *Habermas and the Public Sphere* (Cambridge, Mass.: MIT Press, 1992).

108. Meyer, *Politics of Protest*, 19.

109. Josiah Bartlett Lambert, *"If the Workers Took a Notion"; The Right to Strike and American Political Development* (Ithaca, N.Y.: ILR Press, 2005), 21.

110. Lambert, *"If the Workers Took a Notion,"* 21.

111. Lambert, *"If the Workers Took a Notion,"* 45.

112. Lambert, *"If the Workers Took a Notion,"* 42.

113. Lambert, *"If the Workers Took a Notion,"* 46.

114. Lambert, *"If the Workers Took a Notion,"* 22–23.

115. Geoffrey R. Stone, *Perilous Times: Free Speech in Wartime from the Sedition Act of 1798 to the War on Terrorism* (New York: W.W. Norton, 2004).

116. Darien McWhirter, *Freedom of Speech, Press, and Assembly* (Phoenix, Ariz.: Oryx Press, 1994), 120.

117. Reed, *Art of Protest*; Howard Zinn, *A People's History of the United States: 1492–Present* (New York: HarperCollins, 1999).

118. David Schweingruber, "Mob Sociology and Escalated Force: Sociology's Contribution to Repressive Police Tactics," *Sociological Quarterly* 41, no. 3 (Summer 2000): 379.

119. Such a claim is not meant to ignore the notable exceptions of a handful of large actions such as those in Seattle in 1999 and in Miami in 2003, nor a return of more repressive practices such as routine videotaping of demonstrators and increased surveillance of such groups in the wake of the passage of the U.S. Patriot Act.

120. Schweingruber, "Mob Sociology," 380–381.

121. Joe Mathews, "A Plan for Very Civil Disobedience; Police and Union Will Follow a Script, Which Even Specifies Who will be Arrested, in a March near LAX to Organize Hotel Workers," *Los Angeles Times*, 28 September 2006, A1, retrieved from LexisNexis Academic <http://web.lexis-nexis.com/> (29 Sept. 2006); Joe Mathews, "Labor Protest Targets Airport-Area Hotels; Hundreds are Willingly Arrested After Sit-Ins Shut Century Boulevard; Few Travelers Miss Flights," *Los Angeles Times*, 29 September 2006, B3, retrieved from LexisNexis Academic <http://web.lexis-nexis.com/> (12 Feb. 2007).

122. "Sweatshopper.org," *Sweatshopper* <http://sweatshopper.org/> (7 Apr. 2006).

123. "Sweatshopper.org," *Sweatshopper* <http://sweatshopper.org/podcast_ listen.htm> (7 Apr. 2006).

124. "Main Page—Wikipedia, the Free Encyclopedia," *Wikipedia* <http://en .wikipedia.org/> (16 Feb. 2007). Examples of the many and proliferating video-sharing sites include "YouTube—Broadcast Yourself," *YouTube* <http://www .youtube.com/> (19 Feb. 2007); "Sharkle.com—Free Online Video Sharing Community," *Sharkle* <http://www.sharkle.com/> (16 Feb. 2007); and "Video Activist Network," *Video Activist Network* <http://www.videoactivism.org/> (16 Feb. 2007).

125. Bob Kemper, "Something Wiki This Way Comes in Partisan Politics," *Atlanta Journal-Constitution*, 13 March 2006, 1A; "Current TV // The TV Network Created by the People Who Watch It," *Current TV* <http://www.current.tv/> (1 Mar. 2007).

126. Nato Thompson and Gregory Sholette, eds., *The Interventionists; Users' Manual for the Creative Disruption of Everyday Life* (North Adams, Mass.: MASS MoCA, 2004).

127. A recent example is a promotion in Boston that mobilized the city's police force, which feared it was a terrorist plot. See Louise Story, "A Boston Marketing Stunt That Bombed, or Did It?," *The New York Times*, 2 February 2007, C1.

128. "Surveillance Camera Players," *Surveillance Camera Players* <http://www. factoryschool.org/pubs/scp/> (12 Feb. 2007).

129. "The Yes Men," *The Yes Men* <http://www.theyesmen.org/> (16 Feb. 2007); Jill Lane, "Reverend Billy; Preaching, Protest, and Postindustrial Flânerie," in *The Performance Studies Reader*, ed. Henry Bial (New York: Routledge, 2004), 299; "Reverend Billy and the Church of Stop Shopping," *Reverend Billy and the Church of Stop Shopping* <http://www.revbilly.com/> (16 Feb. 2007). An account of worldwide efforts in-

clude Black Stimson and Gregory Sholette, eds., *Collectivism after Modernism; The Art of Social Imagination after 1945* (Minneapolis, Minn.: University of Minnesota, 2007).

130. "Sniggle.net: The Culture Jammer's Encyclopedia," *Sniggle.net* <http://www.sniggle.net/> (16 Feb. 2007); Glen Tarman, "Digital Activism, the WTO and International Trade Rules," *Digital Opportunity Channel*, 5 September 2003 <http://www.digitalopportunity.org/article/view/67349/1/1134> (9 Feb. 2007); and "WTO | World Trade Organization: WTO / GATT," *World Trade Organization* <http:// gatt.org/> (9 Feb. 2007).

A number of examples can be accessed at "DIYMedia.net—Media Collage Index," *DIYMedia.net* <http://www.diymedia.net/collage/> (16 Feb. 2007); "Illegal-art.org :: A Project of Stay Free! Magazine," *Illegal-art.org* <http://www.illegal-art.org/> (16 Feb. 2007); and "rx" *RX* <http://www.thepartyparty.com/> (16 Feb. 2007).

131. Douglas McMurtrie, "The Photomontage—A New Illustrative Technique," *Printer's Ink*, 26 December 1929, 90.

132. Lears, *Fables of Abundance*, 336; Frank, *Conquest of Cool*; Dettmar and Watt, *Marketing Modernisms*; Catherine Turner, *Marketing Modernism Between the Two World Wars* (Amherst, Mass.: University of Massachusetts Press, 2003); and David Steel, "Surrealism, Literature of Advertising, and the Advertising of Literature in France, 1910–1930," *French Studies* 41, no. 3 (July 1987): 283–97.

133. "AdBusters: Spoof Ads," *AdBusters* <http://adbusters.org/spoofads/> (12 Feb. 2007); Siva Vaidhyanathan, *Copyrights and Copywrongs: The Rise of Intellectual Property and how It Threatens Creativity* (New York: New York University Press, 2001); and Kembrew McLeod, *Owning Culture: Authorship, Ownership, and Intellectual Property Law* (New York: Peter Lang, 2001).

134. Suzanne Vranica, "Best (and Worst) Ads of '06," Associated Press, 22 December 2006, accessed from LexisNexis Academic <http://web.lexis-nexis.com/> (12 Feb. 2007).

135. Eleanor Trickett, "Inside the Mix—Spoof Ads," *PR Week*, 14 November 2005, 10.

136. Barbara Epstein, *Political Protest and Cultural Revolution* (Berkeley, Calif.: University of California Press, 1991), 269.

Part Three

Toward New Formations

Introduction to Part Three

A key point that has been emphasized throughout this book is the indebtedness of market formations of communications as well as those that struggle against them to the formative conditions of capitalism. At the same time, however, while a critical task is to recover the depth of determination of communications projects within such conditions, it is equally important to recognize emergent areas of activity that constitute prefigurative responses which, while less developed, deserve even greater attention. Detecting such emergent formations and practices—much less writing about them—is as difficult as it is necessary. They can only be recognized as suggestive tendencies at best, but they also suggest a practice of democratic communications significantly different from those discussed in Part Two due to the significantly different conditions of their formation.

Chapter 6, "Market Radicalism and the Struggle of Participation," focuses on the recent emergence and institutionalization of so-called consumer-generated (sometimes called user-generated) content. What has made its emergence possible is the long-standing market imperative to engage audiences, which has only become more insistent as markets have become increasingly competitive. While present in advertising and other realms, the institutionalization of consumer-generated content is more developed in the practice of citizen journalism. Granting the corresponding development of more sophisticated means of managing and containing such public contributions so as to coincide with institutional imperatives, the trajectory suggested by the gradual opening of communications institutions to popular contributions suggests in turn the still-open possibility of significant institutional broadening and reformation.

Chapter 7, "Democratic Communications as Critical, Collective Education," examines recent efforts to formulate a practice of democratic communications as direct autonomous composition. The case of migrant labor-camp newspapers of the 1930s Dust Bowl in the United States helps develop a conception of direct autonomous composition as the collective, colloquial production of a social order. This case is contrasted to the recent emergence of programs of media literacy, in which a superficially similar educational project is too often enmeshed in traditional, technicist, authoritarian modes of learning instead of embodying a commitment to open, participatory exploration. At the same time, projects outside of those institutionalized for classroom use demonstrate the open potential of what such a broadened sense of democratic communications—beyond the making of news stories, documentaries, etc.—might also be.

Chapter 6

Market Radicalism and the Struggle of Participation

"Welcome to the revolution in the culture of news production, distribution, and consumption," states the membership-registration information for the Korean-turned-international organization *OhmyNews*. "You are about to join OhmyNews International as a global citizen reporter."[1] Such a status should not be taken lightly, wrote *OhmyNews* founder Yeon-Ho Oh in February 2007 as part of celebrating the seventh anniversary of the founding of *OhmyNews*. Consistent with the adopted motto of "every citizen is a reporter," the brand of citizen journalism practiced by Oh and others is not simply a way to make a few extra dollars or simply to express yourself. Instead, it "concerns how citizen participants create a new journalism and a new world."[2]

When seen in the context of long-standing means of struggling against market formations of communications, this and many additional examples exhibit a fundamental contradiction. They see capitalist business as not repressing, but instead as enabling democratic communications. What makes such a formulation possible is the severing of a critique of capitalism from the struggle for democratic communications. While *OhmyNews* opens its pages to a substantial volume of contributions from people (as "citizen reporters") outside the organization, and in this way is consistent with forms of democratic communications through a critique of the exclusivity of professionalism, it works in concert with capitalism by being a for-profit business that funds itself primarily through advertising.

Not only is this novel formulation increasingly relied upon by self-styled social activists, the severing of a critique of capitalism from the struggle for democratic communications also makes the business model relevant to many established media companies, whose marketshare continues to decline if not hemorrhage. While efforts such as *OhmyNews* see increasing profits as a means of enabling greater civic engagement and social change, other adoptees of this business model invert the formulation, seeing instead greater civic engagement as a means of increasing profits.

Although practiced in opposing ways, the relevance of this media-business model is already widespread. As reported by the authors of an early 2007 marketing report, two-thirds of their sample of the five-hundred fastest growing companies in the United States regarded "social media" as either somewhat important or very important to their business/marketing strategy.[3] Authors of a recent newspaper-industry think-tank report argue that citizen journalism makes it possible to "engage and grow online community in order to build affinity and loyalty to their [users'] brand experience."[4] Citizen journalism and other kinds of so-called social-networking communications projects regard commercial success as a matter not of more effective persuasion and promotion, but of building community and empowering users.[5] By March 2007, even some television stations are "inviting 'citizen journalists' to produce anything from online news footage to complete reports. The coverage could find its way onto local news broadcasts as well." But, for Timur Yarnall, President/CEO of Broadcast Interactive, boosting citizen contributions is a means not of changing the world, but of boosting profits, stating that citizen journalism is, above all, "a revenue-driving initiative."[6]

The emergence of such an assertion—that democratic communications can be seen as aided by the very conditions traditionally seen as evoking its need in the first place—underscores the historical instead of universal nature of such assertions. Prevailing conceptions of democratic communications now can be seen as built upon such conditions as a highly developed consumer society, widespread private ownership of capital and industry, concomitant varying relations between government and industry—in short, a Western capitalist, representative democracy—while also relying on analyses more attuned with the conditions of the 1950s and 1960s than the present.

Recognition of the historical nature of such an assertion also highlights its additional defects. It relies upon simplified theories of mass society, binary understandings of the relationship between capitalism and social change, and a pessimistic assessment of the degree of depoliticization and alienation of people living within such a system. For example, Western conditions that are seen to constitute the danger of depoliticization—such as de-unionization, a tradition of liberal representative government, and corresponding economic and regulatory structures that progressively remove opportunities for meaningful political participation and collective action—may be present in very different forms in other countries, or reduced to the extent they are virtually absent, thus creating a very different political potential.[7] Challenging these "universal" formulas is historical work that suggests that diverse instances such as the commercial bourgeois press, South Wales Miners' Institute cinemas in the1930s (judged at the time to be a "very profitable enterprise"), and popular commercial culture in the twentieth century United States served at the time as important, yet always provisional progressive projects.[8]

This chapter argues that recently emerged conditions have enabled the formation of citizen journalism and other, similar practices, and that these have the potential for being a basis of significantly progressive formations. It argues that the current emergence of what can be described as "market radicalism" is an important prefigurative formation that seeks not simply to broaden the range of content available in media products but to remake the productive relationships that have constituted communications industries in the West for 150 years.[9] At the same time, the severance of democratic communications from a critique of capitalism serves as the primary means by which such progressive projects are reincorporated into the dominant.

Such a formulation seeks to reconceptualize essentialized and uniform domination as an ongoing struggle over productive resources produced in part by capitalist society. In such a conception, capitalism is recognized as a dominant, yet always unstable formation instead of as a uniform description or an exclusive or universal system in the sense of the neoliberal mantra that there is no alternative. By contrast to being uniform, static, or all-encompassing, capitalism is better seen as a highly contextual, contradictory, and historical formation. Such a recognition means replacing a model of mechanical domination (in which there indeed is no alternative) with a conception of conjunctural struggle, in which contrasting modes of theorizing and practice explore the instabilities, cracks, and contradictions of capitalism as well as reclaim and reassert other kinds of social order that respond to these complexities.[10] One must therefore replace demonstrations of essentialist claims with cultural and historical exploration of contexts and specific instances.

To do this, the chapter begins by addressing the conflictual emergence in the United States of market radicalism as a response to an economic and cultural crisis of media industries and, more generally, of rationalist empiricism. Consumer-generated content and citizen journalism constitute communications practices that challenge in some key ways the rationalist-empiricist basis of market formations of communications by hybridizing rationalist empiricism (which valorizes professional expertise as a condition of authoritative knowledge) with the critique of professionalization deriving from critical modernism. As such, these practices exhibit significant contradictions in their attempts to join noncommercial with commercial, democratic with market, popular and professional, personal expression and empirical account.

The chapter continues by recounting the struggle between enabling while at the same time managing and thus containing market radicalism. As will be argued, the primary way this is done is by reasserting the desirability of professional practice, which is equated with quality. The practice of citizen journalism seeks to validate popular participation by demonstrating that the general public is not composed of vegetative, passive readers, but instead are capable with instruction and editorial oversight of writing professional-quality news stories—that is, stories that conform to instead of confront reigning definitions. The use of professional standards to validate the popular is even more explicit in so-called user- or consumer-generated content, which typically takes the form of a contest to see who can make the best advertisement—again, evaluated against existing professional practice. While rational-empiricist determinations in the form of professionalism constitute the current boundaries of practice, the corresponding integral critique of professionalization constitutes a prefigurative potential—albeit, for now, a largely incorporated populist one—for progressive change.

Conceptions of Participation

Practices referred to here as market radicalism developed within specific conditions that determined modes of participation. In order to grasp these determinations, what first must be done is grasp more adequately the complexities of media participation. While a defensible proposal for democratic communications must emphasize the normative goal of greater participation, great care must be taken to specify general claims about participation and to recognize the inadequacies of solely formalist analysis in order to avoid uncritical celebration or eulogy.

Outside of liberal-voluntarist conceptions, one formalist means of defining participation is textually. This has been a hallmark perspective of a variety of critical and cultural perspectives, such as Barthes' distinguishing readerly from writerly texts, Eco's claim (foreshadowing the popularization of reader-response literary theory) that "every reception of a work of art is both an *interpretation* and a *performance* of it [emphasis original]" by the viewer/reader, and the use of these and similar positions to argue that instances of reception are better conceptualized as acts of creative appropriation.[11] However, a theory of semiotic openness, indeterminacy, or polysemy and the resulting textual conception of media participation poses a problem for democratic communications in at least two ways. The first way is theoretical and political, in that such a position can be the basis for suggesting that textual interpretation-as-participation is not the exception but the general rule, with all texts indeterminate and open, at least in principle. To the degree that this is seen as the case, such a position precludes the existence of a particular kind of text that is more or less open, more or less participatory, and thus more or less democratic—indeed, it defines out of existence the topic of and the need for democratic communications itself. In addition, this kind of formalist definition also defines out of relevance historical contexts and determinant conditions by universalizing and thus essentializing textual features, thus placing texts and their effects outside of human practice.

A second formalist means of defining participation—one that is often combined with the textual—is the organizational. A variety of examples suggests that a more horizontal organization (fewer layers of overseeing management and greater rank-and-file autonomy) is comparatively more egalitarian and thus democratic than a vertical organization characterized by rigid and multiple layers of managers and an assembly line, Fordist organization. For example, Brouwer describes a number of HIV/AIDS 'zines of the 1990s as "illustrative of horizontal communication characteristic of alternative media," and as a result showing themselves to be "remarkably inclusive, publishing about seventy different pieces by about forty different authors in addition to their own work."[12] Similarly and elsewhere, Salter suggests that "forms of use that might be considered 'democratic' [include] information sharing, horizontal communication, many-to-many discussions and fully interactive information generation."[13]

Yet, the difficulty of claiming a universally democratic status for particular ways of organizing replicates the problems of essentialism and decontextualization noted above for textual definitions, a problem that is particularly clear when addressing specific cases, as Downing, Atton, Gillett, and others point out.[14] For example, a claim that the BBC's *Video Nation* is a participatory television program based on textual evidence (that people who appear are not paid actors or media employees, and that user-provided footage is intentionally left "rough") and on immediate organizational structure (that "ordinary" people instead of studio executives decide whether their video contributions will air or not) downplays or ignores the determinations of the institutional, social, and cultural contexts that enable the project in the first place. As Carpentier points out, *Video Nation* and many other examples of what he calls "democratainment can never suffice as a tool for media democratization" due to the broader cultural exercise of power and authority that determines participation in multiple ways.[15] Such power and authority emerges not so much from chains of command or personal mandates, but from the deeply institutionalized conventions of the practice, themselves formed in as well as contributing to specific contexts. Ytreberg suggests as much by noting how participation in broadcasting in the West means most generally "mastering a set of performance roles that are given by the production context and by the requirements of the format," and that such mastery, as a condition of participation, means "conform[ing] to the [social and cultural] order of production."[16]

What thus becomes crucial for an analysis of media participation is an historical rather than formalist approach, one that assesses the degree of correspondence between projects and the conditions of their making, but one that also assesses the (re)making of those conditions, and it is in this sense that a second consideration becomes important. In addition to clarifying the contextual and historical nature of participation, one must also reject conceptions of domination as autonomous elites' one-way imposition of their ideas, wishes, and projects on the rest of society. Williams puts the drawbacks of such a view starkly by arguing that "if our social and political and cultural ideas and assumptions and habits were merely the result of a specific manipulation, or a kind of overt training which might be simply ended or withdrawn, then the society would be very much easier to move and to change than in practice it has ever been or is."[17] The positing of domination decontextualizes and dehistoricizes the formation of that dominant and its contextual production, posing it instead as a pre-formed reality and the starting point for rather than the focus of analysis.

For the same reasons, one must also move past conceptions of a struggle between pre-formed, essentialized opposing camps of "producers" and "consumers."[18] To reiterate a point raised in Chapter 3, linear, one-way or essentialized two-party struggles simplify what are in practice complex, constitutive relationships—foregoing a recognition of how, as Shefrin notes, "the producer-consumer dialectic is not solely weighted toward one-sided manipulation [or resistance], but presents a more nuanced set of interactions."[19] While perhaps being too glib, the point must still be made that, without users buying access to products and services, commercial media companies and their ruling elites would very quickly be in all senses out of business.

Such a point should not be taken as a claim that individual choices of what to buy and how to use are in any way more autonomous (an uncritical glorification of consumption-as-resistance) than decisions about what shows to make. Rather, elite domination or popular struggle needs to be retheorized as mutually determining and mutually constituting relationships, with an understanding of such relationships being a task at the core of the development of an emancipatory social theory of democratic communications. The most productive such retheorizations focus on the historical and the conjunctural instead of the essentialized and the automatic, with particular stress placed on probing the determinations of social and historical conditions that enable as well as preclude political possibilities at any single point in time.[20]

Fragmentation and Instability

From such a position, one can clarify the emergence of market radicalism in the United States by recounting key processes related to the emergent instability of capitalist organization of communications. Of primary importance is the contradiction between, on the one hand, the imperative of extending access and, on the other hand, the twin imperative of rationalizing and thus controlling such access. In the United States, and in the context of a rapidly expanding and changing market, this contradiction helped break down an uneasy industry consensus and bring about its crisis (from industry's point of view), the implications of which are still being worked out today.

From the beginning of capitalist markets, a key need has been not to prevent, but to enable greater access and participation in response to the imperative to expand markets. Such efforts include lowering prices while extending distribution, and institutionalizing a genre of commercialized popular culture that concerned the lives of wider swaths of society. Such efforts also included producing new forms of practical involvement—their long-standing presence calling into question how revolutionary many recent efforts of boosting audience involvement are.[21] In contrast to current claims of "paradigmatic shift[s] in producer-consumer affiliations" brought about by ostensibly new forms of audience participation, Griffen-Foley responds that "media producers have been blurring the notion of the passive media consumer [a conception created in the midst of early twentieth-century functionalist fears about the fragmenting effects of mass media on society] for more than a century."[22] Varieties of audience participation in Australia, Britain, and the United States have been practiced in such forms as mass-audience periodicals of the late nineteenth century, confessional magazines, mass-market women's magazines, and "talk-back" radio with its beginnings in the 1930s. To this list one could add the extensions of religious publishing, a number of politically radical as well as popularized commercial newspapers, varieties of live entertainment including vaudeville, and even earlier traditions of the epistolary contributions of correspondents as described in Chapter 1.

And yet, a claim that such extension is de-facto democratizing (a claim relying on a formalist analysis) cannot be taken at face value. "There is, of course," as Griffen-Foley notes further, "a significant contradiction between the [individual empowerment] rhetoric of personalized journalism (for want of a broader term) and the structure of the modern, heavily capitalized mass media."[23] Thus, in addition to acknowledging the extension of participation, one must also acknowledge its rationalization. All contributions do not transparently find a place in the publication. To the degree that the organization requires user participation to coincide with its imperatives, only those contributions—or portions of contributions—that best fit the publication appear, seeking to produce a seamless correspondence of readers' interests with publication and its organization. Thus, recognizable forms of readers' experience are indeed re-represented to readers (an extension of participation), but selectively so, yoked as the selection is to a corresponding effort to consolidate and secure a dependable market (the rationalization of participation). In this way, and at the earliest stages, the emergence of a commercial popular culture depended on a paradoxically selective, intentional, unequal, and necessary collaboration between makers and users while at the same time redefining these relationships into the capitalist, abstract relationships of producers and consumers.[24]

However, such a collaboration has never been a static formula or technique, but instead one that has continually responded to new conditions, with one of the most crucial being the maturation of popular-culture industries and the resulting saturation of popular markets—never simply a stage reached and surpassed, but an ongoing moment in capitalist markets. To the degree that the extension of access to heretofore unaddressed audiences-as-markets becomes largely complete, a new problem is generated. With fewer and fewer new buyers to colonize, greater and greater effort must be directed into taking buyers from other companies, which carries with it the potential not only of the cannibalization of the industry but of clear limits on its collective expansion—anathema to the imperative of limitless capital expansion.

In regards to initial stages of communications industries in the United States, what forestalled such a fate for a time was the spatial reorganization of communications markets. One form of this reorganization has been various means by which sales and commercial revenue were allocated into acceptable, exclusive portions in order to manage direct competition and thus ensure the continued viability of the industry as a whole. For example, oligopolies emerged in such ways as through exclusivity agreements, for example those used by the Associated Press in the United States, with conditions including non-competition clauses.[25] Other means of rationalization included regulatory fiat by virtue of the broadcast-licensing process, which, through the guise of ensuring technical viability, in effect allocated exclusive markets to specific licensees.[26] A second, corresponding form of spatial reorganization was the emergence of hybrid modes of national-local ownership as a dominant form of industry organization. Modes such as syndication and network/affiliate were not simply local nor wholly national, but forms of local ownership that relied upon relationships with larger financial and content-providing services, which placed a premium not on unbridled competition but instead on collaborative agreement to divide up markets in a commercially beneficial way.

The crucial point about the spatial reorganization of markets is that it produced agreed-upon areas of exclusivity that rationalized competition. By the 1940s, the pattern was clear in regards to newspapers. As a sociological study of the day noted, "despite the growing expansiveness of newspaper circulations, newspaper publishing is still an industry the units of which are so diffused geographically that except for those located in the same city or trade area they are not competitive to any effective degree."[27] Newspapers were well suited to using local ownership as a way of rationalizing and thus managing competition due to their geographic specificity and their ability to address needs for more advertising by simply expanding their size, thus obviating conditions conducive to establishing competing companies. The resulting stagnation characteristic of the newspaper industry is clear in the steady decline of the amount of U.S. towns and cities with more than one newspaper, from 502 cities in 1923 to only 19 cities by 1996 with two or more directly competing newspapers.[28]

The management of competition through exclusivity also played a crucial role in broadcasting industries. Although local radio stations and local television affiliates competed with each other for marketshare within localities, the licensing system managed the amount of competitors in any particular locality to ensure as much as possible a sustainable level of support. In addition, the lower costs of radio broadcasting and the institutionalization of narrowly focused programming to secure narrowly defined, local audiences minimized cannibalization. The end result of this spatial reorganization of markets was, for any particular locality, a single newspaper and clearly divided markets for radio and television, and thus a reasonably stable and predictable revenues for those owners who had a slice of the industry and audience pie.

However, what began to upset this structural cartel was the production of new social spaces of commercialization that are part of the transition from Fordism to regimes of flexible accumulation.[29] With the characteristically increasingly mobile capital of these new regimes has come increased pressures on companies for greater and more consistent profitability. Among other responses, this pressure has spurred the search for new products and means of delivery in an intensified effort to develop new markets. This search was intensified further in the United States by the relaxation beginning in the 1980s of long-standing, already minimally obtrusive regulatory frameworks that applied to communications companies.[30]

The key characteristic of these new social spaces of commercialization is that new markets have indeed been produced, but these are increasingly nationalized if not internationalized ones. As such, they overflow their previously manageable local and mutually exclusive boundaries, thus contributing to an intensification of inter-industry competition, as media companies scramble to secure increasingly overlapping if not common markets. This trend had become clear as early as the 1940s, when one study commented that "competition from other media of information, at the present time, is more serious for many newspapers than competition from other newspapers."[31]

Such pressures and changes remade a number of media industries in the United States, with one example being television industry growth at the expense of radio. While annual gross profits averaged 37 percent for the radio industry at its zenith from the late 1930s to early 1940s, network television's takeover in the early 1950s of radio's national audience began to reconfigure radio into a localized medium.[32] Radio-industry revenues declined from being about four times that of television in 1950 to half those of the television industry by 1960. While average radio-industry revenues declined to about three percent during 1951–1976, annual gross profits of the television industry during roughly the same period (1951–1980, excluding prior start-up years) averaged more than twenty-four percent, with the average profits of commercial TV stations as a percentage of net revenues from 1975–2000 averaging 28 percent.[33]

The broadcast-television industry has in turn come under pressure from cable-satellite networks. Their formation was aided both by a gradual relaxation of ownership limits throughout the 1980s and 1990s and the beginning in 1975 of national satellite distribution.[34] Cable was initially a means of extending the signal of existing television stations, thus at the start a part of the existing rationalized system. However, its combination with satellite distribution and the resulting emergence of cable-television networks sowed the seeds of much larger changes in the structure of television broadcasting in the United States. A sense of the change can be seen in audience sizes for evening network news programs, which, for broadcast television, have declined by one million viewers per year since the 1980s to, in 2006, total about 26 million.[35] While revenues for the broadcast-television industry have remained reasonably healthy (varying between a low of thirteen percent in 1991 and a high of 31 percent in 1976, and averaging 24 percent), in the past decade advertiser-supported cable television as an industry exceeded broadcast television in size of viewership, attracting in the 2005–2006 season more viewers than broadcast in the 18–49 age range for the first time and surpassing broadcast networks' audience size for primetime programming.[36] Such challenges and reconfigurations continue with the impending conversion to digital television and integration into television of Internet and of cellular services.[37]

Similar pressures have reshaped the newspaper industry, although in different forms, which evoked different responses. Over time, the management of competition has made the newspaper industry smaller, more exclusive, and steadily profitable, while increasing inter-media competition threatens this stability. The amount of newspapers in the United States declined by more than a quarter from 14,336 in 1937 to 10,521 by 1999, with the amount of daily newspapers falling even further—by close to half—from a high of 2,600 in 1909 to 1,452 by 2005.[38] From a relatively steady proportion early in the twentieth century of one daily newspaper for every three people in the country, a decline set in by 1970, to the point where, in 2005, there were more than five people per daily newspaper circulated.[39] However, this decline made the industry more profitable as it has become more exclusive. As Horwitz summarizes, "operating margins, historically in the 10- to 15-percent range, now [the mid-2000s] range between 20 and 30 percent. This level of profit is two to three times that of the average industrial corporation."[40]

Such profit levels, however, are increasingly challenged by inter-media competition, which are remaking the newspaper industry. In the 1970s newspaper companies engaged in futurist experimentation with home-delivered videotext to retain if not extend reach in local markets primarily through developing new technologies of distribution, a need that has intensified with the emergence of Internet-based classified-advertising services and the resulting challenge to the newspaper industry's core revenue source.[41] Tactics for securing profits in a time of inter-media competition and declining readership include expanding and diversifying through acquisition of additional papers as well as other media properties, and forming or becoming a part of a chain or conglomerate.[42]

Closer to the present, industry representatives and federal regulators claim currently that such pressures have only increased, particularly for companies involved in digital services. Despite some self-serving hyperbole, there is a degree of substance to the claims. For example, in 2003, the Federal Communications Commission (F.C.C.) as part of its justification for relaxing cross-ownership rules claimed that inter-media competition and the maturation of what it termed the "multimedia landscape" had reached the point where inter-media competition—and not government regulation—could serve as the primary means of ensuring localism and system diversity.[43] In recent government testimony, the chief operating officer of a U.S. satellite-radio company noted that satellite radio does not compete only with other satellite-radio services, but also with wireless broadband, mobile-phone streaming, multi-channel high-definition terrestrial radio, and other wireless carriers and services.[44]

Although revised business strategies of the kinds summarized here have been key responses to inter-media competition, the expansion of user participation has only recently joined them. The primary commodity of value, legitimated by rationalist empiricism, was not simply information, but authoritative information, which could be ensured only through professionalization and thus tightly managing user participation in ways that did not affect claims of authority.[45] However, the intensification of the pressures for profitability and of managing an increasingly severe inter-industry competition—particularly with the marketplace consolidation of the 1990s, with takeovers and mergers increasingly financed through speculation and thus requiring immediate and high profitability—meant an intensified search for any distinguishing advantage over competitors that might maintain if not boost profitability.

Paired with and enabling the economic crisis, however, is a deeper crisis based in a general, variable modernist critique of rationalist empiricism that, together with an economic analysis of industry pressures, provides a clearer understanding of why citizen journalism, consumer-created content, and other such instances of social-networking communications have emerged as an important form of market advantage. The current encouragement by commercial communications organizations of a more significant level of user composition can be traced not only to the intensifying imperatives of profitability, but also to a general inability of rationalist empiricism to claim authority in a society confronting extensive challenges to utopian conceptions of modernity.

This crisis of credibility is manifest in the United States in a number of well-known instances. For example, the Commission on Freedom of the Press (known also as the Hutchins Commission) of the late 1940s noted that the failure of the commercial media was not due to the perversion of objectivity by private interests, but instead to the organizational emphasis of the professional ethic of objectivity itself, when defined simply as accuracy. By contrast to a definition of objectivity that amounted to accurate transcription, the commission suggested that facts alone were not enough and had to be placed in a framework which gave them meaning.[46] The failure of the accurate-transcription model of objectivity and of the corresponding organizational emphasis on reporting unfiltered pronouncements of politicians and other institutionalized authorities was particularly clear in the media validation of red-bating charges that fueled McCarthyism of the 1950s, as well as by the 1960s and the U.S. incursion into Vietnam.[47] While existing in various forms since the emergence of newspapers themselves, the generalized critique of claims of neutrality and objectivity emerged fully in the U.S. in the 1960s with the findings by a number of establishment studies and commissions regarding differential coverage of civil rights and of antiwar protests.[48]

Additional challenges to the authority of rationalist empiricism came in terms of form, such as the emergence of the "nonfiction novel" (itself deriving from earlier, critical-modernist experimentation) and other commercially popular experimental writing that challenged distinctions between fact and fiction, which coalesced into what became known as the New Journalism and, later, literary journalism.[49] International pressures mounted as well in such forms as critiques of Western journalism by the Non-Aligned Movement that took place in the United Nations Educational, Scientific and Cultural Organization (U.N.E.S.C.O.) debate in the 1960s and 1970s regarding the necessity of a New World Information and Communications Order to counter Western cultural imperialism; the perceived ill-fit of the practice of Western objectivity for assisting newly emerged nations in nation-building, which was addressed in part by the emergence of development journalism; and efforts to practice participatory journalism, much of which derives from the work of Paolo Friere.[50] Since then, a criticism of the accurate-transcription theory of objectivity has also been adopted by the radical Right, which has built a commercially persuasive foundation upon claims about the bias of the "liberal media."[51]

Pressures since then have only increased. Historically legitimated as "the first rough draft of history," commercial newspapers today are seen in terms of an increasingly popularized critique of news production and a generalized claim that they all are biased. Such a sentiment is suggested by a range of recent public polls. A U.S. survey in 2004 found that the percentage of respondents deeming "that they can believe most of what they read in their daily newspaper" declined from 84 percent in 1985 to 54 percent in 2004.[52] In a 2005 survey, respondents by margins of 3-to-1 or more agreed that "when deciding what stories to report, [major news organizations] care more about attracting the biggest audience rather than about keeping the public informed."[53] Even the C-SPAN cable-television service that simply rebroadcasts speeches, sessions of Congress, and other such official events is viewed as highly believable by only one-third of liberals and only one-in-five conservatives.[54] One-third of working journalists view lack of credibility with the public as one of the top problems facing journalism.[55] The trustworthiness of Cable News Network (the source of television news most trusted by the public) declined from 42 percent in 1998 to 28 percent in 2006.[56]

Advertising as the lynchpin of commercial communications also began to draw widespread popular criticism by the 1950s, which undercut similarly claims of neutral information dissemination or individual instances of limited persuasion.[57] Such criticism began to be institutionalized in the forms of consumer groups and movements. In addition to the Consumers Union, which was established in 1936, a range of groups have emerged that are concerned with interrelations between advertising, commercial media companies, and consumer culture.[58] Such concerns cut across the conventionally separated spheres of journalism and advertising. Even by the late 1980s, an informed estimate by a public-relations scholar put at 40 percent the proportion of newspaper stories that consist "of items churned out by press agencies and public relations bureaus and then regurgitated intact by the 'objective' organs of journalism."[59] Recent concerns continue with the steady incursion of commercially derived publicity into news in ways far beyond conventional practices of advertising and public relations, such as video news releases.[60]

Produced in large part by this extremely complex and ongoing economic and cultural crisis, the most recent emergent competitive advantage of more significant user composition did not come about through industry research and development, but through the commercialization of grassroots developments within transnational digital systems, primarily in the form of the Internet, a process that dates from the mid-1990s. As is well-known, this digital system constitutes a means of enabling users' contributions beyond traditional localized areas into new spaces of globalization. Of great relevance was the development by late 1990s of infrastructure and software that enabled collective, direct publication beyond what had been possible previously. Such developments include the peer-to-peer exchange pioneered by noncommercial collective Independent Media Centers, the development of wiki software and services, and the more recent emergence of simplified means of self-publishing such as weblogs, video-sharing websites, and what have been called social network media. While one must keep in mind the limited extent to which access is available to all who desire it (ranging as of March 2007 from 70 percent of the population in North America to 3.6 percent in Africa), the point remains decisive regarding the comparative expansion.[61]

Indeed, as Strangelove argues and extensively documents, the publishing of noncommercial materials on the Internet is a defining feature with clear progressive possibilities.[62] In his wide-ranging study, he takes issue with conservative and radical claims that the Internet has largely been colonized by commercial interests, and thereby irretrievably incorporated into the dominant. By contrast, and while being well-aware that "the communicative freedom of Internet users is under attack," he argues that presently in process is "a massive shift in the structure of the public sphere" enabled by the present situation of the "production of an alternative symbolic economy and its expansion of the number of contenders that may participate in the normative debate" about the nature of society.[63] While relying at times on an ahistorical conception of the implications of this struggle for the "empire of mind," Strangelove's argument suggests that what has emerged not only with the Internet but with its combination with and concomitant remaking of the conventional media industries is the basis of new kinds of practice, enabled by capitalist extension of media participation and the corresponding and continual crisis it generates.

Advertising and Consumer-Generated Content

With the intensification of pressures to secure nationalized if not transnationalized, transmedia markets, the contradictory imperatives noted above of extending access while managing and channeling it within specific objectives continue to intensify as well. Such pressures affect not only media industries, but the advertising industry as well, which not only competes within itself for clients but teams with media companies in efforts to deliver audiences to business.

Intensified efforts to secure markets and "mindshares" have enabled the formation of user- or consumer-generated content, which is only the most recent form of a central preoccupation with the predilections, preferences, and commercial involvement of buyers. It joins other, long-standing means of gathering this information, which started with the use of comparative sales figures, then expanded into forms of increasingly invasive consumer research including consumer-response tactics such as coupon redemption, statistical survey marketing research originally in the form of polling, and the adoption of psychoanalytic, behavioral, and anthropological methods.

In the wake of the onset by the 1950s of significant popular criticism of such methods of research, the advertising industry has adopted a variety of different practices—for example, such as those signaled by the emergence of the so-called "Creative Revolution" in the 1960s and the formulation of what Frank has called "hip consumerism" as a way of redirecting skepticism about advertising and consumer culture into forms of advertising that paradoxically embrace the rejection of consumer culture.[64] Aided by the recent emergence and embrace of digital communications has been an extended emphasis on "interactivity" as a means of building more durable brand identification through greater participation. Indeed, an early conception of online consumer-generated content was simply as a means of audience feedback—as registering and collecting individual opinions and preferences.[65]

This initial sense of interactivity has been extended in a number of ways. One form of digitally enabled interactivity that is easily managed is personalizing one's own product, such as a sport shoe or car.[66] By contrast, another form quite outside efforts of containment is unmoderated user commentary about and evaluation of products, a concern prompting the emergence of online means of reputation-management for companies.[67]

The intensification of such timid forms of participation has emerged not regarding products but the marketing of those products. The current thinking is that no one knows more about how to effectively advertise products than do members of that very market. Such a view can be seen in books that extol the autonomous power of individual "citizen marketers" to rule the marketplace.[68] User-generated marketing campaigns can be both tightly managed and, in some cases, quite open in unprecedented ways. Forms of rationalization and management include the election, such as contests that ask users to vote for the best advertisement or the best ending to an advertisement, and the collaborative campaign enacted by so-called "guerrilla marketing" projects, "word-of-mouth marketing," and appeals to consumers to join companies' efforts in "building their business."[69] Such marketing efforts often do double duty as marketing research, with one example a recent Internet promotion by an automaker. On the website users would see and read a series of personal goals such as "make an independent film," "tour Europe by car," and "invent something to improve the world." Each goal was set to short, professionally made videos, with visitors to the website asked to comment on the goals and to share their own.[70]

Most recently, user involvement has been extended not just into offering accounts of product uses and preferences that then are refracted through the marketing process, but into actual advertisements. It is this extension—from subjects of marketing to in-effect free creative labor—that characterizes the present cutting edge of extended participation and its management in the advertising industry. Forms in which this takes place include the contest, such as assembling a video ad from video and music segments provided by the company; posting a 15-second video in which the entrant describing what she or he would like to tell the world; and posting an image or a video of a situation or event personally experienced that left one "awed and exhilarated," thus qualifying as a "'wow' moment."[71] An even greater extension of participation is viewer-created ad messages. For example, commercial sponsors of the cable and Internet service Current TV post on the channel's Internet site an assignment for an advertisement, and users whose commercials are judged to be the best are paid up to $1,000.[72]

The extension of user involvement in the making of ads also takes the form of clubs. One such website describes itself as "a marketing service—powered by The Procter & Gamble Company—that develops teen word-of-mouth marketing programs," and does so in part by holding out a promise of being able to put one's own imprint on advertising. It seeks to engage teen-age viewers with claims from its members such as "I wrote a secret message for Vanilla Coke bottles," "I helped The Partnership for a Drug-Free America design their website," and "I chose the design for a Vans Warped Tour t-shirt."[73]

In addition to being potentially more effective, contests and clubs as a way of "cutting through the noise" with more relevant and engaging advertising is also deemed to be "a lot cheaper than spending $1 million on production values via an advertising agency."[74] As an advertising executive recently put it, "the best ideas can come from anywhere—and chances are, your consumers, who love your product, will innovate for you."[75] Of course, the degree to which these remain managed is clear not only in the fact that the company chooses which advertisement to distribute publicly, but more generally in terms of the evaluation of submissions using professional standards.[76]

By contrast to efforts that are rationalized through contests or clubs, efforts that do not rely on official validation and selective distribution enable greater variation and thus can pose problems for companies. For example, part of the capabilities provided through a recent contest to make a video ad for a car was the storage and retrievability of all submissions. Users could view not only ads that were deemed by the company to be the best, but all ads submitted. As a result, a number of ads were created with the music and shots provided in the promotion that critiqued the car's high gasoline consumption and effect on global warming and public health, with public awareness of these critical ads amplified through online email lists and listservs.[77]

The emergence of user-generated content as part of advertising campaigns signals one kind of institutional opening that was unheard of even fifteen years ago. While in one sense asking users to make advertising to more effectively convince themselves to buy or identify with a product has a sinister, almost Orwellian sense about it, in another sense these and other efforts and the critique of professionalism they represent are becoming more widely accepted. As put by the maker of a political advertisement that appeared in March 2007 in which, in a take-off of the Apple Macintosh "1984" ad, U.S. presidential candidate Sen. Hillary Clinton was presented as the menacing, mind-controlling Big Brother, "this ad was not the first citizen ad, and it will not be the last. The game has changed."[78]

OhmyNews as a Global Social Movement

Extending user composition to journalism raises even more direct critiques of professionalism. An initial and still significant such effort emerged in Korea in 2000 with the formation of the Korean citizen-journalist organization *OhmyNews* initiated by its in-part American university-trained founder, Yeon-Ho Oh.[79] *OhmyNews* portrays itself as the leading edge of a media revolution, and a portable model not only for citizen journalism but for a commercial media business. As a self-styled global social movement, it seeks to raze the barriers of professionalist elitism and, by doing so, enable greater public participation in the production of news. However, while it criticizes the social restriction of professionalization, it embraces professionalism in the form of the tenets of good journalism, rooted as it is in rationalist empiricism. *OhmyNews* both critiques and validates rationalist empiricism, and this contradiction is the source of its present instability.

The contradictions and hybridity of *OhmyNews* originated in and can only be understood as a particular response to and enabled by very specific conditions. The practice of citizen journalism by *OhmyNews* has been enabled by very specific social and political conditions of Korea's recent history: social fragmentation, new generations who pursued new values vis-à-vis the vested interests of older generations, skepticism of mainstream media, and significant public investment that produced an enabling and effectively national digital communications system.

Of initial importance is the fragmentation and diversification of society in the wake of the 1987 democratization movement. Civic movements in Korea now raise concerns about previously hidden issues such as the environment, consumerism, gender equality and welfare. Widening generational differences also contribute to this fragmentation, with the most important watersheds being the Korean War and post-war economic reconstruction.[80] The generation born prior to the Korean War and that grew up in the 1960s experienced trying times and poverty, and as a result often takes a strong anti-communist stance. By contrast, the generation born after the 1960s grew up in a comparatively affluent environment and thus tends to adopt a more progressive position. While enjoying a comparative material affluence amid the 1960–1970 economic reconstruction, the post-1960s generation also grew up under military dictatorial regimes in the 1970s and 1980s, formed the main axis of resistance to the military regime in the 1980s, and became by the 1990s a key axis of leadership in Korean society.[81] Many of this younger generation looked askance at Korea's military regimes, which used economic development and anti-communism as a pretext for implementing dictatorial politics. More recent generations have shown even more differentiated social and political inclinations, benefiting as they have from Korea's rapid economic growth, capitalist expansion, and a highly advanced, networked and digitalized society due to widespread technological development.[82]

While this struggle and complexity make the construction of a democratic civil society exceptionally difficult, it also enables the emergence of a coalition of progressive movements that despite its diversity has sought a communication system that matches its openness, responsiveness and level of engagement, unlike those of past years. Such an intention is a response to the pre-1987 media industry landscape and its residual presence after the upheavals of the late 1980s. Before 1987, the state authority controlled the media with rigorous censorship and selective economic support.[83] The media industries gained their independence from state authority amid the expansion of Korean civil society in the wake of the 1987 uprising. However, while state control was considerably weakened, the increase of the power of capital has brought about many of the same problems as those in the West, including increasing centralization of ownership, infringement of editorial rights, and the centralization of media markets, with such problems particularly acute in the newspaper industry. As a result, the newspaper market in Korea is dominated by a few major mainstream newspapers that accumulated their capital under the military governments before 1987, with few media companies becoming quasi-political powers.[84] Such media companies reproduce the interests of capital and conservative groups, with the result that younger generations of reformers demanded media reform. For example, according to the Korea Press Foundation, more than 90 percent of its participants in 1998 voiced support for media reform.[85]

Recent generations in Korea have not only agitated for immense social and political changes, they also use computers and digital technology extensively, which has been made possible by the government-led and funded development of Korea's digital telecommunication infrastructure.[86] As a result of such efforts, Korea established by 2000 a broadband infrastructure that linked 144 cities nationwide.[87] Since the late 1990s, with the rapidly developing telecommunication infrastructure, the number of internet users also has increased greatly, with estimates as of December 2005 that 66 percent of the total population were internet users.[88]

A final, crucial development for the emergence of citizen journalism as a global social movement and business has been the creation and expansion of a consumerist economy.[89] Korea's per-capita gross national income (GNI) in the early 1960s was less than one-hundred dollars, but, by 2003 had expanded to twelve-thousand dollars, with the government proposing a goal of reaching a per-capita GNI of twenty-thousand dollars by 2010.[90] With increasing incomes has come a corresponding increase in the social and economic centrality of leisure and consumption. Corresponding to this level of economic development and consumerism is the growth of Korea's advertising market, with total advertising expenditures from 1981 to 2003 increasing more than twenty-fold from 318.3 billion won to 6.8 trillion won. And, as Internet use has rapidly increased, the internet advertising market has grown concomitantly.

Within these conditions, journalist-turned-activist-turned-online-news-publisher Yeon-Ho Oh conceived of and established in early 2000 an online publication to challenge Korea's entrenched mainstream media companies. Having been a dissident reporter in Korea and learning about American-style commercial, evangelical activist journalism as part of earning a master's degree in the United States from Regent University (funded by evangelist Pat Robertson and his Christian Broadcasting Network), Oh mounted his challenge as a revolutionary business model for twenty-first-century media companies.[91]

Like the practice of consumer-generated content, a key commercial advantage was a dramatic lowering of labor costs paired with the formation of more significant user allegiance. From the beginning, *OhmyNews* quickly built comparatively significant user participation while also stretching scarce start-up capital. Oh started *OhmyNews* with 100 million won (about $85,000), and an initial cadre of 700 citizen reporters who supplemented the efforts of four professional staff.[92] By being an online publication, *OhmyNews* avoided the costs associated with distribution of paper copies while extending its reach nationally and internationally and its ability to accept non-staff contributions. While forms of user involvement include composition and a capability to evaluate and comment on articles, citizen reporting and comment is only one of many means of participation.[93] Additional forms include becoming a member reader, joining a citizen-reporter editing committee, anchoring the *OhMyTV* webcast, writing and posting feedback and commentary on stories and issues, attending a citizens' forum that organizes public discussions of current social issues, undergoing reporters' training, and participating in public, sponsored sporting events such as marathons.

And, yet, this extension of participation is managed by integrating it into a professionalized production process, thus differing in some key ways from those implemented by other sites and services.[94] First, all contributors must register with *OhmyNews*. Second, all submissions are evaluated by staff editors who select, then edit stories, with only a selection of submitted items authorized as *OhmyNews*'s articles.[95] Whether authorized by the publication or not, articles from citizen reporters comprise the bulk of *OhmyNews*'s content, with contributions totaling one-hundred and fifty to two-hundred articles a day.[96] While, by the end of 2006, the *OhmyNews* professional staff has expanded to ninety (sixty-five of them journalists), this small staff is supplemented by a great expansion in the number of registered citizen reporters to, by February 2007, a claimed 50,000 in one-hundred countries.[97] Furthermore, citizen reporters are minimally paid (about US $1.70) for an article. Thus, as labor costs declined, the threshold of profitability correspondingly declined to the point at which *OhmyNews* announced in 2003 that the company had become profitable four years into its operation.[98]

Paired with the cost savings from citizen-journalist labor is the heightened value of the readership to advertisers. The most sizable source of revenue is commercial advertising that seeks to address its sizable user base of young and progressive users. Other sources of revenue include "voluntary charges," a form of self-assigned donation. Together, all sources by the end of 2006 generated about $6 million, with 60 percent coming from advertising.[99] Thus, the model of citizen journalism developed here broadens participation while it generates a loyal readership, greater value to advertisers, and thus greater revenue. As a result, creating a social movement as a commercial business has enabled *OhmyNews* in the span of seven years to become a prominent player not only in Korea but also increasingly in the international sphere of Internet publishing, and it is cited routinely as a model for similar ventures in the West.

However, *OhmyNews* and its brand of citizen journalism as a social movement remains an unstable hybrid. While extolling a critical-modernist challenge to professionalism, at the same time it remains enmeshed in a rationalist-empiricist validation of professionalism. While citizen journalism of this kind may have opened the doors of the profession to nonprofessionals, it has done so by encouraging citizen reporters to aspire to and meet professional standards rather than by remaking or relinquishing the epistemology that supports professional standards in the first place.

This hybridity and instability manifests itself in a number of ways. On the one hand, *OhmyNews* openly critiques exclusionary elite professionalization but, on the other, it exhibits increasingly an insider's progressive integration into the operations and imperatives of other institutions, albeit in somewhat transformed terms. For example, its support of the candidacy of Roh Moo-hyun in 2000—and his election to the presidency—expanded *OhmyNews*'s influence and power.[100] More generally, the size and variety of *OhmyNews* rivals commercial publications. It regularly publishes 15 sections, including politics, economy, society, culture, and four sections on regions of the country. In addition, *OhmyNews* subscribes to major news services and publishes articles related to women, the economy, and other professional issues in alliance with other special interest newspapers. *OhmyNews* was selected as one of fifty-five most influential people and organizations in Korea by *KyungHyang*, one of Korea's mainstream papers.[101] It was ranked tenth in 2000, eighth in 2001 and 2002, and sixth in 2003 and 2004 among "Korea's influential media" in annual surveys by Sisa Journal, a mainstream Korean weekly magazine.[102] Additionally, a recent survey of journalists ranked *OhmyNews* as the fourth most prominent influence on society, with the potential to become the most prominent.[103] Reports from *OhmyNews* are also increasingly reappearing in such traditional papers as the *International Herald Tribune* and used in other Korean media. In August 2006, *OhmyNews* Japan debuted, to follow the launching of an international English-language edition in 2004.[104]

With such expansion has come intensifying pressures. The instability of the global-movement form of citizen journalism has already become apparent in the degree to which its economic viability as a media-business model is being questioned. *OhmyNews* expected to lose money in 2006 after at-best small profits in the preceding few years. Conversely, with increased competition from the continued expansion of other forms of Internet-enabled reporting and commentary, *OhmyNews* finds itself backtracking if not reversing its originating rationale, extolling the value not of breaching of the barriers of professionalization, but of its greater credibility gained from having "professionals screen, edit, and fact-check stories from ordinary folks to filter out inaccuracies and potentially libelous claims."[105] Thus, while severing a critique of professionalization from a critique of capitalism remade the critical challenge to professionalization into allegiance to a particular brand, the intention of broadening participation and viewing such an extension as a form of social change remain progressive contributions.

Community Media and Hyper-Localism

Compared with the Korean experience as seen in *OhmyNews*, citizen journalism in the United States has proceeded in a very different form. In contrast to a challenge to journalism from outside the established industry, citizen journalism in the United States has become largely an industry challenge to itself—as, according to a 2007 industry report, "something that the mainstream media embraced rather than something they saw as a threat."[106] Lacking the extensive social and political upheavals and transformations experienced in Korea during the past twenty years, the most resonant cultural form for U.S. citizen journalism has not been a global, commercialized social movement, but community media as a hyper-localized, commercial public service—"hyper" in the sense of distributing personalized if not mundane accounts of individuals' everyday goings-on.[107] This emphasis on community derives from the long-standing local basis of the U.S. newspaper commercial markets and the need to reverse declines in local readership. As one industry report puts it, newspaper companies that build audiences by using "community platforms" have the best chance of attracting "non-consuming groups your company doesn't reach effectively with its existing products—especially those desirable to advertisers."[108] And a recent marketing-research report listed as a common theme of newspaper business plans for the future the need to "[m]aintain and grow strong local brand focus and build a stronger connection to and community among readers."[109]

Earlier efforts to more fully engage local communities through such projects as public journalism and civic journalism have been critiqued both inside and outside the industry as patrician efforts to legitimize professionalism.[110] Arguments for replacing such projects with citizen journalism not only note the modicum of success of *OhmyNews*, but note as well the shift in advertising media buying increasingly to Internet properties, upon which the brand of citizen journalism advocated by *OhmyNews* depends. In the United States, in 2006 $16.8 billion was spent on online advertising—about 12 percent of total advertising spending but a 34 percent increase over 2005, with expectations of reaching $19.5 billion in 2007.[111] Although still small in comparison to traditional media spending, the steady and accelerating migration of advertising revenue to online properties is the reason why further extensions of user contribution are being tried as a way to gain and maintain buyers. As a recent industry study puts it, "analysts believe consumers will come to demand the ability to [not just talk with but to] interact with the news producers, or they will migrate elsewhere."[112]

The emergence of citizen journalism in the United States as a form of localism has also been encouraged by the seeming possibility of reconciling political activism with commercialism and the traditional recognition of "the importance of a dynamic news business to our civic life, to our educational future, and to our democracy."[113] Such a reconciliation has existed in many forms, beginning perhaps in earnest with the crusading journalism of the late-nineteenth century and muckraking magazine journalism of the early-twentieth century in which the line separating the encouragement of social change and the building of a profitable market becomes extremely difficult if not impossible to locate.[114]

Traditions of localism inform citizen journalism in a number of ways. While some citizen-media start-ups aspire to a deterritorialized reach with an indifference to making money, U.S. industry efforts build upon the local and seek new and profitable business models.[115] Even aspiring national companies see themselves as networks of local projects or as branded franchises that remain local in focus.[116] The legitimacy of a project of community empowerment making money off its community is never questioned, but, then again, in the context of traditions of entrepreneurial small business, the high-tech sheen of the Internet, and the allure of a personal hobby mushrooming into another billion-dollar property along the lines of MySpace, YouTube, or Facebook, neither is the blurring of the distinction between reporting and selling. Indeed, a self-styled expert recommended that local citizen journalists also become local citizen ad salespeople, who could earn an ad commission while reporting a story.[117] Larger citizen-journalism projects are experimenting with syndicating or franchising concepts and contents, and in some cases licensing citizen-journalism software.[118] Present in all current efforts is a clear recognition of the social, political, and economic value of extending user participation, recalling not only the disengagement of the critique of professionalism from a critique of capitalism but the resulting merger of business with social change, defined however as a nostalgic restoration of a past era of widespread civic conscience. Its value not only includes enabling a conversation with local residents but, in doing so, also increasing one's share of local advertising revenue, forestalling competition, generating new sources of revenue, and "reinforc[ing] your leadership position in your community."[119]

A recent report by an American convert to citizen journalism summarizes the task bluntly, that success in securing significant user participation is "the art of making people feel good about themselves by spending their time and money in a way that benefits you or your organization."[120] The way of doing this is to extend not simply to new markets, but to extend the range of material "to dramatically broaden the range of lived experience represented by the news organization's output."[121] Current U.S. industry analysts make the case that "online community strategies are also gaining favor with investors who believe that such initiatives may result in new and sustainable sources of income for media companies."[122] Indeed, the development of what a philanthropic report calls "[commercial] products as participatory community institutions" sets the bar over which the U.S. news industry now feels it has to pass.[123] Here, too, when practiced within the present U.S. commercial media system, the continued extension of media participation from subject to unpaid creative worker is managed not only through the paradoxical critique and valorization of professionalism, but through the reduction of critique from a focus on wider social conditions to simply fulfilling a decontextualized role as a "citizen reporter."

Market Radicalism and Dreams of Change

An industry consultant's advice to media companies about the financial value of expanding access is that "'you have to give up control to gain control'."[124] It is precisely this contradiction that constitutes both the possibility of the reformation as well containment of the framework of commercial media practice in the United States and elsewhere. While remaining largely contained through professionalization and the economic imperatives of capitalism, the steady institutionalization of the critique of professionalization represented by consumer-generated content and citizen journalism embodies the progressive aim of broadening participation and, through it, the possibility of greater democratization.

Other forms of communications organization also struggle with how to concretely embody greater degrees of participation and self-determination, yet through forms that remain entangled in the dominant. In addition to the ongoing project of Indymedia, examples include the recent revival and remaking of the popular documentary movement. As noted in Chapter 3, relatively professionalized organizations such as Brave New Films provide ways for interested publics to collaborate on its films (yet, within the limiting boundaries of professionalization), where those of lesser professional aspirations (although still organized around professional standards) such as Echo Chamber seek to teach the skills of collaborative documentary (instead of critically question "documentary") while producing an "open-source, investigative documentary about how the television news media became an uncritical echo chamber to the Executive Branch leading up to the war in Iraq."[125] Means of socializing distribution include direct selling at very low cost, and the "house party" as a way for interested people in the same geographic area to gather together, view and discuss the film—a form that, however, has also become part of the commercial marketing arsenal.[126]

Although it would be a mistake to assert that consumer-generated content and citizen journalism are uniformly progressive, it would be a mistake of equal magnitude to discount such efforts as simply another instance of the completed containment and incorporation of an emergent practice. For their importance is not in the purported effect of any specific effort, but in the collective furthering and steady institutionalization of the critique of professionalization. Although highly contradictory and uneven, such projects and the provisional openings they assert maintain the possibility of a continued democratic extension and remaking of media practice.

Notes

1. *OhmyNews* International, "Membership Registration," *OhmyNews Inter-national* <http://english.ohmynews.com/member/member_step1.asp?> (7 Mar. 2007).

2. Yeon-Ho Oh, "10 Preconditions for the Value of User-Generated Content," *OhmyNews International*, 26 February 2007 <http://www.ohmynews.com/ articleview/article_view.asp?menu=c10400&no=347268&rel_no=1> (7 Mar. 2007).

3. Eric Mattson and Nora Ganim Barnes, "The Hype Is Real: Social Media Invades the Inc. 500," *Center for Marketing Research, University of Massachusetts Dartmouth* <http://www.umassd.edu/cmr/studies/blogstudy2.cfm> (10 Mar. 2007).

4. Shayne Bowman and Chris Willis, *We Media; How Audiences Are Shaping the Future of News and Information*, The Media Center, American Press Institute, July 2003, 58 <http://www.hypergene.net/wemedia/> (10 Mar. 2007).

5. A recent report identifies seven hundred to eight hundred such efforts currently. See Project for Excellence in Journalism, "Online: Citizen Media," *The State of the News Media 2007* <http://stateofthemedia.org/2007/narrative_online_citizen_media.asp?cat=8&media=4> (22 Mar. 2007).

6. Garth Johnston, "YouNewsTV Turns Viewers Into Stringers," *Broadcasting & Cable*, 5 March 2007 <http://www.broadcastingcable.com/article/CA6421029.html> (12 Mar. 2007).

7. James Curran and Myung-Jin Park, eds., *De-Westernizing Media Studies* (New York: Routledge, 2000); John D.H. Downing, *Internationalizing Media Theory: Transition, Power, Culture* (London: Sage, 1996); James Hamilton and Chris Atton, "Theorizing Anglo-American Alternative Media: Toward a Contextual History and Analysis of U.S. and U.K. Scholarship," *Media History* 7, no. 2 (December 2001): 119–35; Yung-Ho Im, "Critical Communication Studies in South Korea," *Journal of Communication Inquiry* 17, no. 1 (Winter 1993): 108–23; Myungkoo Kang, "There Is No South Korea in South Korean Cultural Studies: Beyond the Colonial Condition of Knowledge Production," *Journal of Communication Inquiry* 28, no. 3 (July 2004): 253–69; Rodney Benson, "Commercialism and Critique; California's Alternative Weeklies," in *Contesting Media Power: Alternative Media in a Networked World*, ed. James Curran and Nick Couldry (Lanham, Md.: Rowman and Littlefield, 2003), 111–27.

8. Raymond Williams, *Politics and Letters* (London: New Left Books, 1979), 310–11; Jürgen Habermas, *The Structural Transformation of the Public Sphere: an Inquiry into a Category of Bourgeois Society*, transl. Thomas Burger and Frederick Lawrence (Cambridge, Mass.: MIT Press, 1989); Robert James, "'A Very Profitable Enterprise': South Wales Miners' Institute Cinemas in the 1930s," *Historical Journal of Film, Radio and Television* 27, no. 1 (March 2007): 27–61; Lizabeth Cohen, *A Consumer's Republic: the Politics of Mass Consumption in Postwar America* (New York: Knopf, 2003).

9. John D.H. Downing, "Radical Media Organization: Two Models," in John D.H. Downing, Tamara Villarreal Ford, Genève Gil, and Laura Stein, *Radical Media; Rebellious Communication and Social Movements* (Thousand Oaks, Calif.: Sage, 2001), 71.

10. Antonio Gramsci, *Prison Notebooks* (New York: Columbia University Press, 1991).

11. Umberto Eco, *The Open Work*, transl. Anna Cancogni (Cambridge, Mass.: Harvard University Press, 1989); and Roland Barthes, *S/Z*, transl. Richard Miller (New York: Hill and Wang, 1974). See also Dick Hebdige, *Subculture, the Meaning of Style* (London:

Methuen, 1979); and Michel de Certeau, *The Practice of Everyday Life*, transl. Steven Rendall (Berkeley, Calif.: University of California Press, 1984).

Such positions have also informed developments in studies of audiences and of popular culture. Key works include Pertti Alasuutari, ed., *Rethinking the Media Audience; The New Agenda* (Thousand Oaks, Calif.: Sage, 1999); Tony Bennett, "Figuring Audiences and Readers," in *The Audience and Its Landscape*, ed. James Hay, Lawrence Grossberg, and Ellen Wartella (Boulder, Colo.: Westview, 1996), 145–59; Roger Dickinson, Ramaswani Harindranath, and Olga Linné, eds., *Approaches to Audiences: A Reader* (London: Arnold, 1998); Ingunn Hagen and Janet Wasko, eds., *Consuming Audiences? Production and Reception in Media Research* (Cresskill, N.J.: Hampton Press, 2000); Roger Silverstone, "Television and Everyday Life: Towards an Anthropology of the Television Audience," in *Public Communication; the New Imperatives*, ed. Marjorie Ferguson (Thousand Oaks, Calif.: Sage, 1990), 173–89; and John Storey, *Cultural Consumption and Everyday Life* (New York: Oxford University Press, 1999).

12. Daniel C. Brouwer, "Counterpublicity and Corporeality in HIV/AIDS Zines," *Critical Studies in Media Communication* 22, no. 5 (December 2005): 353.

13. Lee Salter, "Structure and Forms of Use; A Contribution to Understanding the 'Effects' of the Internet on Deliberative Democracy," *Information, Communication and Society* 7, no. 2 (June 2004): 190.

14. Downing et al., *Radical Media*, 67–74; Chris Atton, *Alternative Media* (Thousand Oaks, Calif.: Sage, 2002), 29; and James Gillett, "The Challenges of Institutionalization for AIDS Media Activism," *Media, Culture and Society* 25, no. 5 (September 2003): 607–24.

15. Nico Carpentier, "The BBC's Video Nation as a Participatory Media Practice: Signifying Everyday Life, Cultural Diversity and Participation in an Online Community," *International Journal of Cultural Studies* 6, no. 4 (December 2003): 442.

16. Espen Ytreberg, "Formatting Participation Within Broadcast Media Production," *Media, Culture and Society* 26, no. 5 (September 2004): 678.

17. Raymond Williams, "Means of Communication as Means of Production," in *Problems in Materialism and Culture* (London: Verso, 1980), 37.

18. James Hamilton, "Remaking Media Participation in Early Modern England," *Journalism: Theory, Practice, Criticism* 4, no. 3 (August 2003): 293–313.

19. Elana Shefrin, "Lord of the Rings, Star Wars, and Participatory Fandom: Mapping New Congruencies Between the Internet and Media Entertainment Culture," *Critical Studies in Media Communication* 21, no. 3 (September 2004): 261–81.

20. Overly zealous variations of de Certeau stand in for the voluntaristic, with the conjunctural relying to varying degrees on Gramsci's conception of hegemony and its historical contingency. Recent attempts in studies of popular culture and user appropriation to do due justice to structure as well as agency include Henry Jenkins, *Textual Poachers: Television Fans and Participatory Culture* (New York: Routledge, 1992).

21. Russel Nye, *The Unembarrassed Muse: The Popular Arts in America* (New York: Dial Press, 1970), 3.

22. Shefrin, "Lord of the Rings," 297; Bridget Griffen-Foley, "From Tit-Bits to Big Brother: A Century of Audience Participation in the Media," *Media, Culture and Society* 26, no. 4 (July 2004): 533–48.

23. Griffen-Foley, "From Tit-Bits to Big Brother," 545.

24. Raymond Williams, *Keywords: A Vocabulary of Culture and Society* (New York: Oxford University Press, 1976), 69.

25. Richard A. Schwarzlose, *The Nation's Newsbrokers, Volume 1: The Formative Years, from Pretelegraph to 1865* (Evanston, Ill.: Northwestern University Press, 1989);

and Richard A. Schwarzlose, *The Nation's Newsbrokers, Volume 2: The Rush to Institution, from 1865 to 1920* (Evanston, Ill.: Northwestern University Press, 1990).

26. Thomas Streeter, *Selling the Air: A Critique of the Policy of Commercial Broadcasting in the United States* (Chicago, Ill.: University of Chicago Press, 1996).

27. Charles V. Kinter, "The Changing Pattern of the Newspaper Publishing Industry," *American Journal of Economics and Sociology* 5, no. 1 (October 1945): 45.

28. Benjamin M. Compaine and Douglas Gomery, *Who Owns the Media? Competition and Concentration in the Mass Communications Industry*, 3rd edition (Mahwah, N.J.: Lawrence Erlbaum, 2000), 8.

29. David Harvey, *The Condition of Postmodernity; An Enquiry into the Origins of Cultural Change* (Cambridge: Mass.: Basil Blackwell, 1989), 141–72.

30. Jeremy Tunstall, *Communications Deregulation: The Unleashing of America's Communication* (New York: Basil Blackwell, 1986).

31. Kinter, "The Changing Pattern of the Newspaper Publishing Industry," 46; Compaine and Gomery, *Who Owns the Media?*, 52–53.

32. Calculated from Alexander J. Field, "Radio—Finances, Employment, and Advertising Expenditures: 1935–1998," Table Dg131-144 in *Historical Statistics of the United States, Earliest Times to the Present: Millennial Edition*, ed. Susan B. Carter, Scott Sigmund Gartner, Michael R. Haines, Alan L. Olmstead, Richard Sutch, and Gavin Wright (New York: Cambridge University Press, 2006), <http://dx.doi.org/10.1017/ISBN-9780511132971.Dg117-18010.1017/ISBN-9780511132971.Dg117-180> (9 Mar. 2007).

33. Christopher H. Sterling, "Television and Radio Broadcasting," in Benjamin Compaine, Christopher H. Sterling, Thomas Guback, and J. Kendrick Noble, Jr., *Who Owns the Media?: Concentration of Ownership in the Mass Communications Industry*, 2nd. edition (White Plains, N.Y.: Knowledge Industry Publications, 1982), 305.

Other comparisons calculated from Alexander J. Field, "Radio—Finances, Employment, and Advertising Expenditures: 1935–1998"; and Alexander J. Field, "Television—Finances, Employment, and Advertising Expenditures: 1945–1998," Table Dg145-161 in *Historical Statistics of the United States*, ed. Carter et al. <http://dx.doi.org/10.1017/ISBN-9780511132971.Dg117-18010.1017/ISBN-9780511132971.Dg117-180> (9 Mar. 2007); and Jonathan Levy, Marcelino Ford-Livene, and Anne Levine, "Broadcast Television: Survivor in a Sea of Competition," OPP Working Paper Series No. 37, Federal Communications Commission, September 2002, 34.

34. Compaine and Gomery, *Who Owns the Media?*, 207.

35. Project for Excellence in Journalism, "Network TV," in *The State of the News Media 2007; An Annual Report on American Journalism*, <http://www. stateofthemedia.org/2007/narrative_networktv_audience.asp?cat=2&media=5> (22 Mar. 2007).

36. Alexander J. Field, "Television—Finances, Employment, and Advertising Expenditures: 1945–1998," Table Dg145-161 in *Historical Statistics of the United States*, ed. Carter et al. <http://dx.doi.org/10.1017/ISBN-9780511132971.Dg117-18010.1017/ISBN-9780511132971.Dg117-180> (9 Mar. 2007); "Ad-Supported Cable Scored a Milestone This 05/06 Official Season," *Cabletelevision Advertising Bureau* <http://www.onetvworld.org/main/cab/research/2006ResearchSnapshots/adsupported-cable-scored.shtml> (26 Mar. 2007); and "Ad-Supported Cable's Highlights & Milestones (Prime & Total Day)," *Cabletelevision Advertising Bureau* <http://www.onetvworld.org/main/cab/research/2007ResearchSnapshots/adsupported-cables-highli-2.shtml> (26 Mar. 2007).

37. Compaine and Gomery, *Who Owns the Media?*, 271–72.

38. Alexander J. Field, " Newspapers and Periodicals—Number and Circulation, By Type: 1850–1967," Table Dg253-266 in *Historical Statistics of the United States*, ed. Carter et al. <http://dx.doi.org/10.1017/ISBN-9780511132971.Dg225-29210.1017/ISBN-9780511132971.Dg225-292> (9 Mar. 2007); Alexander J. Field, " Newspapers—Number and Circulation, By Type: 1920–1999," Table Dg267-274 in *Historical Statistics of the United States*, ed. Carter et al. <http://dx.doi.org/10.1017/ISBN-9780511132971.Dg225-29210.1017/ISBN-9780511132971.Dg225-292> (9 Mar. 2007); Alexander J. Field, "Newspapers and Periodicals—Number, By Type: 1935–1999," Table Dg275-286 in *Historical Statistics of the United States*, ed. Carter et al <http://dx.doi.org/10.1017/ISBN-9780511132971.Dg225-29210.1017/ISBN-9780511132971.Dg225-292> (9 Mar. 2007); and Newspaper Association of America, *The Source; Newspapers by the Numbers 2006*, 22 <http://www.naa.org/thesource/> (9 Mar. 2007).

39. Reprinted in Newspaper Association of America, *The Source; Newspapers by the Numbers 2006*, 24 <http://www.naa.org/thesource/> (9 Mar. 2007). Population figures from "Historical National Population Estimates: July 1, 1900 to July 1, 1999," Population Estimates Program, Population Division, *U.S. Census Bureau* <http://www.census. gov/popest/archives/1990s/popclockest.txt> (9 Mar. 2007); and "Population Estimates Archives: April 1, 2000 to July 1, 2005," Population Estimates Program, Population Division, *U.S. Census Bureau* <http://www.census.gov/popest/national/files/NST_EST2005ALLDATA.csv> (9 Mar. 2007).

40. Robert B. Horwitz, "U.S. Media Policy Then and Now," in *Converging Media, Diverging Politics: a Political Economy of News Media in the United States and Canada*, ed. David Skinner, James R. Compton and Michael Gasher (Lanham, Md.: Lexington Books, 2005), 32.

41. A dystopian, but useful itemization of such trends at the time is John Wicklein, *Electronic Nightmare: The New Communications and Freedom* (New York: Viking, 1981).

42. Compaine and Gomery, *Who Owns the Media?*, 26; Patricia Aufderheide, Erik Barnouw, Richard M. Cohen, Thomas Frank, Todd Gitlin, David Lieberman, Mark Crispin Miller, Gene Roberts, and Thomas Schatz, *Conglomerates and the Media* (New York: New Press, 1997); and "Advertising Age's Media Family Tree; Top Properties of the Nation's 15 Largest Media Companies by Net Revenue in 2005," *Advertising Age* <http://adage.com/images/random/mediafamilytree06.pdf> (28 Mar. 2007).

43. Federal Communications Commission, *Report and Order and Notice of Proposed Rulemaking, In the Matter of 2002 Biennial Regulatory Review*, MB Docket 02-277, 01-235, 01-317, 00-244, and 03-130, 2 June 2003, ¶¶ 86–128.

44. U.S., House, Committee on the Judiciary, Antitrust Task Force, Hearing on Competition and the Future of Digital Music, Testimony of Mel Karmazin, 110th Cong., 1st. sess., 28 February 2007.

45. Auletta notes the "sound business reasons to zealously guard the credibility of journalism. [. . .] Sully the credibility and you damage the brand." Ken Auletta, *Backstory; Inside the Business of News* (New York: Penguin Press, 2003), xvi.

46. Commission on Freedom of the Press, *A Free and Responsible Press; a General Report on Mass Communication: Newspapers, Radio, Motion Pictures, Magazines, and Books* (Chicago, Ill.: University of Chicago Press, 1947).

47. Todd Gitlin, *The Whole World Is Watching; Mass Media in the Making and Unmaking of the New Left* (Berkeley, Calif.: University of California Press, 1980).

48. The long tradition of media criticism in the United States is surveyed in Robert W. McChesney and Ben Scott, eds., *Our Unfree Press: 100 Years of Radical Media Criticism* (New York: New Press, 2004).

49. While some see this development as a quintessentially modern form, others note important premodern precursors. For various positions, see Shelley Fisher Fishkin, *From Fact to Fiction: Journalism and Imaginative Writing in America* (Baltimore, Md.: Johns Hopkins University Press, 1985); Thomas B. Connery, ed., *A Sourcebook of American Literary Journalism: Representative Writers in an Emerging Genre* (New York: Greenwood, 1992); John C. Hartsock, *A History of American Literary Journalism: the Emergence of a Modern Narrative Form* (Amherst, Mass.: University of Massachusetts, 2000); Karen Roggenkamp, *Narrating the News: New Journalism and Literary Genre in Late Nineteenth-Century American Newspapers and Fiction* (Kent, Ohio: Kent State University Press, 2005); and Iona Italia, *The Rise of Literary Journalism in the Eighteenth Century: Anxious Employment* (New York: Routledge, 2005).

50. Paulo Freire, *Pedagogy of the Oppressed*, transl. Myra Bergman Ramos (New York: Herder and Herder, 1970); Richard Shafer, "Comparing Development Journalism and Public Journalism as Interventionist Press Models," *Asian Journal of Communication* 8, no. 1 (March 1998): 31–52; Shirley A. White, ed., *Participatory Video: Images that Transform and Empower* (Thousand Oaks, Calif.: Sage, 2003); and Hemant Shah, "Modernization, Marginalization, and Emancipation: Toward a Normative Model of Journalism and National Development," *Communication Theory* 6, no. 2 (May 1996): 143–66.

51. Claims regarding the "liberal media" continue to be debated in an immense literature. Recent contributions include Eric Alterman, *What Liberal Media?: The Truth About Bias and the News* (New York: Basic Books, 2003); David Edwards and David Cromwell, *Guardians of Power: The Myth of the Liberal Media* (Ann Arbor, Mich.: Pluto Press, 2006); Richard A. Viguerie and David Franke, *America's Right Turn: How Conservatives Used New and Alternative Media to Take Power* (Chicago, Ill.: Bonus Books, 2004); and Brian C. Anderson, *South Park Conservatives: The Revolt against Liberal Media Bias* (Washington, D.C.: Regnery, 2005).

52. Pew Research Center for People and the Press, "Public More Critical of Press, but Goodwill Persists," *Project for Excellence in Journalism*, 26 June 2005 <http://people-press.org/reports/display.php3?ReportID=248> (29 Mar. 2007).

53. Pew Research Center for People and the Press, "Public More Critical of Press," 9.

54. Pew Research Center for the People and the Press, "Media Consumption and Believability Study," *Pew Research Center*, 8 June 2004, 43 <http://people-press.org/reports/pdf/215.pdf> (12 Mar. 2007).

55. Pew Research Center for People and the Press, "Press Going Too Easy on Bush," *The State of the News Media 2007; an Annual Report on American Journalism* <http://www.stateofthemedia.org/2007/journalist_survey_prc1.asp> (22 Mar. 2007).

56. Project for Excellence in Journalism, "Cable TV: Public Attitudes," *The State of the News Media 2007; an Annual Report on American Journalism*, <http://www.stateofthemedia.org/2007/narrative_cabletv_intro.asp?cat=1&media =6> (22 Mar. 2007).

57. A sense of the range is represented by Vance Packard, *The Hidden Persuaders* (New York: D. McKay, 1957); John Kenneth Galbraith, *The Affluent Society* (Boston, Mass.: Houghton Mifflin, 1958); C. Wright Mills, *The Power Elite* (New York: Oxford University Press, 1956); and Herbert Marcuse, *One-Dimensional Man; Studies in the Ideology of Advanced Industrial Society* (Boston, Mass.: Beacon Press, 1964). A useful discussion of social criticism at this time is Daniel Horowitz, *Vance Packard and American Social Criticism* (Chapel Hill, N.C.: University of North Carolina Press, 1994).

58. See Robert N. Mayer, *The Consumer Movement: Guardians of the Marketplace* (Boston, Mass.: Twayne Publishers, 1989); and Kathy M. Newman, *Radio Active: Ad-*

vertising and Consumer Activism, 1935–1947 (Berkeley, Calif.: University of California Press, 2004).

59. Quoted in Christopher Lasch, "Journalism, Publicity, and the Lost Art of Argument," *Media Studies Journal* 9, no. 1 (Winter 1995): 91.

60. Doug Underwood, *When MBAs Rule the Newsroom* (New York: Columbia University Press, 1993); Richard McCord, *The Chain Gang: One Newspaper Versus the Gannett Empire* (Columbia, Mo.: University of Missouri Press, 1996); Kristina Borjesson, ed., *Into the Buzzsaw: Leading Journalists Expose the Myth of a Free Press* (Amherst, N.Y.: Prometheus Books, 2002); Eric Klinenberg, *Fighting for Air: The Battle to Control America's Media* (New York: Metropolitan, 2007); and Trudy Lieberman, "The Epidemic," *Columbia Journalism Review*, March/April 2007 <http://www.cjr.org/issues/2007/2/Leiberman.asp> (14 Mar. 2007).

61. "Usage and Population Statistics," *Internet World Stats* <http://www. internetworldstats.com/stats.htm> (26 Mar. 2007). See also Organisation for Economic Co-Operation and Development, "OECD Broadband Statistics to December 2006," *OECD* <http://www.oecd.org/sti/ict/broadband/> (30 Apr. 2007).

62. Michael Strangelove, *The Empire of Mind; Digital Piracy and the Anti-Capitalist Movement* (Toronto: University of Toronto Press, 2005).

63. Strangelove, *Empire of Mind*, 218–19.

64. Thomas Frank, *The Conquest of Cool; Business Culture, Counterculture, and the Rise of Hip Consumerism* (Chicago, Ill.: University of Chicago Press, 1997; Raymond Williams, "Advertising: the Magic System," in *Problems in Materialism and Culture* (London: Verso, 1980), 170–95.

65. "OpinionWare Establishes New Benchmark for Online Opinion-Expression," *PR Newswire*, 6 September 2000 <http://web.lexis-nexis.com/> (22 Feb. 2007).

66. Examples include Nike athletic shoes and the Toyota Scion.

67. "Factiva Launches Most Sophisticated Reputation Intelligence Tool Available," *PR Newswire*, 4 August 2005 <http://web.lexis-nexis.com/> (22 Feb. 2007).

68. Jackie Huba and Ben McConnell, *Citizen Marketers; When People Are the Message* (Chicago, Ill.: Kaplan, 2007); Jackie Huba and Ben McConnell, *Creating Customer Evangelists: How Loyal Customers Become a Volunteer Sales Force* (Chicago, Ill.: Dearborn, 2003).

69. "Word of Mouth Marketing Association," *Word of Mouth Marketing Association* <http://www.womma.org/> (10 Mar. 2007); "Web Analytics Pioneer Brett Hurt Starts New Company To Help Online Brands Build Their Business Through Word of Mouth," *Business Wire*, 11 January 2006 <http://web.lexis-nexis.com/> (22 Feb. 2007).

70. "Home Page: My Dream," *Lincoln (Ford Motor Company)* <http://mydream. tv> (14 Mar. 2007).

71. Edward Cone, "Customer-Created Content; Ready for Prime Time?," *CIO Insight*, January 2007, 21; "Dallas Cowboys Cheerleaders—Making the Commercial," *Country Music Channel* <http:/www.makingthecommercial.cmt.com/> (16 Oct. 2006); "15 Seconds," *CBS.com* <http://www.cbs.com/originals/15_seconds/> (22 Jan. 2007); "Windows Vista – Show Us Your 'Wow'," *Microsoft Corp.* <http:// showusyourwow.msn.com/> (22 Jan. 2007).

72. "Current TV // Make Video // V-CAM = Viewer Created Ad Message," *Current TV* <http://www.current.tv/make/vc2/vcam> (21 Mar. 2007).

73. "Tremor – Step 1," *Tremor* <http://www.tremor.com/teenscreener/ Page1.aspx> (14 Mar. 2007); "Welcome to Tremor," *Tremor* <http://www.tremor.com/login/about_us.aspx> (14 Mar. 2007).

74. Jake Coyle, "TV Networks Embrace the YouTube Model," Associated Press, 6 March 2007 <http://web.lexis-nexis.com/> (26 Mar. 2007).

75. Elizabeth Ross, "To Win, Don't Hide Behind False Brand Pretenses," *Advertising Age*, 26 February 2007, 18.

76. Lisa Sanders, "Creative License; Consumer-Generated Content Is the Future, but as Chevy Discovered, It's Also a Scary Prospect," *Advertising Age*, 17 April 2006, 12.

77. Sanders, "Creative License."

78. Phil De Vellis, "I Made the 'Vote Different' Ad," *The Huffington Post* <http://www.huffingtonpost.com/phil-de-vellis-aka-parkridge/i-made-the-vote-differen_b_43989.html> (22 Mar. 2007).

79. Steve Outing, "The 11 Layers of Citizen Journalism," *Poynter Online*, 13 June 2005, <http://www.poynter.org/content/content_view.asp?id=83126> (10 Mar. 2007; *OhmyNews*, "Biography of Oh Yeon Ho, CEO and Founder of *OhmyNews*," in *OhmyNews; Every Citizen is a Reporter* (Seoul: Seoul Selection, 2004), 5.

80. Sung-Nam Cho and Sook-Mi Park, "The Concept of Generation and the Social Change in Korea," *The Research of Social Science* 9, no. 1 (December 2002): 39–68.

81. Dae-Youp Cho, "The Generation of Social Movement in Korea: 386," *Ideology* 14, no. 3 (September 2002): 125–47.

82. Hong-Sub Jung, "Gap in Generations and the Role of Education," *Education Science* 8, no. 1 (February 2003): 131–48.

83. Myung-Jin Park, Chang-Nam Kim and Byung-Woo Sohn, "Modernization, Globalization, and the Powerful State: The Korean Media," in *De-Westernizing Media Studies*, ed. James Curran and Myung-Jin Park (New York: Routledge, 2000), 111–23.

84. Seung-Su Kim, *Media Reform for People* (Seoul: Segyesa, 2002); and Hae-Sik Kim, "A Sociological Study on the Transformation of the Korean Press Since 1960s" (unpublished Ph.D. diss., Seoul University, 1993).

85. Korea Press Foundation, *A Survey of Audience* (Seoul: Korea Press Foundation, 1998).

86. Ministry of Information and Communication, *The Strategy of Informationization in Korea* (Seoul: Ministry of Information and Communication, 2003).

87. Ministry of Information and Communication, *Strategy of Informationization in Korea*; International Telecommunication Union, *ITU Internet Reports 2003—Birth of Broadband* (Geneva: International Telecommunication Union, 2003).

88. "Top 35 Countries With the Highest Internet Penetration Rate," *Internet World Stats* <http://www.internetworldstats.com/top25.htm> (26 Mar. 2007).

89. Dal-Ho Cho, *In the Age of US $10,000 Income, the Meaning and Task* (Seoul: Samsung Economic Research Institute, 1996).

90. Ministry of Finance and Economy, *Economic Indicator, 2004–6* (Seoul: Ministry of Finance and Economy, 2004).

91. Joe Miracle, "Regent Journalism Alum Making Waves in Korea; Yeon-Ho Oh Makes Participatory Journalism Work with Internet Newpaper OhmyNews," *News@ Regent University*, 5 January 2005 <http://www.regent.edu/news/ohmynews.html> (26 Mar. 2007).

92. Yeon-Ho Oh, *The Speciality of Korea: OhmyNews* (Seoul: Humanist, 2004), 55; "OhmyNews Runs to the Third Generation Internet Newspaper," *OhmyNews*, 23 February 2004 <http://www.ohmynews.com/articleview/article_view.asp?menu=0446&no=153582&rel_no=1> (10 Jun. 2004); Yong-Cheol Cheon, "Internet Newspapers as Alternative Media: OhmyNews in South Korea," *Media Development* 51, no. 1 (2004): 28.

93. Yeon-Ho Oh, "Internet Media and Korean Society," speech at the Korea Institute of Future Strategy, 18 May 2002.

94. Chris Atton, *An Alternative Internet: Radical Media, Politics and Creativity* (Edinburgh: Edinburgh University Press, 2004), 25–60.

95. "On These Sites, We're All Reporters," *Joins.com*, 5 March 2001 <http://www.factiva.com/> (18 Mar. 2004).

96. Oh, *Specialty of Korea*, 33.

97. Moon Ihlwan, "OhmyNews' Oh My Biz Problem," *BusinessWeek.com*, 1 November 2006 <http://www.businessweek.com/globalbiz/content/nov2006/gb20061101_539412.htm> (10 Mar. 2007); Oh, "Ten Preconditions."

98. "OhmyNews Goes into the Black," *OhmyNews*, 31 December 2003 <http://www.ohmynews/board/bbs_view.asp> (20 Jun. 2004).

99. Ihlwan, "OhmyNews."

100. Mark Clifford and Moon Ihlwan, "The Web Site that Elected a President," *Business Week*, 24 February 2003, 30; Yun-Su Jeong, "Brave News World," *Joins.com*, 30 March 2003 <http://global.factiva.com/> (10 Apr. 2004); "Blue House Ready to Open Door Wider for Media Reps," *Korea Times*, 10 January 2003 <http://global.factiva.com/> (10 Apr. 2004); Dong-Min Lee, "Internet Campaigns Race Ahead of Election Laws," *Korea Herald*, 15 February 2002 <http://global.factiva.com/> (10 Apr. 2004).

101. "Fifty-Five People in Korea," *KyungHyang Newspaper*, 8 October 2001, 42–47.

102. "Korea's Influential Media," *Sisa Journal*, 18 October 2000; "Korea's Influential Media," *Sisa Journal*, 25 October 2001; "Korea's Influential Media," *Sisa Journal*, 22 October 2002; "Korea's Influential Media," *Sisa Journal*, 30 October 2003; and "Korea's Influential Media," *Sisa Journal*, 28 October 2004.

103. "Internet Newspaper, Development Possibility Is the Biggest," *Media Today*, 7 January 2004 <http://www.mediatoday.co.kr/> (20 Jun. 2004).

104. Bobbie Johnson and Michael Fitzpatrick, "Korean Website's 40,000 Citizen Journalists Get Shot at Byline in Herald Tribune," *Guardian Unlimited*, 31 May 2006 <http://business.guardian.co.uk/story/0,,1786356,00.html> (26 Mar. 2007); "*OhmyNews Japan* Debuts," *OhmyNews International*, 28 August 2006 <http://english.ohmynews.com/articleview/article_view.asp?article_class=8&no=313808&rel_no=1> (10 Mar. 2007); Moon Ihlwan and Kenji Hall, "*OhmyNews*: Voices From the Street," *BusinessWeek.com*, 4 May 2006 <http://www.businessweek.com/globalbiz/content/may2006/gb20060504_995206.htm> (10 Mar. 2007).

105. Ihlwan, "OhmyNews."

106. Project for Excellence in Journalism, "Online: Citizen Media," in *The State of the News Media 2007; An Annual Report on American Journalism* <http://www.stateofthemedia.org/2007/narrative_online_citizen_media.asp?cat=8&media=4> (13 Mar. 2007).

107. Steve Outing, "CitJ's National Networks: Will They Bloom?," *Poynter Online*, 6 April 2006 <http://www.poynter.org/content/content_view.asp?id= 96566> (8 Mar. 2007).

108. "Newspaper Next: Blueprint for Transformation," *American Press Institute*, 44 <http://www.newspapernext.org/> (10 Mar. 2007).

109. Barbara Cohen and Craig Kaczorowski, "Newspapers' Business Model Transformation," *Newspaper Association of America*, 2004 <http://www.naa.org/ horizon/Biz_model/Biz_model_2.html> (14 Mar. 2007).

110. Civic journalism continues to be supported by philanthropies, such as the Pew Center for Civic Journalism (http://www.pewcenter.org/) and the Kettering Foundation (http://www.kettering.org/). Foundational studies and critiques include Jay Rosen, *Getting the Connections Right: Public Journalism and the Troubles in the Press* (New York:

Twentieth Century Fund, 1996); Jay Black, ed., *Mixed News: The Public/Civic/Commun-0itarian Journalism Debate* (Mahwah, N.J.: Lawrence Erlbaum, 1997); Hanno Hardt, "Conflicts of Interest: Newsworkers, Media, and Patronage Journalism," in *Media Power, Professionals and Policies*, ed. Howard Tumber (New York: Routledge, 2000), 209–24; Anthony J. Eksterowicz and Robert N. Roberts, eds., *Public Journalism and Political Knowledge* (Lanham, Md.: Rowman & Littlefield, 2000); and Tanni Haas and Linda Steiner, "Public Journalism; A Reply to the Critics," *Journalism: Theory/Practice/Criticism* 7, no. 2 (May 2006): 238–54.

From the point of view of citizen-journalism advocates in the newspaper industry, civic journalism means patrician control rather than user empowerment. For example, see Shayne Bowman and Chris Willis, *We Media; How Audiences Are Shaping the Future of News and Information*, The Media Center, *American Press Institute*, July 2003 <http://www.hypergene.net/wemedia/> (10 Mar. 2007).

111. Juan Carlos Perez, "Online Ad Spending had Healthy Growth in 2006," *InfoWorld*, 7 March 2007 <http://www.infoworld.com/article/07/03/07/ Hnonlineadspending_1.html> (10 Mar. 2007); Bradley Johnson, "Adwatch," *Advertising Age*, 12 February 2007, 8; "Online Ad Spending to Total $19.5 Billion in 2007," *eMarketer.com*, 28 February 2007 <http://www.emarketer.com/Article.aspx ?1004635> (10 Mar. 2007).

112. Project for Excellence in Journalism, "Online: Citizen Media."

113. Merrill Brown, "Abandoning the News," *Carnegie Reporter* 3, no. 2 (Spring 2005) <http://www.carnegie.org/reporter/10/news/> (10 Mar. 2007).

114. W. Joseph Campbell, *The Year that Defined American Journalism: 1897 and the Clash of Paradigms* (New York: Routledge, 2006).

115. Steve Outing, "Citizen-Media Start-ups Ply Innovative Business Models," *Poynter Online*, 6 April 2006 <http://www.poynter.org/content/content_view.asp?id =97601> (8 Mar. 2007).

116. Outing, "CitJ's National Networks."

117. Steve Outing, "Innovative Thoughts from CitJ's Gurus," *Poynter Online*, 6 April 2006 <http://www.poynter.org/content/content_view.asp?id=97298> (8 Mar. 2007).

118. Outing, "CitJ's National Networks"; Fran Willis, "Citizen Journalism, the Next Generation of Local News and Advertising," presentation to Growing Audience Seminar, *Newspaper Association of America*, 5 June 2006 <http://www.naa.org/ conferences/gas/program.html> (9 Mar. 2007).

119. Willis, "Citizen Journalism."

120. Lisa Williams, Dan Gillmor, and Jane MacKay, "Frontiers of Innovation in Community Engagement: News Organizations Forge New Relationships with Communities," *Center for Citizen Media*, <http://citmedia.org/frontiers/> (8 Mar. 2007).

121. Williams et al., "Frontiers of Innovation," 10.

122. Williams et al., "Frontiers of Innovation," 18. See also Steve Outing, "Independent CitJ: Web sites and Networks," *Poynter Online*, 6 April 2006 <http://www.poynter.org/content/content_view.asp?id=96563> (8 Mar. 2007).

123. Merrill Brown, "Abandoning the News," *Carnegie Reporter* 3, no. 2 (Spring 2005) <http://www.carnegie.org/reporter/10/news/> (10 Mar. 2007).

124. Shayne Bowman and Chris Willis, "The Future Is Here, But Do News Media Companies See It?," *Nieman Reports* 59, no. 4 (Winter 2005), 6.

125. "Brave New Films," <http://www.bravenewfilms.org/> (10 Mar. 2007); Robert S. Boynton, "How to Make a Guerrilla Documentary," *The New York Times Magazine*, 11 July 2004, 20–3; "Echo Chamber Project," *Echo Chamber Project* <http://www. echochamberproject.com/> (10 Mar. 2007).

126. Lianne Stewart, "Promo Partner Personals: House Party Raises the Roof on In-Home Marketing," *Marketing*, 1 January 2006, 33.

Chapter 7

Democratic Communications as Critical, Collective Education

Early in 2000, writes Raj Jayadev, "we, the young temps of Silicon Valley, were invisible." In the epicenter of the high-tech economy and amidst the stereotype of the 21-year-old dotcom multimillionaire, "the vast majority of the work of the new high-tech economy, it turns out, was actually old-fashioned assembling, packaging, and shipping" or "service industries set up to support high-tech workers." Not only was the work low-paying, much of it was temporary and without benefits. Jayadev concludes that "the people who were really turning the engines of Silicon Valley were nowhere in the public's perception."[1]

Instead of taking the well-worn steps of proto social movements and mounting an organized publicity campaign and series of protest marches to garner attention, Jayadev and a small group of friends simply "started meeting after work at a Vietnamese café in downtown San Jose, trying to outdo each other with stories of maniacal supervisors." When a call from a San Francisco Bay-area monthly magazine for "stories about the 'digital divide'" spurred them to "put our stories down on paper," they found that the process of writing turned into a process of collective self-recognition that helped them "reflect on our lives, the world around us, and how we related to it. We wrote about what we really wanted to say to the C.E.O., why people do drugs at work, and what it feels like to be so close to unattainable wealth." Their stories were published, and they handed out the magazine issue that contained their stories to "working people at Silicon Valley bus stops, light-rail stations, and fast-food joints," who "read our stories and for the first time saw their experiences reflected on paper. It was a confirmation that they weren't alone."[2]

This informal, collective project of recognition and exploration has since expanded into a wide range of projects, including a bilingual, bimonthly 'zine (*De-Bug*), a weekly local television show, and a twice-a-month radio discussion of culture and politics plus local music. It also consists of projects typically placed far outside the "media" realm, including conducting discussion and writing workshops for incarcerated women in transitional living centers, aiding a formation of a similar magazine project for high-school students in Central and West Fresno, helping communities organize "against police brutality, unfair terminations at the workplace, and discrimination of immigrants," staging "writing, art, and discussion sessions for high schools and community organizations," and serving as a home for photographers and those interested in photography who discuss and show their work as well as produce photo essays and a photo 'zine series.[3]

Despite the range and complexity of these efforts, the collective still works in a distinctly noninstitutional way, grounded in weekly meetings/discussions to which anyone off the street can attend and that build upon individuals' experiences which, upon their airing and discussion, turn out to be not so individual. Stories suggested or written in this process "provide insights not obvious to outsiders," ones that go much deeper than observations about bad pay and no benefits for the majority of young high-tech workers, to insights about "how the temp economy assaults one's faith—in oneself, in others, and in the possibility of change." Jayadev sees its emphasis on open, collaborative creation as a better means of political organization. Such a project is able to "go places where traditional organizing does not go, because media creation encourages people to think and analyze for themselves. Traditionally, it has always been the organizers who identify the key issues and the strategies to address them, whereas the workers simply hold the signs and chant in picket lines. Writing and art, however, allow people to develop their own analysis and strategies based on their experiences."[4]

The emphasis on the value of open, creative collaboration and what the contributions to democratic communications might be and might become is the topic of this chapter. For what Jayadev and others advocate is a collective practice of education—more precisely, self-education conducted not in a classroom, but through a collective process of composition in the fullest sense. Along with the institutional struggle that continues to challenge and reshape professional media practice (as discussed in Chapter 6), democratic communications can also be recognized as a process of colloquial, collective education. Seeing it this way helps de-emphasize a preoccupation with technical, organizational, and capital necessities of message dissemination, and emphasize instead the much more difficult and yet much more crucial task of producing new, open forms of collaboration and collective composition through which new kinds of social movements and new kinds of institutions and polities can arise.

The chapter begins by contrasting two responses to the drawbacks of market systems in the United States. While a distinctly undemocratic response is to form a ruling, expert bureaucracy in sole charge of generating information, a second democratic response is what Raymond Williams called direct autonomous composition. The situation of U.S. migrant labor-camp newspapers in the Dust-Bowl years of the 1930s is used to develop a grounded conception of direct autonomous composition as the fluid, colloquial, collective practice of discovering forms of social organization, thus not only extending the notion of democratic communications beyond matters of media to matters of social life, but also suggesting the extent to which democratic communications can be thought about as a process of self-education and production of new forms of collective action. The recent emergence of programs of media literacy are then examined as a kind of educational attempt related to democratic communications but much different than this social and cultural sense. While its practice replicates many of the drawbacks of traditional approaches to education, the continued attention to media literacy within the context of the formation of self-managed projects such as *De-Bug*, Independent Media Centers, and others holds great potential for developing new forms of collective, creative composition not only of media products, but of the very bases of social life.

Education Matters

As seen from within liberal political theory, democratic communications depends upon the presence and expansion of the marketplace of ideas. Constituting this marketplace are ideas that emerge from individual lives and their wants, needs, and views, with the partial truths contained in them forming a composite truth upon which agreement can be reached and action taken.[5] Such a process is seen to produce a well-informed citizenry that can make intelligent decisions about affairs of the day, thus enabling not only a surviving but a well-working democracy.

Setting aside for the moment the immense complexity and global scale of problems that now confront the world (which increase by many times the challenge of organizing an adequate response to them), one can identify other difficulties with this normative conception. First, while it valuably enshrines freedom from legal constraint, it does not address the need for enabling conditions for the exercise of communications, thus leaving it open to the familiar critique that, in a market system, "freedom of the press is guaranteed only to those who own one."[6] In addition, as Keane argues, a system informed by such a conception makes little headway against the "unaccountable and arcane power of state institutions"—a point of particular concern in the post-9/11 United States informed by the egregiously misnamed Patriot Act and related federal and politically sanctioned actions to reign in public expression. Forms in which such power is exercised include evoking emergency powers, secrecy, lying, government advertising and public relations, and the growth of corporatism through government-industry alliances that move decisions outside of democratic processes.[7]

At least two responses to these difficulties are possible, with one being to restrict, manage, and professionalize it further in the hopes of making it more efficient and dependable. In the wake of the Progressive-era push for efficiency and professionalism in government as a way of eliminating cronyism, graft, and corruption, it was not only the defects of the press but the power of propaganda in the midst of a diversifying United States, the rising visibility and power of the labor movement, and fears of political revolution that prompted newspaperman Walter Lippmann to focus his critique of market communications not only on its inefficiencies, but its dangers as well. By contrast to the liberal market model, he advocated a professional if not vanguard elitism. Dismissing what he calls the "original dogma of democracy [. . .] that the knowledge needed for the management of human affairs comes up spontaneously from the human heart," Lippmann argued that "we cannot rely upon intuition, conscience, or the accidents of casual opinion if we are to deal with the world beyond our reach" for the very reason that "the outsider [. . .] has neither time nor attention, nor interest, nor the equipment for specific judgment. It is on the men inside, working under conditions that are sound, that the daily administrations of society must rest."[8] Thus, instead of thrusting the public into the middle of all kinds of complex issues, Lippmann advocated the formation of a cadre of experts who would provide the dependable, authoritative information needed to administer a complex society. In Lippmann's terms, to boil off the "large fatty mass of stereotyped phrases" in which is embedded the "intricate series of facts" necessary for informed action, "expert reporters" would provide "a valid picture of the environment" as a means of grounding an expert-mediated hearing and resulting decision.[9] The difficulty with such a position is how it works against rather than enables a democratic, participatory extension of involvement and accountability.

A second, very different response to the shortcomings of market systems of communications emphasizes the necessity of direct popular involvement. This response does not simply mean greater popular participation (which, as was discussed in Chapter 6, has been a long-standing imperative in the U.S. market system), but participation that is more direct and autonomous in addition to being more generally available. The most valuable kinds aspire to what Raymond Williams calls direct autonomous composition. As Jones summarizes, the educational project embodied in this term was "not the paternalist one of [English literary and cultural critic F.R.] Leavis or others," but an "empowering radical-democratic one." Paraphrasing Williams's argument, Jones emphasizes that "Williams never shifted from his earliest contention that a popular culture worthy of the name required democratic organization of the cultural means and skills of cultural composition so that all societal members might be 'direct autonomous composers'."[10] Exemplifying a "longstanding advocacy of the teaching of skills to critical analysis and its extension to compositional practice," Williams emphasized in a late essay the need to join the teaching of critical analysis to its "extension to compositional practice."[11]

> [Critical studies of] reification will have to be distinguished from the open, conscious composition of works, or the only results will be negative, as in some contemporary semiotic tendencies, which demystify the practice by calling all such practice into question, and then predictably fall back on ideas of universal (inherent and unsurpassable) alienation, within the terms of a pessimistic and universal psychology. The critical demystification has indeed to continue, but always in association with practice: regular practice, as part of a normal education, in this transforming labour process itself: practice in the production of alternative "images" of the "same event"; practice in processes of basic editing and the making of sequences; practice, following this, in direct autonomous composition.[12]

One can valuably contrast the term "direct autonomous composition" with its implicit opposites to make its emphases clearer. It is seen as direct as opposed to representative, thus constituting a critical–modernist critique of professionalization (a critique that, importantly, has nothing to do with attained levels of skill—only with the degree of social restriction of the practice). It is autonomous as opposed to managed or controlled, thus also distancing itself from professionalization. And the term "composition" emphasizes formulating and making rather than solely interpreting and demystifying. Williams ultimately regarded this creative activity as much more than composing written, visual, or aural works, as important as those capabilities are. Instead, he never wavered in his view that the production of new kinds of association were not only as creative as the narrowly construed artistic ones, but equally if not more crucial. In 1958, for example, Williams already considered the production of the "collective democratic institution, whether in the trade unions, the cooperative movement or a political party [. . .] as a very remarkable creative achievement."[13] Close to twenty years after that, he wrote that "we have to learn and to teach each other the connections between a political and economic formation, a cultural and educational formation, and, perhaps hardest of all, the formations of feeling and relationship which are our immediate resources in any struggle."[14]

Importantly, the normative goal of direct autonomous composition should not be mistaken as a liberal call for simply setting free ostensibly self-sufficient individuals. If misread through liberalism, direct autonomous composition seems to champion an original, ahistorical individual creativity the likes of which would be at home in liberal claims of essential human autonomy—a conception against which Williams argued throughout his lifetime, not in pessimistic terms of the inescapability of rigid determination, but in critical terms of the defects in liberal individualism as a means of solidarity and ultimately—in these days of nuclear threat and accelerating ecological catastrophes—of enabling the survival of the species. When pressed to respond to critiques of his earlier positions as seeming to rely on a distinctly uncritical liberal humanism, Williams clarified his position in a reprinted revision of an earlier essay (the additions consist largely of the parenthetical material).

> [N]o mode of production, and therefore no dominant society or order of society, and therefore no dominant culture, in reality exhausts the full range of human practice, human energy, human intention (this range is not the inventory of some original "human nature" but, on the contrary, is that extraordinary range of variations, both practiced and imagined, of which human beings are and have shown themselves to be capable).[15]

Such a position makes it possible for Williams to consistently reject any kind of essentialized authenticity as well as totalized claims of a complete and rigid determination of all human activity. As but one example, he wrote in some of his last work of "the intense vitality of some kinds of popular music, always reached for by the market and often grasped and tamed, but repeatedly renewing its impulses in new and vigorous forms."[16]

In contrast to this liberal reduction, Williams sees direct autonomous composition as related theoretically and politically to a particular conception of communications and its role in the formation and extension of participatory democracy. This conception emerges from the centrality Williams attributed to communications. Instead of seeing it in instrumental terms as information about the world upon which to base decisions, Williams regards communications materially and dialectically, as the making of meanings, values, and practices that in turn constitute the world, society, and ourselves. In this way, individuals are not simply original and self-contained, but continually made through human practice, which at the same time makes the conditions of this making. In this way and for this reason, as Jones notes, communications becomes "a central element of the 'expanding culture' of his [Williams's] desired educated and participating democracy."[17] For Williams, the education that produces such a democracy enables people to recognize the deep determinations of their present circumstance, the open possibilities for change, and the means to try to work out collectively how to make new conditions and practices. What Williams means by "participation" is direct, reflexive involvement in this dialectical process of making while being made. It is also what he means by "composition," the dual meaning of the word (both as substance and as the act of representation) recalling Williams's emphasis on the materiality of culture, thus connecting direct autonomous composition not only to a normative goal of participatory democratic process but to its theoretical basis. "Composition" refers not simply to the making of messages, but in the most general social and historical scale to the continual (re)making of our world and ourselves.

In a conception such as that of Williams's cultural materialism, what is at stake in democratic communications is not simply opinions, ideas, or pictures of the world separate from any "real" substance, but quite literally the material shape and social substance of our world. It is for these reasons and in these ways that education, which was for Williams in the most general sense the activity of enabling the broadest and fullest possible making of a social order, is of central importance. Ultimately, education matters for Williams quite deeply by being something much broader than mastering a body of thought or training in a technical skill to be individually employed for personal gain. Instead, education is at once a precondition of, a productive process contributing to, and a means of participating in an open, expanding, fully democratic culture. It is from such a position that Williams regarded free adult education of the kind he was involved in during 1946–1961 as a Staff Tutor for the Oxford Delegacy for Extra-Mural Studies, which organized and taught classes in collaboration with the Workers' Educational Association, as always a processual and uneven, yet valuable embodiment and realization of this normative goal of greater access to a democratic form of education grounded in specific, conflictual life experience and thus always open to mutual reformulation and change.[18]

Social Composition in Migrant Labor Camp Newspapers of the 1930s

Among the many possibilities for investigating the complexities of direct autonomous composition in the United States is one particularly useful and underused project. The general situation is well-known. During the 1930s, more than 500,000 people moved from the south-central drought-stricken states of Oklahoma, Arkansas, Texas, and Missouri, with more than 300,000 of them making their way to California, only to find infrequent, low-paying work amidst widespread persecution and poverty.[19] Although a large migratory labor force had been roaming California since the 1870s, what made the 1930s influx so much more visible and such a problem was the inability of the Depression-slowed California economy to absorb these new arrivals.[20]

To respond to what had rapidly become a crisis in size and in magnitude of suffering, the federal Resettlement Administration (R.A.) established and operated migrant-labor camps by virtue of its broad mandate to control and manage population movements.[21] New Dealers intended the camps as one of many means of addressing through rational planning the general economic and humanitarian crisis of the Depression.[22] To start the federal migrant-labor camp program, the R.A. in late 1935 took over California state camps at Marysville and Arvin, thus beginning a camp project that by 1941 expanded into fifty-three camps in eleven states from California to Florida, Washington to Texas.[23] The federal migrant-camp program ran for a number of years, providing shelter and other assistance until the economic expansion as part of the effort of World War Two reached the point where the excess labor pool that had been working as farm laborers was largely absorbed by the expanding wartime economy.[24]

More contentious than this outline is its interpretation. One view sees Dust-Bowl migrants as unable to help or improve themselves and as paradigmatic victims of forces beyond human control. Paired with this view is an interpretation of the situation as the rescue and social reintegration of people undeservedly set adrift by inhuman forces. A second, very different view emphasizes migrants as gritty, determined embodiments of the power of the American Dream, thus demonstrating individual, personal resilience, and depth of commitment to improving one's life and chances for the future.[25]

Despite differences, what all the interpretations share is a basis outside direct knowledge. Just as folk-singer Woody Guthrie's often strident criticism of capitalism is excised from these popular accounts, and just as the cheerful, plucky ending of the popular feature film "The Grapes of Wrath" belies the quite different conclusion of the novel on which it was based, a broader situation exists from which these views have been selected. Restoring the breadth of the situation by addressing migrants' own definitions of self and their actions provides a way of addressing more specifically the immense complexities of struggling for direct autonomous composition as a cultural practice of democratic communications.[26]

What makes it possible to approach such an issue through this situation and in such a way is the existence of newspapers published in migrant-labor camps in the 1930s and 1940s. A pair of such newspapers titled the *Weedpatch Cultivator* (later named the *Tow-Sack Tattler*) appeared from 1938 to 1942 in the federally run Arvin Migratory Labor Camp seventeen miles southeast of Bakersfield, Calif. at the southern entrance to the San Joaquin Valley.[27] The Camp Arvin newspaper was only one of many such newspaper projects. By 1940, Farm Security Administration (F.S.A.) personnel in San Francisco noted that camp newspapers were being produced at camps in Visalia, Gridley, Indio, Brawley, Shafter, Marysville, and Yuba City in addition to Arvin.[28] Archival holdings identify newspaper projects in sixteen different camps.[29]

Studies such as Gitlin's work on media and the New Left suggest the depth of contextual analysis required to reach a corresponding depth of understanding—one that in this case also drew upon his own personal experience inside the movement he wrote about.[30] Distinguishing the project of migrant-camp newspapers from more recent and current situations is the deeply sedimented documentary record, which allows one to extend an analysis beyond the textual and to formative historical relationships that display the complexities of direct autonomous composition as collective education.

Institutions and Formations

One way to conceive direct autonomous composition is as direct access to technical means of making media products, which treats such access as indications of autonomy of control and of expression. The newspaper at Camp Arvin fits such a conception in that it was directly produced by camp residents. It began in 1938 as a double-sided single sheet, eight inches wide and fifteen inches deep, with copies produced by mimeograph. The masthead was hand-drawn and stories typewritten in vertical columns. While camp management oversaw its operation during periods between editors, camp residents were intended to fill the role of newspaper editors.[31]

However, despite this direct access, the newspaper project was many times removed from an essentialized autonomy, embedded as it was in formations that provided its resources, intention, design, rationale, and situation. Understanding the multiple levels of determination from local and immediate decisions to national and historical trends provides a clearer understanding of the contextual determination of the newspaper project.

At the most local and immediate level, camp management provided labor (with the camp secretary in the absence of the editor as the person who accepted items, transcribed items when necessary, and likely typed the newspaper), instruction about the practice (institutionalized at Camp Arvin in 1941 in the form of a Journalism Club), supplies, and means of making and distributing copies.[32] In addition, the impetus for the newspaper project came not from migrant farmworkers themselves but from F.S.A. personnel, who saw it as part of a comprehensive social-reform education program meant to rehabilitate migrants from (ostensibly) rootless wanderers to responsible, wage-earning productive members of society.[33] In 1936, as part of proposing the camp project, F.S.A. Region IX director Jonathon Garst emphasized the humanitarian aid, but also the "great opportunity for Adult Education" that these camps represented, and that this opportunity "should be seized. Lectures and demonstration classes in hygiene, baby care, cooking, preserving, dressmaking would be eagerly welcomed by the migrants."[34] The details of the program (deemed to be "absolutely standard forming" by one assessment) were improvised by Thomas Collins, the initial manager of Camp Arvin.[35] Collins had been the manager of the California state camp at nearby Marysville, a guide to author John Steinbeck through the maze of migrant culture, and the model for "Jim Rawley," the "Wheatpatch" camp manager in Steinbeck's book *The Grapes of Wrath.*[36] The adult-education program's range of projects—from clubs and activities to town meetings and sports teams—sought to functionally replicate a prototypical American small-town culture.[37]

At a more expansive level of determination, the educational project was indebted in turn to a broader conception of education as social intervention and social reform, a position that had emerged within the immense social changes of the nineteenth century of industrialization, urbanization, immigration and, as the century wore on, abolition.[38] Antebellum educator Horace Mann's project of a nationwide system of taxpayer-supported common schools as a means of social improvement embodied his "grand conception of education as the transformation of the individual's moral and intellectual character, a virtual rebirth that would be comparable in effect to a religious conversion in a revival."[39] Noninstitutionalized efforts were equally important. For example, African-American experience suggests the extent to which education as social intervention consisted of learning and sharing clandestine tactics and strategies that enslaved people employed to "gain some control over their own lives."[40] The confluence of education and social reform/intervention received perhaps its most important exponent by the end of the century in the work of John Dewey, whose combination of sociology, political theory, and education in the form of pragmatism emphasized the importance of education as a means and subset of socialization, as both "a form of community life" and "a continuing reconstruction of experience."[41]

Such a reformulation of education—from abstract book learning to a way of improving society—had its analog in adult education, which had its start as job-skills training. In the midst of an expanding and changing economy and society in the late-nineteenth and early twentieth centuries, "young working adults [. . .] flocked in astonishing numbers to proprietary schools that offered a mixture of classroom and correspondence instruction," and "that promised immediate dividends in the form of job-specific skills rather than any broad integration of culture and experience."[42] A more expansive effort was that by Booker T. Washington and his version of adult education that addressed the social and economic situation of African-American freedpeople.[43] Through efforts such as the Tuskegee Institution, Washington developed "strong occupation- and career-oriented programs [that] would give economic stability to families and communities," with these programs also seeking to promote "personal development that aided self-realization, physical health, sound morals, and spiritual values."[44] In ways similar to Washington, the settlement-house project pioneered by Jane Addams (discussed in Chapter 3) was another effort to establish adult education as a means of addressing not simply individual desires for a better job but widespread social problems.[45]

Such an expansive practice of adult education became only more crucial by the time of the Great Depression. Events such as marches by Army veterans on Washington to demand immediate payment of war bonuses, highway blockades by farmers in the Midwest to prevent the shipment of food and thus help boost commodity prices, forced evictions that were blocked by black renters in Harlem, and marches by thousands of unemployed workers on a Ford plant in Michigan to demand jobs and relief underscored the extent of discontent. By 1933, close to one-quarter of the nation's workforce was unemployed.[46] As a response to this desperate situation, education became not only a form of social intervention, but a means of preventing greater social unrest. As Grant paraphrases, reformers of the day felt that "the unemployed and underemployed had to be reoriented, retrained, and reinvigorated through jobs and vocational and citizenship training before they became radicalized, and therefore disruptive."[47] Such a tradition and practice established the foundation for the migrant-labor-camp education project's design, and it also provided the basis for justifying it as of crucial importance not only to migrants but to all of society.[48]

While the political importance of adult education provided the deeper rationale, the migrant education project also sought more immediate political benefits as a means of deflecting objections both from state and national critics, who opposed the camp project for its ostensible harboring of radicals.[49] F.S.A. directors therefore spent a good deal of resources to document activities and to build public approval and political support for the camp program as humanitarian aid and social reform, often using the newspapers themselves as evidence of its success in integrating migrants into middle-class American society.[50]

Yet, education as social intervention was as much a means of building solidarity and resistance as it was a means of social integration. What Denning refers to as the Cultural Front of the 1930s sought to use the existing and immense productive ability of capitalist, commercial entertainment as well as smaller-scale noncommercial efforts to consolidate a broad-based social and educational movement.[51] Other educational efforts included the Social Reconstructionist Movement, which "was among the largest radical dissenting educational movements in the twentieth century," and the proto-Pan-Africanist pedagogical practice of W.E.B. DuBois, who "upheld the notion that schooling was both personally and socially emancipatory."[52]

Not only did the F.S.A. see its education program as a key social intervention, labor organizations saw their own efforts at organizing the diffuse difficulty agricultural workforce in similar educational terms.[53] While earlier unions included "Japanese and Mexican sugar beet workers around Oxnard in 1903" and a Mexican labor movement in California agriculture from the late 1920s to the mid-1930s, institutions of organized labor felt in the 1930s that the potential for organizing had reached a crucial point.[54] Although migrant laborers had worked California fields since the later 1800s, attempts to organize them faced difficulties ranging from practical problems from mobility to deeper differences of race and nationality. As Shindo notes, foreign-born migratory farm laborers from China, Japan, India, the Philippines, and Mexico had long been a part of the Californian agriculture industry, and had endured for some time the conditions that had only in the late 1930s elicited general outrage. A key difference, of course, was that the generic "migrants" of the 1930s referred to a specific segment that were Anglo, Christian, and born in the United States.[55]

Organized labor in the form of the Congress of Industrial Organization (C.I.O.), under the leadership of John L. Lewis and his seeming tolerance of Communist Party of the United States of America (C.P.U.S.A.) members and activities, renewed earlier efforts by others to organize farmworkers and began to commit resources through the C.I.O.-affiliated United Cannery, Agricultural, Packing and Allied Workers of America (U.C.A.P.A.W.A.).[56] Kern County, in which Camp Arvin was located, constituted at the time what Weber calls "a relatively hospitable atmosphere for Anglo organizing." Remnants of earlier labor, populist, and socialist movements persisted, as did the Communist Party (which "found enough members there to become the strongest branch in the

[San Joaquin] Valley"), and the Socialist Party.[57] The decade of the 1930s was a time of the largest series of agricultural strikes in U.S. history, all occurring in California.[58] Labor education and organizing in migrant-labor camps was possible largely because of the sympathy most F.S.A. personnel (especially those in the field) had for the goals of organized labor. Many F.S.A. workers, including the camp managers (mostly liberals, some socialists) personally supported efforts by the migrants to organize.[59] At least five camps (Arvin, Gridley, Marysville, Shafter, and Visalia) had active locals of the U.C.A.P.A.W.A., the Worker's Alliance of America (a national pressure group aligned with the U.C.A.P.A.W.A.), or both, and during strikes the U.C.A.P.A.W.A. used several F.S.A. camps as strike headquarters without interference from government employees.[60]

When restoring the context of the camp educational project, it becomes clear that the depth and degree of determination it exerted establishes that migrant-labor involvement in the camp newspaper was not the result of migrants' personal initiative or individual needs or wants, but the outcome of a long series of deep traditions, developments, and decisions external to them, thus also seeming to undercut any claims that the newspaper embodied direct autonomous composition.[61]

Cultural Forms of Association

However, the implicit conceptual framework that informs polarizing questions (did the newspaper exemplify direct autonomous composition or not?) needs to be itself critically investigated. The normative yardstick that informs the question—the desirability if not the possibility of a migrant culture exercising direct autonomous composition—posits an essentialized migrant culture that practices such an agency. But, as further contextualization shows, the migrant formation was not only fluid, variable, and contradictory, it also worked through and with many formations, the recognition of which calls into question the degree to which migrants can even be isolated as essential and thus turned into a self-directed and unified agency in the first place.[62]

What is referred to as the Anglo migrant farmworker culture of the 1930s consisted of a complex and contradictory range of anti-monopoly and citizen-producer ideas and practices, agrarian and working-class radicalism, evangelism, and nationalist and sometimes sexist and racist attempts to preserve the country's white male Protestant dominance.[63] Migrants professed patriotic pride as an American (inferring an acceptance if not admiration for things as they were) but also advocated radical change.[64] "If they cut that relief off in California they will have a revolution in California. They'll [migrants will] fight fer it, they always have. [. . .] And by-god I ain't no Communist, but I may sound like one though."[65] Not only was the formation itself striated by other formations, its complexities were subject to other kinds of determinations, such as generational. A 1936 interview of a migrant farmworker noted substantial disagreement within his family about the best action to take in the situation, indicating differing generational, gender and political alignments. "His father: 'There kaint [*sic*] be any recovery until the workingman gets paid enough so he can buy what there is to sell.' His mother: 'It's been worser [*sic*] than this in Oklahoma. There's been times when we'd been glad to work for ten cents a day.' His brother: 'Blast their God damn fields with dynamite.'"[66] What made this situation particularly complex was the fact that points of agreement were also polarizing differences. Migrants chafed within the patriarchal F.S.A. educational program, which not only devalued their ability to make their own choices but also addressed symptoms rather than causes of the migrants' plight.[67] And, despite the Popular Front strategy of the C.P.U.S.A. which called for collaboration with trade unions, the U.C.A.P.A.W.A.'s revolutionary rhetoric offended many migrants' sense of patriotic allegiance to the United States while confirming their equally deep fear of "creeping" communism.[68]

Thus, just as the cultural work of the newspaper project was far from autonomous (embedded as it was in formational and historical relationships), claims of unified agency also cannot be substantiated. As a result, to the degree that all cultural work is similarly embedded and determined, and to the degree that all claims of unified agency lack a basis, the possibility of practicing direct autonomous composition (and, to the degree it relies on such a premise, of democratic communications worthy of the name) of the kind Williams and others advocate seems to be emptied of its promise if not also its practicality. At this point, the possibilities of democratic communication as direct autonomous composition appear bleak indeed.

A way through this conceptual roadblock is to historicize direct autonomous composition rather than essentialize it. Instead of aspiring to escape context, history, and thus human existence, direct autonomous composition can be seen as social and historical—as creative, open practice within specific conditions and their determinations. Such a formulation recalls a materialist conception of people making their own history in conditions not of their own making. In doing so, it also redefines the practice of democratic communications not as requiring an escape from the social process, but as a struggle to remake the social process with means at hand as well as with means that may yet be newly formed from those at hand.

Direct autonomous composition in the way conceived here is not solely technical, but in its shared nature more broadly social, thus emphasizing the need to inform social and historical analysis with that broadly literary and artistic forms and styles—the combination what Williams calls the analysis of "a special kind of material practice: that of human sociality."[69] Thus, the activity of the newspaper project and the cultural forms that work through it become crucial as evidence and means of such a struggle, not simply as the transparent expression of a unified culture outside history nor simply as evidence of their subjugation to institutional imperatives. The variety of cultural forms embodied by the newspaper's letters, recipes, opinions, notices, reminiscences, jokes, gossip, and stories thus helps constitute the formation itself in a struggle for the communal as a process instead of as an ideal state that never can be reached.

Furthermore, this continual activity is best seen as educational in the long-standing sense of embodying efforts at social reform, improvement, and change. But, compared with the institutional projects of in this case the F.S.A. and organized labor, this effort can be more broadly seen as the social struggle to renew existing and formulate new means of sociality. Editorial requests for contributions recognized the reciprocity in the experience and representation of camp residents. "If you've been moved either to laugh or cry by something that's happened to you or around you, it's pretty certain that some of your neighbors would be moved in the same way if they saw the story in print."[70] An analysis of the camp-newspaper project as efforts to produce new means of sociality seeks to understand the social and material form of this reciprocity.

For its part, camp management sought to build its version of sociality through the form and obligations of mutual help. Official notices and reports in the newspaper invited migrant farmworkers' solidarity by embodying, demonstrating, and practicing mutual help. Eyewitness accounts of work conditions, and personal evaluations of whether opportunities were better packing lemons or picking prunes or peaches in nearby areas were offered as advice to readers, thus demonstrating not only what other camp residents did for readers but also what readers could do in return for others.[71]

Yet, limits existed to the resonance of appeals of mutual help in the camp context. The confluence between camp residents and F.S.A. camp management was at best a provisional rapprochement. When appeals to mutual help fell short, rational arguments called for camp residents to exercise middle-class individual responsibility. Reports of labor conditions in the area embodied the implicit expectation that camp residents were not to pick and choose, but to pursue all legitimate opportunities for working. The camp manager explained the basis of decisions to underscore the need to abide by them ("there are many families now on W.P.A. which we hate to see leave Camp. Nevertheless, rules are rules").[72] Notices in later issues urge parents to mind their children, to clean up trash on lots, to visit the camp clinic for smallpox and diphtheria inoculations, and to have visitors sign in with the camp manager upon arrival.[73] On behalf of the camp management and its education program, the newspaper also reprinted meeting schedules of various groups and services that included at various times sports, church meetings, a mothers' club, the Well-Baby Clinic, Sunday School, the sewing room, the Library, the Agriculture Workers Health and Medical Association clinic, a girls' club, a boys' club called "Young Pioneers," a Drama Club, a Craft Shop, and a Good Neighbors' Club. The newspaper also contained minutes from meetings, as well as the occasional overtly pedagogical item, such as one that outlined a proper nutritional diet and another that instructed migrants about personal hygiene.[74] All such appeals to individual responsibility were in effect assertions of unquestionable authority rather than a means of egalitarian collaboration.

Similar to camp management, labor organizing embraced the centrality of education as a means of solidarity and social change and, due to its being institutionally outside the migrant-labor camp project, had a greater potential of overcoming the divisive authority lurking within recommendations from camp management. Yet, many labor-aligned items were offered in forms just as alienating as those by camp management, despite being penned by migrant-farmworker writers such as U.C.A.P.A.W.A. organizer Sam Birkhimer who edited the paper during the 1939 strike season.[75] In addition to explanatory essays about the purposes of the U.C.A.P.A.W.A. and the W.A. of A., Birkhimer penned accounts of how to organize as well as pep talks to maintain flagging interest and support.[76] Didactic poems and song lyrics urged migrant laborers to "get off the row" and join the C.I.O.[77] Reprinted lyrics to songs sung on the picket lines as well as those penned by Woody Guthrie appeared often to assuage the fears of those yet to join as well as to teach tactics to those who had.[78] One example consisted of the head and shoulders of Woody Guthrie, with a caption: "The Dust Bowl Kid says: Prices is *High* wages *Low* / A man that would pick / 80¢ cotton is a slave / and nothing more! — Woody [emphasis original]."[79]

By contrast to overtly pedagogical or assertive forms by camp management and organized labor, camp residents engaged in other, much broader and open forms of sociality. For example, while concerned with the labor situation, many items offered personal positions that relied on personal testimony. By refraining from asserting required actions and offering instead personal views, they created room for loyal disagreement. One migrant farmworker simply stated that he and his family would not return "to the cotton fields until this strike is over."[80] Another concluded that, "as far as I'm concerned, the work is about over in the cotton fields. This will just about finish the fruit and cotton work for 1938. We hope next year will be different."[81]

By contrast to institutional efforts, such forms based their communality on the provisional and individual. Plaintive, aphoristic verse rehearsed well-known and broadly acknowledged lessons in order to renew their currency. The good-natured unpretentiousness of many of these verses encouraged an approachability (that one should not be prejudged when meeting another) while it depoliticized migrants' position in suggesting the legitimacy of existing society and one's place in it: "If you're just a little tadpole / Don't try to be a frog / If you're just a tail / Don't try to wag the dog."[82] However, such a stance often became a fatalism underscored by religious resignation, such as in a poem that concluded "It is not for us to understand / Just leave it all in jessus [*sic*] hand."[83] Reluctance to appear "uppity" by telling others what to do informs a poem that poked fun at "grumblers," who complained about everything. The advice given to people who were confronted with grumblers was to "turn a deaf ear, and pretend you can't hear."[84]

In these and other ways, migrant efforts at sociality operated at a level much more fundamental than the overtly pedagogical or assertive. In contrast, primary forms of autonomous association outside institutional imperatives included gossip, jokes, and teasing as tentative means of a noncommittal deepening of social relationships. Some jokes displayed a self-conscious literariness and thus were likely appropriated from generally available popular reading material, with specific campers' names replacing those in the original. Allusions to quinine and strychnine, to men being more "zoological" or "as transparent as cellophane, and as hard to remove from your mind, once you get caught up in them" pose in structure and diction a middle-class genteelism that in many was outside migrant farmworker everyday experience.[85]

In contrast to genteel, bookish middle-class jokes that typically were built around a humorous surprise generated through recontextualization, attempts at humor in more indigenous jokes relied on exaggeration. The volume and variety of what were typically very short items suggests the extent to which the form was widely available as a means of cultural production in the new, uncertain context of the camp, surrounded as people were by others unknown but yet intimately known by virtue of their common experience. "Andy chased his catfish over in Red River. The fish pushed up all the water ahead of him, and he kicked up so much dust behind, that Andy couldn't see him."[86] "John Durham said he had counted his weeds this week, but thought there was still as many as there were, and thought Carter's weeds were coming over to visit him."[87] Such indigenous jokes were also present in verse, as in an example by a poem written by camp resident 13-year-old George Tapp titled "The Skinny Man": "For exercise he jumped through a hole in a nutmeg grater / He's so skinny he looks like a sweet potato / [. . .] And when he died, he died an awful death / He fell through a hole in the seat of his britches / And choked hisself [*sic*] to death."[88] Anonymous teasing was a means of playful public recognition and social acknowledgment. "The other day out in the Cotton Patch I saw Andy Mathis holding a girl's hand. Oh! Boy if only his wife had seen him."[89] "Heyden Davis has a new girl friend in Camp. Wonder if she could be Mary Taylor?"[90] "Has anybody noticed Arley Terry lately. The reason he looks so sad, is that he lost his best girl friend. She don't like cotton pickers."[91] "Wonder if Bertie and Gladys are looking for a house? They sure are doing lots of walking lately. They say they are just walking around and looking at the trees, but we all think it might be the boys."[92] "Ask Wayne Grimmett when he comes up this Saturday who his new girlfriend is."[93] "Why does it make Mrs. Ona Moss mad to call her slim, ask her why."[94]

When compared to explanations of principles of organized labor and calls to support strikes, the importance of jokes, gossip, and teasing can seem slight indeed. From within the imperative of teaching migrants to build a community through forms of mutual help, individual obligation, and solidarity, the casual form in which such association took place did not correspond at all with the weighty civic lessons to be learned. However, in the context of the migrant camp, such activity was a leading edge of a fluid (instead of orthodox), general (instead of elite-driven), everyday (instead of special) social project of association.[95] While the F.S.A. and labor efforts regarded effectiveness of their educational efforts in vanguard terms (the extent to which migrants accepted the lessons put to them), colloquial associational activity by migrants suggests a redefinition of effectiveness as the cultivation of resonant, generally available forms of association, with its recognition and extension an educational process not in the sense of learning new information, but of increasingly recognizing resources at hand—cultural (jokes, gossip, teasing) as well as technical (the camp newspaper)—and putting them to use as best as one can.

Despite this many faceted, intensive effort, the institutional realization of migrant self-determination never materialized. Upon the end of the 1939 growing season, the onset of World War Two, the collapse of the labor-organizing effort, the wartime improvement in the nation's economy (which meant large numbers of new war-related jobs for unskilled workers in southern California), and increased nationalism all undercut oppositional positions, with the institutional milieu changing substantially.[96] By 1941, the newspaper displayed little of the freewheeling, broad-based association in earlier issues and none of the labor consciousness. Although a group of camp residents was credited with putting out the camp newspaper, and a column of "Camp Chatter" emerged (later to become titled and personalized as "Snoopy Sue; She Knows All and Tells Everything") to try to resuscitate gossip and teasing, the newspaper project embodied to its end an aloof, professional form and style suitable for an aspiring "proper" newspaper, even to the extent of carrying columns of hand-drawn advertising for local commercial businesses.

Overall, however, the migrant-camp newspaper project of 1930s and 1940s California vividly illustrates how a cultural and historical mode of analysis helps overcome theoretical dead ends in the conceptualization and practice of democratic communications as direct autonomous composition. Instead of evaluating specific projects by using a definition of autonomy that aspires to a state as far outside human history and experience as it is unattainable, thus dooming it to well-intentioned but unattainable dream, a cultural and historical conception of autonomy regrounds the practice within human history, experience, and capability. Direct autonomous composition describes not simply the availability of technical means of mechanical reproduction, but of the deployment of cultural forms of association as an always available and ever-present capacity, changing as contexts change. And, as a social and historical process of recognition, realization, extension, and adaptation to specific conditions, direct autonomous composition is best understood as the kind of colloquial self-education that fits well within a participatory democratic practice.

Media Literacy

The contradictions between the professed intention of aiding the development of democratic self-determination and the patrician practice of the F.S.A. and organized-labor educational projects of the 1930s persist in much institutionalized educational practice today, with similar limits and pressures exerted on democratic communications. Patrician practice today consists of mastering the arcane techniques of textual analysis, thus reducing direct autonomous composition to learning a set of standard techniques and proper responses instead of enabling transformative, exploratory, and participatory cultural work.

The retreat of this kind of media-literacy training to within the walls of classrooms is all the more surprising, considering the wellspring to which the project of media literacy in the West is routinely traced. Among those cited as direct inspirations is Brazilian educator, philosopher, and activist Paulo Freire—demonstrating here as in the case in Chapter 6 about *OhmyNews* the extent to which critical conceptions and practices of democratic communications current in the United States emerged outside the Anglo-American context.[97] The case of Freire also makes clear the degree to which his project of critical literacy engaged issues of skill but only as they in turn engaged concrete and specific situations and conditions. Working from 1946 to 1964 in adult literacy education among poor rural Brazilians, Freire's project developed not a set of unchanging, abstract, universal techniques meant to be used anywhere, but was instead a response to specific conditions in Brazil, immersed as it was in uneven and polarizing processes of modernization and an increasingly authoritarian political establishment.[98] As Giroux puts it, Freire's theory of education was hardly a set of classroom techniques, but instead a practical engagement with colonial power in the world, a practice that "takes seriously the relationship between radical critical theory and the imperatives of radical commitment and struggle," what Freire referred to as "the relationship between social class and knowledge."[99] Mayo concurs by emphasizing the way in which "the issue concerning the very poor distribution of resources in Brazil and the rest of Latin America rendered discussions regarding social class an important feature of such works [by Freire] as *Pedagogy of the Oppressed*."[100]

Given this formative source, it is all the more surprising that the teaching of media literacy it is said to inspire retreats from this breadth of engagement into the enclave of the classroom. What enables such a retreat is its rendition in the form of detailed textual analysis and a too-often vanguard-directed ideology critique. A handbook directed at teachers of kindergarten-to-twelfth-grade students displays such a view, citing Freire as laying the groundwork but regarding "critical literacy [. . .] as the active and often resistant engagement with texts."[101] From within this basis, media literacy means being able to "decode, evaluate, analyze, and produce both print and electronic media."[102] Another influential proponent sees media literacy as entirely a technical skill that is "primarily investigative. It does not seek to impose specific cultural values."[103] The many notions of media literacy sketched by Bazalgette are all skills, ranging from mastery of semiotic interpretive techniques to the ability to produce a "film, video, audio tape, or photo sequence"; the ability to deploy a range of critical social theories in relation to media; historical knowledge of media forms; in-depth knowledge of specific texts and "their aesthetic, dramatic, or moral values"; knowledge of the industrial and economic structures of media industries; and an awareness of the functions of media texts and of stereotypes and bias.[104] Such a practice essentializes and homogenizes "the media," foregrounds textual attention often at the price of contextual and historical understanding, while it also places the practice within institutionalized and thus containable formal education. Such a reduction also makes possible a practice not only vanguard, but orthodox. As one recent item suggests, and despite differences between media-literacy programs, they are said to "have one objective in common, and that is to foster the *right* critical stance toward the media [emphasis added]."[105]

Kellner and Share characterize the core of this dominant conception of media literacy as a set of neutral interpretive skills, validating Williams's observation noted earlier about the limitations of such an approach.[106] The concepts of "non-transparency" (that all media messages are constructed), of "codes and conventions" (that "media messages are constructed using a language with its own rules"), of "audience decoding" (that "different people experience the same media message differently"), and "content and message" (that "media have embedded values and points of view") all derive from a semiotic-structuralist tradition of ideological critique. By contrast, only one aspect of this core (that "media are organized gain profit and/or power") poses questions about intention and purpose, and the homogenization of the claim (citing "the media" as a monolith) is striking.[107]

The main difficulty with a textualist, skills-based conception of media literacy is that it is uncomfortably close to attempts to create a clerisy of specially enlightened, discerning viewers. Seen in this light, the project shares much with early twentieth-century Arnoldian-Leavisite attempts to inoculate people against the deleterious effects of degraded mass culture by developing in them special powers of discernment and insight.[108] Indeed, in their review essay, Kellner and Share identify and characterize iterations of media-literacy practice with such goals. Whether the traditionalist-protectionist approach (one that attempts "to 'inoculate' young people against the effects of media addiction and manipulation by cultivating a taste for book literacy, high culture, and the values of truth and beauty") or a media-literacy movement that "attempts to teach students to read, analyze, and decode media texts in a fashion parallel to the advancement of print literacy," both are modeled on the Arnoldian-Leavisite project of skills of close reading and the building of discernment as a technical skill of appreciation.[109] The mediacentric solution that runs through this kind of mediacentric definition of the problem is as simple as it is simplistic: What is needed to change society is different texts with different (better) meanings. As many people have noted, such a practice not only divorces such texts, codes, and conventions from traditions and institutions that make them meaningful in the first place, in doing so they also tend to isolate and reify the psychological dimension of individual meaning-making while putting students in the quite uncollaborative position of needing to undergo such remedial treatment because the educator has already decided a priori they need it.

A starker contrast to the project Williams and likeminded educators practice and envision would be hard to find. Williams commented often on the masked elitism of such a view. McIlroy notes how Williams (who was involved for fifteen years in adult education prior to the critical and theoretical work for which he is more widely known) came to see the "failures in communication" typical of most kinds of education "as characteristic of industrial society itself."[110] Williams made such a point in the late 1950s. "The failure is due to an arrogant preoccupation with transmission which rests on the assumption that the common answers have been found and need only to be applied. But people will (damn them do you say?) learn only by experience and this normally is uneven and slow."[111]

By contrast, the steady commitment to democratic forms of education as collective, collaborative discovery can be seen in an address Williams gave in 1986 regarding the future of cultural studies. Referring to the Open University, and despite its seeking to make education more accessible (and to a good degree accomplishing this), Williams noted that

> it lacks to this day that crucial process of interchange and encounter between the people offering the intellectual disciplines and those using them, who have far more than a right to be tested to see if they are following them or if they are being put in a form which is convenient—when in fact they have this more basic right to define the questions. These people were, after all, in a practical position to say, "well, if you tell me that question goes outside your discipline, then bring me someone whose discipline *will* cover it, or bloody well get outside of the discipline and answer it yourself [emphasis original]."[112]

As part of their critical discussion of conceptions of media literacy, Kellner and Share identify directions that diverge from textualist/protectionist versions and that display instead a much greater affinity to conceptions shared by Williams and others. What they call "critical media literacy" draws from textual criticism (as does Williams's own critical practice), but does not stop there. Instead, it acknowledges the "importance of learning to use the media" as means of expression and more broadly as a means of collaboration, thus moving more fully into the direction of the normative goal of direct autonomous composition.[113] The call by Kellner and Share is at its most compelling when it advocates "a comprehensive approach that would teach critical skills and how to use media as instruments of social communication and change," to "help invigorate democratic debate and participation" and, most directly, when encouraging a practice that "enables people to create their own meanings and identities and to shape and transform the material and social conditions of their culture and society."[114] Such goals require not only technical skills, but searching, collaborative, and creative analysis of exactly what kinds of social change and specifically what kinds of transformations are worth struggling for in the first place.

The degree to which an open, enabling approach ill-fits current conceptions of media education is clear in a number of ways and instances. A pedagogy of knowledge transmission informs efforts such as "The Citizen's Guide to the Airwaves," a poster and 52-page report that documents the value of the electromagnetic spectrum in order to establish the necessity for the need to lease it to media companies instead of give it away. The effort at regulatory reform is admirable, but the unquestioned assumption of the perpetuation of a market system severely limits the range of change proposed, while the form of the effort itself does little to build open, collaborative participation.[115] A set of twelve ideas for lessons that incorporate material generated through Project Censored includes classroom exercises for recognizing the degree to which news is selected, understanding industry market concentration, and reflecting on students' own usage of news reports. While limiting itself to news (which ignores the immense range of other kinds of communications practices that students engage in), a proposal for critical viewing of a news program asks teachers to have students "apply A.C.M.E.'s 'Questioning Media' principles" instead of suggesting that students develop their own or develop an entirely different approach to the topic and issue.[116] Its welcome proposals for engaging in composition unfortunately limit such engagement to writing letters to the local newspaper, or producing a radio or a video news story that a local radio or television station might be able to use, thus ignoring drawbacks of aspirations to meet the expectations of professional forms rather than enhancing the means of exploring new ones.[117]

Other efforts that seek to dispense with the dominant models of media literacy focus valuably on composition, but in a limited sense as the individual acquisition of technical skills, thus divorcing intention from the practice of composition. A particularly notable effort is a free guide intended primarily for community groups in England that wish to create and produce videos.[118] While enabling in a number of ways, what fills the normative vacuum created by isolating skills from intentions are the expectations of professionalism in ways that citizen-journalism projects already struggle with (and as discussed in Chapter 6).

What anchors to the limitations of these models of textualist, skills-based, and mediacentric pedagogy is the often implicit and unquestioning acceptance of a market system of communications. This can be seen in a variety of efforts outside the United States, such as the Media Literacy component of the Audiovisual and Media Policies group of the European Commission, which in its emphasis on accessing, analyzing, evaluating, and using already-existing "media on a personal basis" does not question the current system.[119] Recent speeches and announcements by U.S. senators, and testimony given at Senate hearings also relies on what Kellner and Sharp identify as the "inoculation" paradigm of media literacy: that such programs should allow the few enlightened ones to inoculate the ignorant rest from the bad effects of media.[120] A comprehensive curriculum in media literacy proposed by the Center for Media Literacy defines media literacy as providing "a framework to access, analyze, evaluate, and create messages in a variety of forms" as well as building "an understanding of the role of media in society, as well as essential skills of inquiry and self-expression necessary for citizens in a democracy." Although admirably inclusive in its way, such a definition not only focuses on a generic "media" as a pale stand-in for the immense range of specific, human practices of communications, it also relies on the "five core concepts" of the textualist dominant of media literacy identified by Kellner and Sharp while it promotes a program of skill mastery to the point of habitual acceptance, as the promotional pitch by the organization's president makes clear: "It is our dream that by the time they graduate from high school, all students will be able to apply the Five Key Questions almost without thinking."[121]

While efforts at enabling direct autonomous composition are limited by professionalism and by traditional pedagogies, other efforts continue to struggle with developing a sense of what such a process might mean and how it might be approached. Experiments in organizing communications serve well as attempts to explore their broader implications. Along with the *De-Bug* effort described at the beginning of this chapter, Independent Media Centers emerge here too as an immensely important project not simply of media production, but of the working out of practical matters of large-scale organization that does not rely on a rigidly authoritarian chain of command, in order to enable the open work of conceiving responses to neoliberalism. Garcelon identifies the cultural and technical attempt to overcome the opposition between corporate and anarchist forms in the development of peer-to-peer exchange as a distributed rather than centralized or individualized project.[122] Despite persistent difficulties and challenges, this is a significant advance toward new kinds of social organization that have the potential of supporting the kind of direct autonomous composition envisioned by Williams and others, not only in terms of media, but as a social process of communications as the collective formation of new forms of engagement and action.

Democratic Communications as Open, Social Composition

Technological achievements are not easy to accomplish, and coordinating large-scale protest at a specific time and place is even more difficult. Yet, when recognizing the necessity of building generalized struggles against such amorphous, delocalized sources of oppression variously located in capitalism, racism, and patriarchy, the need to provide a means of engaging immense, dispersed populations—not just radical elites for their rank and file—remains paramount, with the Internet and any other technical means only in service of a more fundamental conception of democratic communications not as radical content or new techniques, but as a critical pedagogy of open, social composition.

Projects that embody open, social composition have the greatest possibility not only for enabling previously unrecognized critique, but for exploring new, contextually resonant responses to ever-changing situation. They can enable the proposition, debate, practice, and revision of new kinds of collaboration and organization—an emphasis sorely lacking in institutionalized educational projects but increasingly a potential if not also and on occasion an inspired actuality. As a long-time Indymedia activist Sheri Herndon puts it, "'Ultimately, it's not enough for us to talk about what we are against. We have to articulate what we are for. It's not enough to slow the rate of destruction. We have to increase the rate of creation."[123]

Democratic communications today is at its most compelling when increasing the rate of creation. Instead of seen as an essentialized and universalized technology or product, the argument throughout this book puts forth a sense of democratic communications as the social and historical creation, adoption, and constant revision of public forms of representing, critiquing, and responding to the current crisis of neoliberalism. Writing in 1930, Brecht put well the dilemma and the ever-present potential and possibility. "The technique for all such projects has still to be developed; but it will be directed towards the prime task of ensuring that the public is not only taught but must also itself teach."[124]

Notes

1. Raj Jayadev, "Making Media Make Change with Young Temps in Silicon Valley," *Media Alliance*, 14 May 2004 <http://www.media-alliance.org/article.php?story=20040514190157379> (12 Jan. 2006).

2. Jayadev, "Making Media."

3. "About Us," *Silicon Valley De-Bug* <http://siliconvalleydebug.org> (30 Mar. 2007).

4. Jayadev, "Making Media."

5. For the philosophical bases for this position and responses to it, see J. Herbert Altschull, *From Milton to McLuhan; The Ideas Behind American Journalism* (White Plains, N.Y.: Longman, 1990).

6. A.J. Liebling, quoted in Richard Kluger, *The Paper: The Life and Death of the New York Herald Tribune* (New York: Knopf, 1986), 341.

7. John Keane, *The Media and Democracy* (Cambridge, Mass.: Basil Blackwell, 1989), 94–114.

8. Walter Lippmann, *Public Opinion* (New York: Free Press, 1965), 158, 251.

9. Lippmann, *Public Opinion*, 252–55.

10. Paul Jones, *Raymond Williams's Sociology of Culture; A Critical Reconstruction* (New York: Palgrave Macmillan, 2004), 174–75.

11. Jones, *Raymond Williams's Sociology of Culture*, 173.

12. Raymond Williams, "Means of Communication as Means of Production," in *Problems in Materialism and Culture* (London: Verso, 1983), 62.

13. Raymond Williams, *Culture and Society, 1780–1950* (New York: Columbia University Press, 1958), 327.

14. Raymond Williams, "'You're a Marxist, Aren't You?'," in *Resources of Hope; Culture, Democracy, Socialism*, ed. Robin Gale (London: Verso, 1989), 76.

15. Raymond Williams, "Base and Superstructure in Marxist Cultural Theory," in *Problems in Materialism and Culture* (London: Verso, 1983), 43. The significance of the differences between the 1973 version and the 1980 version is discussed in Jones, *Raymond Williams's Sociology of Culture*, 104.

16. Raymond Williams, *The Year 2000* (New York: Pantheon, 1983), 146.

17. Jones, *Raymond Williams's Sociology of Culture*, 157.

18. John McIlroy and Sallie Westwood, eds., *Border Country; Raymond Williams in Adult Education* (Leicester: National Institute of Adult Continuing Education, 1993), 3; and Raymond Williams, "The Future of Cultural Studies," in *The Politics of Modernism; Against the New Conformists*, ed. Tony Pinkney (London: Verso, 1989), 151–62.

19. Carey McWilliams, *Factories in the Field; The Story of Migratory Farm Labor in California* (Boston, Mass.: Little, Brown, 1939); James Gregory, *American Exodus; The Dust Bowl Migration and Okie Culture in California* (New York: Oxford University Press, 1989), 3–35; U.S., Congress, Senate, Committee on Education and Labor, *Violations of Free Speech and Rights of Labor*, reprint ed. (New York: Arno Press, 1975); Walter J. Stein, "A New Deal Experiment with Guided Democracy: The F.S.A. Migrant Camps in California," *Canadian Historical Papers 1970*, 132–46.

Photographic records include Dorothea Lange and Paul S. Taylor, *An American Exodus; A Record of Human Erosion in the Thirties*, revised edition (New Haven, Conn.: Yale University Press, 1969); Carl Fleischhauer and Beverly W. Brannan, eds., *Documenting America, 1935–1943* (Berkeley, Calif.: University of California Press, 1988); and Richard Steven Street, *Photographing Farmworkers in California* (Stanford, Calif.: Stanford University Press, 2004). A useful bibliographic reference is Penelope Dixon, *Photographers of the Farm Security Administration: An Annotated Bibliography, 1930–1980* (New York: Garland, 1983).

20. Linda C. Majka and Theo J. Majka, *Farmworkers, Agribusiness, and the State* (Philadelphia, Pa.: Temple University Press, 1982), 105–8; McWilliams, *Factories*, 7–8, 316–23; Irving Bernstein, *Turbulent Years; A History of the American Worker, 1933–1941* (Boston, Mass.: Houghton Mifflin, 1969), 147.

21. A much more controversial project was the creation of entire communities meant for long-term residence. See Paul Conkin, *Tomorrow a New World: The New Deal Community Program* (Ithaca, N.Y.: Cornell University Press, 1959).

22. Arthur Ekirch, *Ideologies and Utopias; The Impact of the New Deal on American Thought* (Chicago, Ill.: Quadrangle, 1969), 110.

23. Carl C. Taylor and Walter E. Packard, memorandum, 27 March 1936, file RF-CF-25-430, "Arvin: accounting," Coded Administrative Camp Files, 1933–45— Arvin; Farm Security Administration, San Francisco/Berkeley, Record Group 96, Records of the Farm Security Administration, National Archives–Pacific Sierra Region, San Bruno, Calif. [hereafter cited as "Arvin Camp Records, N.A.P.S.R."]; U.S. Department of Agriculture, Farm Security Administration, *Report of the Farm Security Administration, 1941* (Washington D.C.: U.S. Department of Agriculture, Farm Security Administration, 1941), 38; "Community and Family Services Activities Described in Narrative Reports from Regions VI and XII," file 934, "Jan. 1935–1939 Dec. inclusive [1]," General Correspondence, 1940–42; Farm Security Administration, San Francisco/Berkeley, Record Group 96, Records of the Farm Security Administration, National Archives–Pacific Sierra Region, San Bruno, Calif., 2 [hereafter cited as "General Correspondence, N.A.P.S.R."].

24. Gregory, *American Exodus*.

25. In addition to works already cited, see Charles J. Shindo, *Dust Bowl Migrants in the American Imagination* (Lawrence, Kan.: University Press of Kansas, 1997); and Keith Windschuttle, "Steinbeck's Myth of the Okies," *The New Criterion* 20, no. 10 (June 2002): 24–32.

26. Woody Guthrie, *Pastures of Plenty; A Self-Portrait*, ed. Dave Marsh and Harold Leventhal (New York: HarperPerennial, 1992). Novek's work addresses in a different context the political issues surrounding newspapers in prisons. See Eleanor M. Novek, "'The Devil's Bargain'; Censorship, Identity and the Promise of Empowerment in a Prison Newspaper," *Journalism: Theory, Practice, Criticism* 6, no. 1 (February 2005): 5–23; and Eleanor M. Novek, "'Heaven, Hell, and Here': Understanding the Impact of Incarceration through a Prison Newspaper," *Critical Studies in Media Communication* 22, no. 4 (October 2005): 281–301.

27. The run of the Arvin camp newspaper is in places very sparse, due to uneven publication and somewhat haphazard preservation. Largely complimentary holdings of surviving issues are held at the National Archives-Pacific Sierra Region in San Bruno, California and at the University of California at Berkeley. No end-date for the newspaper is listed in *The National Union Catalog Pre-1956 Imprints*, v. 617 (Chicago, Ill.: American Library Association, 1979), 661. The most recent issue that can be located is dated February 5, 1942.

28. Jerome Wilcox to Frederick Soule, 9 April 1940, folder 163-01 "Genl [Jan. to June 1940] [1]," box 33-C, General Correspondence, N.A.P.S.R. Reference to the spread of newspaper activity to other camps is in "Community and Family Services."

29. "Migratory Labor Camp Papers," *University of California at Berkeley Library* <http://www.lib.berkeley.edu/doemoff/govinfo/federal/gov_migrat.html> (17 Apr. 2007).

30. Todd Gitlin, *The Whole World is Watching; Mass Media in the Making and Unmaking of the New Left* (Berkeley, Calif.: University of California Press, 1980).

31. "Mrs. Greene, Editor," *Weedpatch Cultivator*, 23 December 1938, 1; Fred W. Ross, Arvin Camp Weekly Narrative Report, 28 September 1939, folder RF-CF-16-918 "Arvin; Reports prior 7-1-40," box 22, Arvin Camp Records, N.A.P.S.R.

32. "Camp to Have a Weekly Paper," *Weedpatch Cultivator*, 2 September 1938, 1; Katharine Dietz, "Organization of a Community Council," file 934 "[Jan. 1940–May 1940]," General Correspondence, N.A.P.S.R.; "Journalism Club," *Tow-Sack Tattler*, 28 February 1941, 6.

33. Taylor and Packard, memorandum; "Camp to Have a Weekly Paper," *Weed-Patch Cultivator*, 2 September 1938, 1; Stein, "New Deal Experiment," 132-46; Katharine Dietz, "Organization of a Community Council," file 934, "[Jan. 1940–May 1940]," 1, General Correspondence, N.A.P.S.R.

34. Jonathan Garst, "California Migrants' Camp; Exhibit 'A': Social and Economic Justification," file RR-CF-16 [703], "California Migrant Camps, State—California, Reg. IX, RR. Project No.," Arvin Camp Records, N.A.P.S.R.

35. After a year at Arvin, the F.S.A. put Collins to work opening many other camps in California. Eric Thomsen, "Preliminary Report on Arvin Migratory Camp," 3 August 1936, file RF-CF-16-918, "Arvin, reports prior 7–1–40," 10, Arvin Camp Records, N.A.P.S.R.; Thomas Collins, "Kern Migratory Labor Camp, Report for Week Ending March 21, 1936," file RF-CF-26-918-01, "Arvin [report] [March 1936]," 7, Arvin Camp Records, N.A.P.S.R.

36. Jackson Benson, "'To Tom, Who Lived It'; John Steinbeck and the Man from Weedpatch," and "An Afterward and an Introduction," *Journal of Modern Literature* 5, no. 2 (April 1976): 151–210; Charles Wollenberg, "Introduction," in John Steinbeck, *The Harvest Gypsies* (Berkeley, Calif.: Heyday Books, 1988); and Robert DeMott, *Working Days: The Journals of The Grapes of Wrath 1938–1941* (New York: Viking, 1989), xxvii.

37. Collins, "Kern Migratory Labor Camp, Report for Week Ending March 21, 1936," 4; Thomas Collins, "Kern Migratory Labor Camp, Report for Week Ending April 18, 1936," 7, file RF-CF-26-918-01, "Arvin [report] [April 1936]," Arvin Camp Records, N.A.P.S.R.; Arthur A. Lundin, "Observations Made During Stay at the Arvin Migratory Labor Camp, July 10 to July 17, 1936," 7, file RF-CF-25-911-01, "Arvin Migratory Labor Camp, 7–1–40 6–30–41 [blueprint]," Arvin Camp Records, N.A.P.S.R.; Thomas Collins, "Kern Migratory Labor Camp, Report for Week Ending February 22, 1936," 4–7, file RF-CF-26-918-01 [A], "Arvin Feb. 1936," Arvin Camp Records, N.A.P.S.R.

38. Useful recent, general accounts include John L. Rury, *Education and Social Change; Themes in the History of American Schooling* (Mahwah, N.J.: Lawrence Erlbaum, 2005); William J. Reese, *America's Public Schools; From the Common School to "No Child Left Behind"* (Baltimore, Md.: Johns Hopkins University Press, 2005); and William H. Jeynes, *American Educational History: Schools, Society, and the Common Good* (Thousand Oaks, Calif: Sage, 2007).

39. Horace Mann, "Prospectus of the Common-School Journal," in *Lectures and Annual Reports on Education* (Cambridge, Mass.: no named publisher, 1867), 3. Quotation from Joseph F. Kett, *The Pursuit of Knowledge Under Difficulties; From Self-Improvement to Adult Education in America, 1750–1990* (Stanford, Calif.: Stanford University Press, 1994), 71.

40. Heather Andrea Williams, *Self-Taught; African-American Education in Slavery and Freedom* (Chapel Hill, N.C.: University of North Carolina Press, 2005), 8. See also C.G. Woodson, *The Education of the Negro Prior to 1861; a History of the Education of the Colored People of the United States from the Beginning of Slavery to the Civil War* (New York: G.P. Putnam's, 1915), 307–35; Henry Allen Bullock, *A History of Negro Education in the South; From 1619 to the Present* (Cambridge, Mass.: Harvard University Press, 1967); L.H. Whiteaker, "Adult Education within the Slave Community," in *Education of the African American Adult; An Historical Overview*, ed. Harvey G. Neufeldt and Leo McGee (Westport, Conn.: Greenwood, 1990), 3–10; and Adam Fairclough, *A Class of Their Own; Black Teachers in the Segregated South* (Cambridge, Mass.: Belknap Press, 2007).

41. John Dewey, *My Pedagogical Creed* (Washington, D.C.: Progressive Education Association, 1929), 8, 12 (originally published in 1897). For a Progressive evaluation of Dewey's work, see Merle Curti, *The Social Ideas of American Educators* (New York: Scribner's, 1935), 499–541.

42. Kett, *Pursuit of Knowledge*, 224–25.

43. Huey B. Long, *Early Innovators in Adult Education* (New York: Routledge, 1991).

44. Virginia Lantz Denton, *Booker T. Washington and the Adult Education Movement* (Gainesville, Fla.: University Press of Florida, 1993), 107. See also Peter M. Ascoli, *Julius Rosenwald; The Man Who Built Sears, Roebuck and Advanced the Cause of Black Education in the American South* (Bloomington, Ind.: Indiana University Press, 2006).

45. Kett, *Pursuit of Knowledge*, 297.

46. Nancy L. Grant, "Adult Education for Blacks during the New Deal and World War II: The Federal Program," in *Education of the African American Adult; An Historical Overview*, ed. Harvey G. Neufeldt and Leo McGee (Westport, Conn.: Greewood, 1990), 212.

47. Grant, "Adult Education," 212.

48. Eric Thomsen, "Why Plan Security for the Migratory Laborer?," paper presented to the California Conference of Social Work, San Jose, 12 May 1937, National Agricultural Library, Bethesda, Md. The global case is summarized in Eric Hobsbawm, *The Age of Extremes* (New York: Pantheon, 1994), 85–108.

49. Stein, "New Deal Experiment," 133; McWilliams, *Factories*, 152–263; Majka and Majka, 74; Charles L. Todd, "Trampling Out the Vintage," *Common Sense*, July 1939, 7–8, 30.

50. Sidney Baldwin, *Poverty and Politics; The Rise and Decline of the Farm Security Administration* (Chapel Hill, N.C.: University of North Carolina Press, 1968); and Frederick Soule to Fred Ross, 5 October 1939, file 163-01, "Newspaper and Magazines, Articles and Press Releases, Jan. 1939–Dec. 1939," General Correspondence, N.A.P.S.R.

The government-funded camp newspaper was only a recent example of the long-standing government practice of self-promotion, to which substantial financial resources had long been channeled and that were at a high level in the 1930s. See James L. McCamy, *Government Publicity; Its Practice in Federal Administration* (Chicago, Ill.: University of Chicago Press, 1939). Other F.S.A. public-relations efforts had this purpose in mind. For example, Pare Lorenz' documentary about the Dust Bowl "The Plow that Broke the Plains" was financed and produced in concert with the Resettlement Administration, and was used by R.A. director Rexford Guy Tugwell to help drum up congressional support for the program. See Charles C. Alexander, *Nationalism in American Thought 1930–45* (Chicago, Ill.: Rand McNally, 1969), 91; and R. Douglas Hurt, *The Dust Bowl: An Agricultural and Social History* (Chicago, Ill.: Nelson-Hall, 1981), 61–63.

51. Michael Denning, *The Cultural Front; The Laboring of American Culture in the Twentieth Century* (New York: Verso, 1996), especially 68–74.

52. William H. Watkins, "A Marxian and Radical Reconstructionist Critique of American Education," in *Black Protest Thought and Education*, ed. William H. Watkins (New York: Peter Lang, 2005), 125–26.

53. McWilliams, *Factories*; Paul S. Taylor and Tom Vasey, "Historical Background of California Farm Labor," *Rural Sociology* 1, no. 2 (June 1936): 281–95; Paul S. Taylor and Tom Vasey, "Contemporary Background of California Farm Labor," *Rural Sociology* 1, no. 4 (December 1936): 401, 404; Alan L. Olmstead and Paul Rhode, "An Overview of California Agricultural Mechanization, 1870-1930," *Agricultural History* 62, no. 3 (Summer 1988): 86–112.

54. In addition to Majka and Majka, *Farm Workers*, see Mark Reisler, "Mexican Unionization in California Agriculture, 1927–1936," *Labor History* 14, no. 4 (Fall 1973): 562–79; U.S., Department of Labor, Bureau of Labor Statistics, *Labor Unionism in American Agriculture*, by Stuart M. Jamieson, Bulletin No. 836 (Washington: GPO, 1945; reprint New York: Arno Press, 1975); Cletus E. Daniel, *Bitter Harvest; A History of California Farmworkers, 1870–1941* (Ithaca, N.Y.: Cornell University Press, 1981), Devra Weber, *Dark Sweat, White Gold; California Farm Workers, Cotton, and the New Deal* (Berkeley, Calif.: University of California Press, 1994); Camille Guerin-Gonzales, *Mexican Workers and American Dreams; Immigration, Repatriation, and California Farm Labor, 1910–1939* (New Brunswick, N.J.: Rutgers University Press, 1994); Richard Steven Street, *Beasts of the Field; A Narrative History of California Farmworkers, 1769–1913* (Stanford, Calif.: Stanford University Press, 2004); and Richard Steven Street, "Poverty in the Valley of Plenty: The National Farm Labor Union, DiGiorgio Farms, and Suppression of Documentary Photography in California, 1947–66," *Labor History* 48, no. 1 (February 2007): 25–48.

55. Charles J. Shindo, "The Dust Bowl Myth," *Wilson Quarterly* 24, no. 4 (Autumn 2000): 25–30.

56. Majka and Majka, *Farm Workers*, 74, 85; and Vicki L. Ruiz, *Cannery Women, Cannery Lives; Mexican Women, Unionization, and the California Food Processing Industry, 1930–1950* (Albuquerque, N.M.: University of New Mexico Press, 1987). Klehr points out that, despite the important alliance of Comintern with the C.I.O. during the 1930s, its role could hardly be described as dominant or even unproblematic. See Harvey Klehr, *The Heyday of American Communism; The Depression Decade* (New York: Basic Books, 1984), 136–46, 223–51. See also Harvey A. Levenstein, *Communism, Anti-Communism, and the C.I.O.* (Westport: Greenwood, 1981.)

57. Weber, *Dark Sweat*, 153–61. In addition to McWilliams, *Factories*, Daniel, *Bitter Harvest*, and Majka and Majka, *Farm Workers*, see Dick Meister and Anne Loftus, *A Long Time Coming: The Struggle to Unionize America's Farm Workers* (New York: Macmillan, 1977).

58. Weber, *Dark Sweat*, 183; Majka and Majka, *Farm Workers*, 74, 128–29; Bernstein, *Turbulent Years*, 153.

59. Majka and Majka, *Farm Workers*, 111, 127–29; Weber, *Dark Sweat*, 164–65. Examples of privately held sentiments suppressed publicly include Collins' refusal to review McWilliams' book *Factories in the Field*. See Thomas Collins to L.W. Harvison, 15 September 1939, file 160, "Public Relations, General, Jan. 1939-Dec. 1939," General Correspondence, N.A.P.S.R. Unofficially, however, *Factories in the Field* and *The Grapes of Wrath* were highly regarded by the F.S.A. See Frederick Soule to John Fischer, 2 August 1939; and John Fischer to Frederick Soule, 1 June 1939, 8 August 1939, and 8 September 1939, File 160, "Public Relations, General, Jan. 1939–Dec. 1939," General Correspondence, N.A.P.S.R.

60. Majka and Majka, *Farm Workers*, 111, 127–29; Weber, *Dark Sweat*, 164–65.

61. Gregory, *American Exodus*, 139–71.

62. A similar point about a similar case is developed in Ranu Samantrai, *AlterNatives; Black Feminism in the Postimperial Nation* (Stanford, Calif.: Stanford University Press, 2002).

63. Gregory, *American Exodus*, 141–2. See also Weber, *Dark Sweat*, 137–51.

64. James Bright Wilson, "Social Attitudes of Certain Migratory Agricultural Workers in Kern County, California" (unpublished M.A. thesis, University of Southern California, 1942).

65. Wilson, "Social Attitudes," 277.

66. "Shafter-Wasco potatoe [*sic*] district, Kern Co. 5-6/36," folder "History of AFL Agricultural Unions," carton 6 "F.S.A.," Simon J. Lubin Society Papers, Bancroft Library, University of California, Berkeley, Calif.

67. Majka and Majka, *Farm Workers*, 130–2; Sheila Goldring Manes, "Depression Pioneers: The Conclusion of an American Odyssey; Oklahoma to California, 1930-1950, A Reinterpretation," (unpublished Ph.D. dissertation, University of California, Los Angeles, 1982), 3; and Wilson, "Social Attitudes."

68. Wilson, "Social Attitudes," 332–33; John Diggins, *The Rise and Fall of the American Left* (New York: Norton, 1992), 173–74; Levenstein, *Communism*, 36.

69. Raymond Williams, *Marxism and Literature* (Oxford: Oxford University Press, 1977), 165. A recent work that develops a valuable argument for placing such cultural work at the center of social activism and challenge is T.V. Reed, *The Art of Protest; Culture and Activism from the Civil Rights Movement to the Streets of Seattle* (Minneapolis, Minn.: University of Minnesota Press, 2005).

70. "Prize for Best Poem or Idea," *Tow-Sack Tattler*, 24 August 1939, 1.

71. "Work Conditions," *Weedpatch Cultivator*, 2 September 1938, 1.

72. "Editorial," "Camp News," "W.P.A. Workers," *Weedpatch Cultivator*, 2 September 1938, 2.

73. "Notice," *Weedpatch Cultivator*, 21 April 1939, 2.

74. "Food for the Family," *Weedpatch Cultivator*, 4 August 1939, 2; "Clinic News; Body Cleanliness," *Tow-Sack Tattler*, 17 November 1939, 8.

75. Walter Stein, *California and the Dust Bowl Migration* (Westport, Conn.: Greenwood Press, 1973), 252. Birkhimer and his family had been in California for three years, and he had started a chapter of the Workers' Alliance at Camp Indio in 1938. "Wage Hearing is Held," *Tow-Sack Tattler*, 29 September 1939, 4; "This and That," *Tow-Sack Tattler*, 6 October 1939, 9.

76. Sam Birkhimer, "Weekly Letter from the Editor," *Tow-Sack Tattler*, 13 October 1939, 4; Sam Birkhimer, "Our Strike," *Tow-Sack Tattler*, 28 October 1939, 4–5; Birkhimer, "Our Strike," 17 November 1939, 5.

77. Untitled, *Tow-Sack Tattler*, 20 October 1939, 3. Woody Guthrie, who noted that he had "made the Arvin Camp lots of times with the old trusty guitar, and listened to the Campers sing in their churches and at their dances, and pie suppers and speakins," later set this verse to music. In a published collection of songs in which it was included, Guthrie mentioned hearing "a little fourteen year old boy's poem called 'I'd Ruther To Die on My Feet than Live on My Knees . . .' Can you beat that? No, you can't. It leapt out of this boy's mind like a young mountain lion, and the road was lined with cops in their big black sedans, laughing, grunting, and talking, and a listening to jazz music on their radios." The boy—George Tapp—also authored the cited poem. See Alan Lomax, ed., *Hard-Hitting Songs for Hard-Hit People* (New York: Oak Publications, 1967), 225.

78. "Join the Union," *Tow-Sack Tattler*, 28 October 1939, 16; "'Greenback Dollar' (streamlined)," *Tow-Sack Tattler*, 11 November 1939, 4.

79. Untitled, *Tow-Sack Tattler*, 11 November 1939, 10. As Guthrie was an accomplished illustrator in the homey style of this illustration, it is likely that Guthrie drew it as well as signed it. At least one notice appeared of an impending visit of Guthrie and Hollywood actor Will Geer to the Arvin Camp. See "Woody and Gear [*sic*] to Entertain," *Tow-Sack Tattler*, 22 September 1939, 2.

80. "To Them This May Concern," *Weedpatch Cultivator*, 21 October 1938, 1.

81. "Work Conditions," *Weedpatch Cultivator*, 25 November 1938, 1.

82. "Be What You Is," *Weed Patch Cultivator*, 11 November 1938, 1. This poem appeared widely in other migrant labor camp newspapers. See Gregory, *American Exodus*, 152.

83. "Leave It In Jesus's Hand," *Weed Patch Cultivator*, 12 May 1939, 2. Early in the camp's existence, manager Tom Collins also commented in a weekly report on the religious core of migrant fatalism: "The campers 'Trust in the Lord.' That is good of course, [. . .] However we cannot encourage them to become dependent with the hope that the ravens will feed them or that Jonah will come along with his whale and swallow all their troubles." See Thomas Collins, "Kern Migratory Labor Camp, Report for week ending March 7, 1936," 7, file RF CF 26 918-01, "Arvin [Report] [March 1936]," Arvin Camp Records, N.A.P.S.R.

84. "'A Grumbler'," *Weed Patch Cultivator*, 11 November 1938, 2.

85. "Letters from Campers," Weedpatch Cultivator, 30 September 1938, 2; "By Campers," Weedpatch Cultivator, 25 November 1938, 2.

86. "From Campers," *Weedpatch Cultivator*, 28 October 1938, 2.

87. "Garden Home News," *Weedpatch Cultivator*, 11 November 1938, 1.

88. George Tapp, "The Skinny Man," *Tow-Sack Tattler*, 28 October 1939, 10.

89. "By Campers," *Weedpatch Cultivator*, 11 November 1938, 2.

90. "By Campers," *Weedpatch Cultivator*, 11 November 1938, 2.

91. "By Campers," *Weedpatch Cultivator*, 11 November 1938, 2.

92. "By Campers," *Weedpatch Cultivator*, 20 January 1939, 3.

93. "Jokes," *Tow-Sack Tattler*, 13 October 1939, 11.

94. "Fun Fun Fun," *Tow-Sack Tattler*, 20 October 1939, 10.

95. The relation of "mundane" activities to matters of democratic communications is broached in Chris Atton, "The Mundane and Its Reproduction in Alternative Media," *Journal of Mundane Behavior* 2, no. 1 (February 2001): 122–37.

96. Levenstein, *Communism* (68) notes that Lewis lent little support to the U.C.A.P.A.W.A.. In January 1938, he stopped C.I.O. aid. See also Bert Cochran, *Labor and Communism; The Conflict that Shaped American Unions* (Princeton, N.J.: Princeton University Press, 1977), 143–44. An able overview is Anthony J. Badger, *The New Deal; The Depression Years, 1933–1940* (New York: Noonday, 1989), 286–90.

97. Stephanie A. Flores-Koulish, *Teacher Education for Critical Consumption of Mass Media and Popular Culture* (New York: RoutledgeFalmer, 2005), 15.

98. Paulo Freire, "Education as the Practice of Freedom," in *Education for Critical Consciousness*, ed. and transl. Myra Bergman Ramos (New York: Seabury, 1973), 1–84; Paul V. Taylor, *The Texts of Paulo Freire* (Philadelphia, Pa.: Open University Press, 1993), 1–33; Moacir Gadotti, *Reading Paulo Freire; His Life and Work*, transl. John Milton (Albany, N.Y.: State University of New York Press, 1994), 5–35.

99. Paulo Freire, *The Politics of Education; Culture, Power, and Liberation* (South Hadley, Mass.: Bergin and Garvey, 1985), xii.

100. Peter Mayo, *Liberating Praxis; Paulo Freire's Legacy for Radical Education and Politics* (Westport, Conn.: Praeger, 2004), 6. The work referred to is Paulo Freire, *Pedagogy of the Oppressed*, transl. Myra Bergman Ramos (New York: Herder and Herder, 1970).

101. Lisa Patel Stevens and Thomas W. Bean, *Critical Literacy; Context, Research, and Practice in the K–12 Classroom* (Thousand Oaks, Calif.: Sage, 2007), 5.

102. Patricia Aufderheide, "Media Literacy: From a Report of the National Leadership Conference on Media Literacy," in *Media Literacy in the Information Age*, ed. Robert Kubey (New Brunswick, N.J.: Transaction, 1997), 79–86.

103. Len Masterman, "A Rationale for Media Education," in *Media Literacy in the Information Age*, ed. Robert Kubey (New Brunswick, N.J.: Transaction, 1997), 41.

104. Cary Bazalgette, "An Agenda for the Second Phase of Media Literacy Development," in *Media Literacy in the Information Age*, ed. Robert Kubey (New Brunswick, N.J.: Transaction, 1997), 73–74.

105. Jacques Piette and Luc Giroux, "The Theoretical Foundations of Media Education Programs," in *Media Literacy in the Information Age*, ed. Robert Kubey (New Brunswick, N.J.: Transaction, 1997), 90.

106. Williams, "Means of Communication," 62.

107. Douglas Kellner and Jeff Share, "Toward Critical Media Literacy: Core Concepts, Debates, Organizations, and Policy," *Discourse: Studies in the Cultural Politics of Education* 26, no. 3 (September 2005): 374–77. See also David Sholle and Stan Denski, *Media Education and the (Re)Production of Culture* (Westport, Conn.: Bergin and Garvey, 1994); Sut Jhally and Justin Lewis, "The Struggle for Media Literacy," *Journal of Communication* 48, no. 1 (Winter 1998): 109–20, and Sut Jhally and Jeremy Earp, "Empowering Literacy: Media Education as a Democratic Imperative," in Sut Jhally, *The Spectacle of Accumulation; Essays in Culture, Media, and Politics* (New York: Peter Lang, 2006), 239-67.

108. Jones, *Raymond Williams's Sociology of Culture*, 1–36; Richard Lee, *Life and Times of Cultural Studies; The Politics and Transformation of the Structures of Knowledge* (Durham, N.C.: Duke University Press, 2003), 35–69.

109. Kellner and Share, "Toward Critical Media Literacy," 372.

110. McIlroy and Westwood, *Border Country*, 6.

111. Williams, *Culture and Society*, 314.

112. Raymond Williams, "The Future of Cultural Studies," in *The Politics of Modernism; Against the New Conformists* (London: Verso, 1989), 157.

113. Kellner and Share, "Toward Critical Media Literacy," 372.

114. Kellner and Share, "Toward Critical Media Literacy," 373, 381.

115. *Citizen's Guide To The Airwaves; A Graphic Depiction of the Uses—and Misuses—of the Radiofrequency Spectrum* (Washington, D.C.: Spectrum Policy Program, New America Foundation, 2003); J.H. Snider, *An Explanation of the Citizen's Guide to the Airwaves* (Washington, D.C.: Spectrum Policy Program, New America Foundation, 2003).

116. "Questioning Media; Ten Basic Principles of Media Literacy Education," *Action Coalition for Media Education* <http://www.acmecoalition.org/page.cfm?ID =102> (25 Apr. 2007).

117. "Challenging Big Media News And Censorship; A.C.M.E. Activities Guide," *Project Censored* <http://www.projectcensored.org/resources/teacherguide .htm> (15 Mar. 2007).

118. "YOUR Video—A Practical Guide to Planning and Creating Effective Videos," *Video Toolkit* <http://www.ourvideo.org/toolkit/> (25 Apr. 2007).

119. "Audiovisual and Media Policies – Media Literacy | Europa," *European Commission* <http://ec.europa.eu/avpolicy/media_literacy/> (25 Apr. 2007).

120. Joseph Lieberman, "Prepared Statement of Senator Joe Lieberman, Media Research Forum," news release, 9 April 2003 <http://lieberman.senate.gov/newsroom/ release.cfm?id=207585> (25 Apr. 2007); Hillary Clinton, speech to Kaiser Family Foundation, 5 March 2005 <http://clinton.senate.gov/news/statements/details.cfm ?id=233740> (25 Apr. 2007); U.S Senate, Committee on the Judiciary, Subcommittee on the Constitution, Civil Rights and Property Rights, hearing on "Why the Government Should Care About Pornography: The State Interest in Protecting Children and Families," testimony of

Leslie Harris, 109th Cong., 1st sess., 10 November 2005 (Washington, D.C.: GPO, 2006), 28–34; U.S Senate, Committee on the Judiciary, Subcommittee on the Constitution, Civil Rights and Property Rights, hearing on "What's in a Game? Regulation of Violent Video Games and the First Amendment," testimony of David S. Bickham and Elizabeth K. Carll, 109th Cong., 2nd sess., 29 March 2006 (Washington, D.C.: GPO, 2006), 66–71, 79–90; and U.S. Senate, Committee on Health, Education, Labor, and Pensions, hearing on "Elementary and Secondary Education Act Reauthorization: Improving NCLB To Close The Achievement Gap," testimony of Reg Weaver, 13 March 2007, 109th Cong., 1st sess. <http://help.senate.gov/Hearings/2007_03_13/Weaver.pdf> (25 Apr. 2007).

121. "What Is Media Literacy?," *Center for Media Literacy* <http://www. medialit.org/gifs/MML_404.jpg> (25 Apr. 2007); "CML Media Kit," *Center for Media Literacy* <http://www.medialit.org/mlk_c.html> (25 Apr. 2007); Elizabeth Thoman and Jessica Jolls, "Literacy for the 21st Century; An Overview and Orientation Guide to Media Literacy Education," *Center for Media Literacy* <http://www.medialit.org/pdf/mlk/01_MLKorientation.pdf> (25 Apr. 2007); Jeff Share, Tessa Jolls, and Elizabeth Thoman, "Five Key Questions That Can Change the World; Classroom Activities for Media Literacy," *Center for Media Literacy* <http://www.medialit.org/pdf/mlk/02_5KQ_ClassroomGuide.pdf> (25 Apr. 2007); "Five Key Questions That Can Change the World," *Center for Media Literacy* <http://www.medialit.org/reading_room/article661 .html> (25 Apr. 2007).

122. Marc Garcelon, "The 'Indymedia' Experiment: The Internet as Movement Facilitator Against Institutional Control," *Convergence: The International Journal of Research into New Media Technologies* 12, no. 1 (February 2006): 55–82. See also Victor W. Pickard, "Assessing the Radical Democracy of Indymedia: Discursive, Technical, and Institutional Constructions," *Critical Studies in Media Communication* 23, no. 1 (March 2006): 19–38; and Victor W. Pickard, "United Yet Autonomous: Indymedia and the Struggle to Sustain a Radical Democratic Network," *Media, Culture and Society* 28, no. 3 (May 2006): 315–36.

123. "Sheri Herndon," *Olympic Movement for Justice and Peace* <http://www. omjp.org/herndon.html> (25 Apr. 2007). Quotation in Gal Beckerman, "Emerging Alternatives; Edging Away from Anarchy; Inside the Indymedia Collective, Passion vs. Pragmatism," *Columbia Journalism Review*, September/October 2003 <http://www.cjr.org/issues/2003/5/anarchy-beckerman .asp> (19 Apr. 2007).

124. Bertolt Brecht, "Radio as a Means of Communication: A Talk of the Function of Radio," in *Communication and Class Struggle*, Vol. 2: *Liberation, Socialism*, ed. Armand Mattelart and Seth Siegelaub (New York: International General, 1983), 169–71.

Afterword: Utopia and Inspiration

As this book has discussed in detail, common approaches to democratic communications emphasize technical matters such as choosing the best technology, selecting an optimal mode of organization, and finding the right content mix. The hope is that, with the right choices, the project will have the desired effect on society. However, the preoccupation with vanguard direction and instrumental effects works completely contrary to open, participatory democratic practice. Thus, the fundamental task for democratic communications is to remake it into a fully democratic practice.

One way of overcoming the contradiction between democratic intentions and too-often undemocratic practice, argues Bourdieu, is to rethink the role of vanguards in relation to social movements as well as to rethink the kinds of cultural work represented by democratic communications. He writes that scholars' (and, by extension, everyone's) objective in projects of social change should be "not only to invent responses, but to invent a way of inventing responses, to invent a new form of organization of the work of contestation and of organization of contestation, of the task of activism."[1] Put in terms more closely aligned with this book, Bourdieu places great importance on what are in practice two inseparable needs: generating creative responses to current situations, but also creatively remaking conditions that enable the generation of creative responses in the first place.

In concert with Bourdieu's proposal, this book has argued that projects of democratic communications are best seen not only as the task of conveying information, but also as the project of inventing new forms of collaborative composition, interpretation, and social action. Such a sense recasts democratic communication from being a closed, universal formula or technique to find and treasure, to being an open, ongoing, historical problem in constant need of attention. While matters of technology, organization, details of form and composition, and their implications remain important, the practice of democratic communications must also include the continual rethinking of foundational, essentialist premises, not the least of which is a belief in the self-sufficiency of matters of technology, organization, and form. Today, such a rethinking must focus on replacing elite, instrumentalist, and undemocratic forms of practice with those that are more thoroughly popular, creative, and democratic.

To conclude, I want to consider some ways of continuing and extending the critical project represented by this book. The speculative nature of what follows fits my belief that such an effort is an exercise as much of imagination as it is of theory.

To do this, I want to suggest the relevance of and thus to rehabilitate utopia and inspiration, two currently quite-unfashionable terms in social theory. While utopia has a long, acknowledged tradition in critical social thought, inspiration has an as-yet unrecognized relevance.[2] Both together suggest a way of addressing further the need for a reconstituted practice of alternative media and of democratic communication.

In view of immense and pressing problems confronted by social movements and the necessity of confronting these problems with any and all means necessary, earnest appeals for utopian-sounding projects appear as admirable in the abstract as they do unworkable in the trenches. As the thinking goes, with the house burning down, should the primary effort really be to make sure that everyone has an equal say in determining who gets to throw onto the blaze the first bucket of water?

Pressures for immediate action (damn the details) are quite real, but the formulation above relies on an assumed trade-off between, on the one hand, efficiency and results and, on the other, ideals and principles. In corresponding terms, the choice is between acting based on reality (the "real world") or on utopia (the "unreal"). Given such a choice, of course, immediate and seemingly common-sense action always makes the most sense.

While many activists in the trenches express little patience for pie-in-the-sky utopian proposals, one must still recognize (as, for example, Brecht did) that the airy abstraction of utopia is actually quite grounded, in the sense of its being relational, comparative, and historical. Items taken today as commonplace were a short time ago the stuff of fantasy (utopia?). Consider for example how cellular telephones and their global reach have emerged as disposable commodities casually holstered by grade-school children in the West, or the broad, coalitional protests against the World Trade Organization in Seattle in 1999 (and others following it) when it was thought that large-scale social protests were as passé as bell bottoms and mutton-chop sideburns.

More generally, a judgment that a project is utopian and thus dismissible asserts its unspoken opposite: a particular conception of reality. For it is only with the acceptance of present conditions as natural, stable, and unchangeable reality (despite volumes of evidence to the contrary) that makes it possible to dismiss projects as utopian in the first place.[3] As it is asserted, because this is simply the way things are, attempts at social change beyond those conceivable in the current circumstances are by definition unrealistic and therefore a waste of time.

Acceptance of the inevitability of the real not only prevents recognition of what often becomes real. It also prevents a consideration of what might yet be made the real. Rehabilitating utopia as a relevant and critical concept means regarding the real as a problem instead of as the inviolable limit of action. To be worthy of attention, proposals should not situate themselves within the real nor favor solely what is doable today. Of course, what is possible indeed depends on what is available, but only in part. Proposals for social change must also embrace the utopian critique of the real by creatively projecting—given specific and changing conditions—how a bridge might be forged between the real and what could be made real.

Instead of dismissing projects due to being utopian, just the opposite should be the case. The more unrealistic they are, the more worthy they are of sustained discussion and debate. As argued throughout this book, democratic communications as the social, historical, and creative project of making new kinds of societies is a clearly utopian conception, but one that deserves continued attention for precisely that reason. And perhaps it is not that utopian after all, considering the recent and uneven but unmistakable presence of progressive coalitions composed of feminism, peace, labor, and ecology movements and their articulation with those for racial and human global justice as a basis for a new kind of politics.[4]

The rehabilitation of the critical value of utopia suggests in turn new ways of thinking about exactly what kind of difference democratic communications should seek to bring about. As this book has argued in detail, the common answer to this question—get people to accept our ideas—is informed by behaviorism, technological determinism, and functionalism, all of which close off their explanations from history, context, and possibility of change. But, if the difference that democratic communications makes is not conceived as a media effect with all its instrumental, undemocratic implications, then how could it be differently seen?

Placed in more general terms, addressing such a question recalls the dilemma of social life. Contrary to pluralist formulations of this dilemma as how to balance the needs of society with the needs of the individual, or mechanically dialectical formulations of how to achieve a new synthesis from the clash between the social and the individual, the key task of a more humane social order is not how to balance or synthesize, but to make possible a fluid complex of irreducible but different kinds of relationships.

To address this general dilemma in terms of arguments about democratic communications, one can sketch in a provisional way here common conceptions of effects of democratic communications and where they fall along this social continuum. Each illustrates how common conceptions have yet to move past the opposition of the social and the individual and to provide a basis for mutually constituting social collaboration without enforcing uniformity.[5]

Emphasizing the goal of uniformity is the culture of truth of providentialism and its desired effect of conversion, in which unbelievers join instigators to become one body. Corporations seek similarly unifying effects as part of building brand identification and brand-loyal consumers. Still other efforts that privilege the production of unity are statist projects and their practice of agitprop, which validates a vanguard as instigator and director while seeking to form a corresponding mass of followers.

By contrast, projects that emphasize the goal of differentiation include those informed by the culture of truth of rationalist empiricism. Its rationalist inflection emphasizes the authority of self-expression, but with one's voice seen as essentially separate from others, with the most isolating forms becoming versions of self-contained, solipsistic psychotherapy. By contrast, its empiricist inflection emphasizes the authority of witnessing or of a statistical compilation, both of which in their empiricism separate the witness from the thing seen or the situation established. Finally, the critical-modernist conception of performance sees the audience as playing a constitutive instead of subservient role, with the gauge of value as artistic merit encouraging a communal sense but too-often of an isolating aesthetic appreciation.

While many variations exist, these formations and their corresponding conceptions of the effects of democratic communications make urgent a return to the original questions. To the extent that democratic communications can be seen in the affirmative and utopian sense as the collective composition of new cultural forms of democratic association, the question remains: How might one rethink the difference that democratic communications might make? How might one collaborate without conformity? And innovate without chaos?

In a way similar to recalling the critical value of utopia, one might rehabilitate inspiration by drawing upon but also distancing it from its religious and Romantic usages and traditions in order to suggest a ways of rethinking the difference that democratic communications makes.

Within the usages of "inspiration" resides a productive ambiguity rooted in its various traditions, which, like those of "culture," makes it particularly open to such a rehabilitation.[6] Early in its English-language and largely religious usage, inspiration suggests predominantly the reliance on an outside source or, in a less prominent usage, the use of an already existing individual capability.[7] Where the first usage emphasized being acted upon or infused by an outside force such as the Holy Spirit, the second emphasized acting by virtue of an inherent feeling or impulse. While each of these emphasizes either wholly external or internal sources, a third focuses on inspiration as the practical, life-enabling action of inhaling or absorbing air. Thus, in this third usage, the two oppositions (internal and external) become inseparably related without a clear hierarchy of power. Air make life possible, but it must also be actively used by living beings.

The distinction between external and internal sources of inspiration persists into the nineteenth century. By the late nineteenth-century and the elite fear of the rise of mass culture brought about in part by mass-circulation publications and the extension of literacy, the supernatural external source of inspiration became—particularly in the form of a political leader—an earthly external source who inculcates his or her views among the passive, susceptible public. Also at this time, the usage of inspiration as an internal source emerges as the inherent feeling or impulse of Romantic creativity, sometimes inspired by the artistic muse, but that also refers to the definitional capability of an individual artistic genius.

Alongside the replication of the familiar dualism (external or internal) was the persistence of the third, relational usage of inspiration, but in new forms. In some religious contexts, it became a redirection or, in a stronger sense, a transformation, one that was seen as both an outcome of an interaction with an external source, but with the result not the simple replication of a prescribed action to take, but the varied attainment of a new level of capability and purpose, with its form an open one. Like the usages that emphasize the external and internal, such a merged usage also remains common today in such ways as claims of being inspired by an act, person, or situation to take on or renew a commitment, but with its specific form open and to be determined.

Inspiration in this merged, social, and creative sense is a promising way to rethink the difference that democratic communications might make. It renders such a difference as not one of one person or group achieving control of others, nor as an isolated, individualized preference or urge, but as a collectively made and open resource of strength through which individual and collective creative action can emerge.

A way to further utopian thinking about democratic communications is to see it as the social production of inspiration. Whether this conception is realistic remains to be seen.

Notes

1. Pierre Bourdieu, "Social Scientists, Economic Science and the Social Movement," in *Acts of Resistance; Against the Tyranny of the Market*, transl. Richard Nice (New York: The New Press, 1998), 58.

2. A recent installment in the debate in critical social theory about utopia is Fredric Jameson, *Archaeologies of the Future; The Desire Called Utopia and Other Science Fictions* (London: Verso, 2005).

3. Bertolt Brecht, "Radio as a Means of Communication: A Talk of the Function of Radio," in *Communication and Class Struggle, Vol. 2: Liberation, Socialism*, ed. Armand Mattelart and Seth Siegelaub (New York: International General, 1983), 169.

4. Raymond Williams, *The Year 2000* (New York: Pantheon, 1983), 249; Van Jones, "Beyond Eco-Apartheid; Is the Green Movement Too White?," *Common Ground*, April 2007 <http://www.commongroundmag.com/2007/04/eco-apartheid0704.html> (27 Apr. 2007).

5. Recent works that focus on such complexities in varying ways include Ranu Samantrai, *AlterNatives; Black Feminism in the Postimperial Nation* (Stanford, Calif.: Stanford University Press, 2002); and John Holloway, *Change the World Without Taking Power* (London: Pluto Press, 2002).

6. Raymond Williams, *Keywords; A Vocabulary of Culture and Society* (New York: Oxford University Press, 1976), 76–82.

7. *Oxford English Dictionary*, 2nd edition, s.v. "inspiration."

Bibliography

Books and Articles

"15 Seconds." *CBS.com.* <http://www.cbs.com/originals/15_seconds/> (22 Jan. 2007).

"A Prairie Home Companion for Jan. 27, 2007 from American Public Media." *A Prairie Home Companion.* <http://prairiehome.publicradio.org/programs/ 2007/01/27/> (5 Feb. 2007).

Abelove, Henry, Betsy Blackmar, Peter Dimock, and Jonathan Schneer, eds. *Visions of History.* New York: Pantheon, 1984.

Abelson, Elaine S. *When Ladies Go A-Thieving: Middle-Class Shoplifters in the Victorian Department Store.* New York: Oxford University Press, 1989.

"About AAN." *Association of Alternative Newsweeklies.* <http://aan.org/gyrobase/Aan/ViewPage?oid=2086> (2 Dec. 2005).

"About Common Dreams." *Common Dreams.* <http://www.commondreams.org/about.htm> (19 Oct. 2006).

"About GNN." *Guerrilla News Network.* <http://www.guerrillanews.com/profile/about.html> (13 May 2002).

"About the Brave New Foundation." *Brave New Foundation.* <http://www.bravenewfoundation.org/about.php> (19 Oct. 2006).

"About the Radio Round-Table: Opinions of Radio Leaders Regarding the Past, Present and Future of Broadcasting." *Scientific American*, December 1922, 379.

"About Us." *Silicon Valley De-Bug* <http://siliconvalleydebug.org> (30 Mar. 2007).

"About Us." *Youth Radio.* <http://www.youthradio.org/index.shtml> (20 Sept. 2006).

Achinstein, Sharon. *Literature and Dissent in Milton's England.* Cambridge: Cambridge University Press, 2003.

"ACM: About the Alliance." *Alliance for Community Media.* <http://www.alliancecm.org/index.php?page_id=3> (27 Nov. 2006).

Adams, Edward E. and Gerald J. Baldasty. "Syndicated Service Dependence and a Lack of Commitment to Localism: Scripps Newspapers and Market Subordination." *Journalism and Mass Communication Quarterly* 78, no. 3 (Autumn 2001): 519–532.

Adams, Katherine. *Progressive Politics and the Training of America's Persuaders.* Mahwah, N.J.: Lawrence Erlbaum Associates, 1999.

"AdBusters : Spoof Ads." *AdBusters.* <http://adbusters.org/spoofads/> (12 Feb. 2007).

Ades, Dawn. *Photomontage.* New York: Pantheon, 1976.

"Ad-Supported Cable Scored a Milestone This 05/06 Official Season." *Cabletelevision Advertising Bureau.* <http://www.onetvworld.org/main/cab/research/2006ResearchSnapshots/adsupported-cable-scored-.shtml> (26 Mar. 2007).

"Ad-Supported Cable's Highlights & Milestones (Prime & Total Day)." *Cabletelevision Advertising Bureau.* <http://www.onetvworld.org/main/cab/research/2007Research Snapshots/adsupported-cables-highli-2.shtml> (26 Mar. 2007).

"Advertising Age's Media Family Tree; Top Properties of the Nation's 15 Largest Media Companies by Net Revenue in 2005." *Advertising Age.* <http://adage.com/images/random/mediafamilytree06.pdf> (28 Mar. 2007).

Agnew, Elizabeth N. *From Charity to Social Work; Mary E. Richmond and the Creation of an American Profession.* Urbana, Ill.: University of Illinois Press, 2004.

Ahrens, Frank. "FCC Releases Its New Media Ownership Rules." *Washington Post,* 3 July 2003, E01.

Aitken, Hugh G.J. *Scientific Management in Action: Taylorism at Watertown Arsenal, 1908–1915.* Princeton, N.J.: Princeton University Press, 1985.

———. *The Continuous Wave: Technology and American Radio, 1900–1932.* Princeton: Princeton University Press, 1985.

Alasuutari, Pertti, ed. *Rethinking the Media Audience; The New Agenda.* Thousand Oaks, Calif.: Sage, 1999.

Albert, Michael. "Mass Media: Theirs and Ours." *Z Magazine.* <http://zena.secureforum.com/Znet/ZMag/articles/oldalbert6.htm> (19 Jan. 2006).

Alexander, Charles C. *Nationalism in American Thought 1930–45.* Chicago, Ill.: Rand McNally, 1969.

Alterman, Eric. *What Liberal Media?: The Truth about Bias and the News.* New York: Basic Books, 2003.

Altschull, J. Herbert. *From Milton to McLuhan; The Ideas Behind American Journalism.* White Plains, N.Y.: Longman, 1990.

American Society of Newspaper Editors. "ASNE Statement of Principles." <http://www.asne.org/kiosk/archive/principl.htm> (14 July 2006).

Ames, William E. "Federal Patronage and the Washington D.C. Press." *Journalism Quarterly* 49, no. 1. (Spring 1972): 22–30.

———. *A History of the National Intelligencer.* Chapel Hill, N.C.: University of North Carolina Press, 1972.

Amos, William. *The Theory and Practice of the Drill Husbandry.* 1802.

Anderson, Benedict. *Imagined Communities; Reflections on the Origin and Spread of Nationalism.* Revised edition. London: Verso, 1991.

Anderson, Brian C. *South Park Conservatives: the Revolt against Liberal Media Bias.* Washington, D.C.: Regnery, 2005.

Anderson, Perry. *In the Tracks of Historical Materialism.* London: Verso, 1983.

Andreski, Stanislav, ed. and transl. *Max Weber on Capitalism, Bureaucracy, and Religion: a Selection of Texts.* Boston, Mass.: Allen & Unwin, 1983.

Anstruther, John. *Remarks on the Drill Husbandry.* London: T. Egerton, 1796.

Antonio, Robert J. and Ronald M. Glassman, eds. *A Weber-Marx Dialogue.* Lawrence, Ks.: University Press of Kansas, 1985.

Appadurai, Arjun, ed. *Globalization.* Durham, N.C.: Duke University Press, 2001.

———, ed. *Modernity at Large: Cultural Dimensions of Globalization.* Minneapolis, Minn.: University of Minnesota Press, 1996.

Armstrong, David. *A Trumpet to Arms: Alternative Media in America.* Boston, Mass.: South End Press, 1981.

Arnove, Robert F. *Philanthropy and Cultural Imperialism: The Foundations at Home and Abroad.* Boston, Mass.: G.K. Hall, 1980.

Aronowitz, Stanley. *The Death and Rebirth of American Radicalism.* New York: Routledge, 1996.

Ascoli, Peter M. *Julius Rosenwald; The Man Who Built Sears, Roebuck and Advanced the Cause of Black Education in the American South*. Bloomington, Ind.: Indiana University Press, 2006.

Ash, Eric H. *Power, Knowledge, and Expertise in Elizabethan England*. Baltimore, Md.: Johns Hopkins University Press, 2004.

Ashton, Owen, Robert Fyson, and Stephen Roberts, eds. *The Chartist Legacy*. Woodbridge, Suffolk: Merlin Press, 1999.

Associated Press. "The Associated Press Statement of News Values." *American Society of Newspaper Editors*. <http://www.asne.org/index.cfm?ID=6120> (14 July 2006).

Atton, Chris. "The Mundane and Its Reproduction in Alternative Media." *Journal of Mundane Behavior* 2, no. 1 (February 2001): 122–37.

———. *Alternative Media*. Thousand Oaks, Calif.: Sage, 2002.

———. *An Alternative Internet: Radical Media, Politics, and Creativity*. Edinburgh: Edinburgh University Press, 2004.

"Audiovisual and Media Policies—Media Literacy | Europa." *European Commission* <http://ec.europa.eu/avpolicy/media_literacy/> (25 Apr. 2007).

Aufderheide, Patricia, Erik Barnouw, Richard M. Cohen, Thomas Frank, Todd Gitlin, David Lieberman, Mark Crispin Miller, Gene Roberts, and Thomas Schatz. *Conglomerates and the Media*. New York: New Press, 1997.

———. "In the Battle for Reality: Social Documentaries in the U.S." Center for Social Media, School of Communication, American University, December 2003. <http://www.centerforsocialmedia.org/battle/Battle_for_Reality3.pdf> (29 Jan. 2004).

———. *Communications Policy and the Public Interest; The Telecommunications Act of 1996*. New York: Guilford, 1999.

Auletta, Ken. *Backstory; Inside the Business of News*. New York: Penguin Press, 2003.

Bacon, Francis. *The Philosophy of Francis Bacon*. Translated by Benjamin Farrington. Chicago, Ill.: University of Chicago Press, 1966.

Badger, Anthony J. *The New Deal; The Depression Years, 1933–1940*. New York: Noonday, 1989.

Bagdikian, Ben. *The Media Monopoly*, 6th ed. Boston, Mass.: Beacon Press, 2000.

Bakhtin, Mikhail. *Rabelais and His World*. Translated by Hélène Iswolsky. Bloomington, Ind.: Indiana University Press, 1984.

Baldasty, Gerald J. "The Press and Politics in the Age of Jackson." *Journalism Monographs*, no. 89 (August 1984).

———. *The Commercialization of News in the Nineteenth Century*. Madison, Wisc.: University of Wisconsin Press, 1992.

Baldwin, Sidney. *Poverty and Politics; The Rise and Decline of the Farm Security Administration*. Chapel Hill, N.C.: University of North Carolina Press, 1968.

Barbuto, Domenica M. *American Settlement Houses and Progressive Social Reform; An Encyclopedia of the American Settlement Movement*. Phoenix, Ariz.: Oryx Press, 1999.

Barlow, William. "Community Radio in the US: The Struggle for a Democratic Medium." *Media, Culture & Society* 10, no. 1 (January 1988): 81–105.

Barnhurst, Kevin and John Nerone. *The Form of News; A History*. New York: Guilford, 2001.

Barnouw, Erik. *A History of Broadcasting in the United States*. 3 vols. New York: Oxford University Press, 1966–70.

———. *Documentary: A History of the Non-Fiction Film*. new York: Oxford University Press, 1974.

Barrett, James R. *William Z. Foster and the Tragedies of American Radicalism*. Urbana, Ill.: University of Illinois Press, 1999.

Barsam, Richard. *Nonfiction Film: A Critical History*. Bloomington, Ind.: Indiana University Press, 1992.

Barsamian, David. *The Decline and Fall of Public Broadcasting*. Cambridge, Mass.: South End Press, 2001.

Barth, Gunther. *City People; The Rise of Modern City Culture in Nineteenth-Century America*. New York: Oxford University Press, 1980.

Barthes, Roland. *S/Z*. Translated by Richard Miller. New York: Hill and Wang, 1974.

Bauman, Zygmunt. *Modernity and Ambivalence*. Ithaca, N.Y.: Cornell University Press, 1991.

Beal, Peter. *In Praise of Scribes; Manuscripts and Their Makers in Seventeenth-Century England*. Oxford: Clarendon Press, 1998.

Beard, Charles A. *An Economic Interpretation of the Constitution of the United States*. New York: Macmillan, 1913.

———. *Economic Origins of Jeffersonian Democracy*. New York: Macmillan, 1915.

Beckerman, Gal. "Emerging Alternatives; Edging Away from Anarchy; Inside the Indymedia Collective, Passion vs. Pragmatism." *Columbia Journalism Review*, September/October 2003 <http://www.cjr.org/issues/2003/5/anarchy-beckerman.asp> (19 Apr. 2007).

Bekken, Jon. "'These Great and Dangerous Powers': Postal Censorship of the Press." *Journal of Communication* 15, no. 1 (Winter 1991): 55–71.

———. "The Working Class Press at the Turn of the Century." Pp. 151–75 in *Ruthless Criticism: New Perspectives in U.S. Communications History*, edited by William S. Solomon and Robert McChesney. Minneapolis, Minn.: University of Minnesota Press, 1993.

Benjamin, Walter. *Illuminations; Essays and Reflections*. Edited by Hannah Arendt. Translated by Harry Zohn. New York: Schocken, 1968.

Bennett, Tony. "Figuring Audiences and Readers." Pp. 145–59 in *The Audience and its Landscape*, by James Hay, Lawrence Grossberg, and Ellen Wartella. Boulder, Colo.: Westview, 1996.

———. "Theories of the Media, Theories of Society." Pp. 30–55 in *Culture, Society and the Media*, edited by Michael Gurevitch, Tony Bennett, James Curran, and Janet Woollacott. Reprint edition. London: Routledge, 1995.

Benson, Jackson. "'To Tom, Who Lived It'; John Steinbeck and the Man from Weedpatch." and "An Afterward and an Introduction." *Journal of Modern Literature* 5, no. 2 (April 1976): 151–210.

Benson, Rodney. "Commercialism and Critique; California's Alternative Weeklies." Pp. 111–127 in *Contesting Media Power: Alternative Media in a Networked World*, edited by James Curran and Nick Couldry. Lanham, Md.: Rowman and Littlefield, 2003.

Berger, John. *The Sense of Sight; Writings by John Berger*. New York: Vintage International, 1993.

Berman, Edward H. *The Influence of the Carnegie, Ford, and Rockefeller Foundations on American Foreign Policy: The Ideology of Philanthropy*. Albany, N.Y.: State University of New York Press, 1983.

Bernstein, Irving. *Turbulent Years; A History of the American Worker, 1933–1941*. Boston, Mass.: Houghton Mifflin, 1969.

Bernstein, Jay. "Critical Theory—The Very Idea (Reflections on Nihilism and Domination)." Pp. 3–38 in *The Frankfurt School: Critical Assessments*, Volume 1. Edited by Jay Bernstein. London: Routledge, 1994.

Berry, Helen. *Gender, Society and Print Culture in Late-Stuart England; The Cultural World of the* Athenian Mercury. Burlington, Vt.: Ashgate, 2003.

Bestor, Arthur. *Backwoods Utopias; The Sectarian and Owenite Phases of Communitarian Socialism in America: 1663–1829*. Philadelphia, Pa.: University of Pennsylvania Press, 1950.

"Biography of Oh Yeon Ho, CEO and Founder of OhmyNews." P. 5 in *OhmyNews; Every Citizen is a Reporter*. Seoul: Seoul Selection, 2004.

Black, Jay, ed. *Mixed News: The Public/Civic/Communitarian Journalism Debate*. Mahwah, N.J.: Lawrence Erlbaum, 1997.

Bloom, Allan. *The Closing of the American Mind*. New York: Simon and Schuster, 1987.

Blue House Ready to Open Door Wider for Media Reps." *Korea Times*, 10 January 2003, accessed from Factiva at http://global.factiva.com/> (10 Apr. 2004).

Bodnar, John. *The Transplanted; a History of Immigrants in Urban America*. Bloomington, Ind.: Indiana University Press, 1985.

Bold, Christine. *The WPA Guides: Mapping America*. Jackson, Miss.: University Press of Mississippi, 1999.

Borjesson, Kristina, ed. *Into the Buzzsaw: Leading Journalists Expose the Myth of a Free Press*. Amherst, N.Y.: Prometheus Books, 2002.

Boucheron, Pierre. "News and Music from the Air." *Scientific American*, December 1921, 104–5.

Bourdieu, Pierre. *Acts of Resistance; Against the Tyranny of the Market*. Translated by Richard Nice. New York: The New Press, 1998.

———. *Language and Symbolic Power*. Cambridge: Polity, 1991.

Bowman, Shayne and Chris Willis. "The Future Is Here, But Do News Media Companies See It?" *Nieman Reports*, Winter 2005, 6–10.

———. *We Media; How Audiences are Shaping the Future of News and Information*. The Media Center, American Press Institute, July 2003 <http://www.hypergene.net/wemedia/> (10 Mar. 2007).

Boyle, Deirdre. *Subject to Change: Guerrilla Television Revisited*. New York: Oxford University Press, 1997.

Boynton, Robert S. "How to Make a Guerrilla Documentary." *The New York Times Magazine*, 11 July 2004, 20–23.

Branner, Amy C. "Culture Work: the Production of Radical Feminism and Underground Print Media at *Off Our Backs*." Unpublished M.A. thesis, University of Missouri, 1996.

Brantlinger, Patrick. *Bread & Circuses; Theories of Mass Culture as Social Decay*. Ithaca, N.Y.: Cornell University Press, 1983.

"Brave New Films." *Brave New Films*. <http://www.bravenewfilms.org/> (10 Mar. 2007).

Braverman, Harry. *Labor and Monopoly Capital; The Degradation of Work in the Twentieth Century*. New York: Monthly Review Press, 1975.

Brecht, Bertolt. "Radio as a Means of Communication: A Talk of the Function of Radio." Pp. 169–171 in *Communication and Class Struggle*, Vol. 2: *Liberation, Socialism*, edited by Armand Mattelart and Seth Siegelaub. New York: International General, 1983.

Bremer, Francis. *The Puritan Experiment: New England Society from Bradford to Edwards*. Revised edition. Hanover, N.H.: University Press of New England, 1995.

Bremner, Robert H. *Giving; Charity and Philanthropy in History*. New Brunswick, N.J.: Transaction Publishers, 1994.

Breton, Andre. *Manifestoes of Surrealism*. Translated by Richard Seaver and Helen R. Lane. Ann Arbor, Mich.: University of Michigan Press, 1972.

Briggs, Asa. *A Social History of England*. 3rd ed. London: Penguin, 1999.

———. *The BBC: The First Fifty Years*. Oxford: Oxford University Press, 1985.

Brison, Jeffrey D. *Rockefeller, Carnegie, and Canada: American Philanthropy and the Arts and Letters in Canada*. Montréal: McGill-Queen's University Press, 2005.

Brissenden, Paul. *The I.W.W.; A Study of American Syndicalism*. New York: Columbia University Press, 1919.

Bristol, Michael. "Shamelessness in Arden: Early Modern Theater and the Obsolescence of Popular Theatricality." Pp. 279–306 in *Print, Manuscript & Performance; the Changing Relations of the Media in Early Modern England*, edited by Arthur F. Marotti and Michael D. Bristol. Columbus, Ohio: Ohio State University Press, 2000.

Brody, David. "Labor History in the 1970s: Toward a History of the American Worker." Pp. 252–269 in *The Past Before Us; Contemporary Historical Writing in the United States*, edited by Michael Kammen. Ithaca, N.Y.: Cornell University Press, 1980.

Bromley, Claude M. *Crazy Camera; Secrets of Photomontage*. New York: Focal Press, 1943.

Bronner, Simon J., ed. *Consuming Visions; Accumulation and Display of Goods in America, 1880–1920*. New York: W.W. Norton, 1989.

Brooker, Peter and Andrew Thacker, eds. *Geographies of Modernism: Literatures, Cultures, Spaces*. New York: Routledge, 2005.

Brooks, John Graham. *American Syndicalism; the I.W.W.* New York: Macmillan, 1913.

Brouwer, Daniel C. "Counterpublicity and Corporeality in HIV/AIDS Zines." *Critical Studies in Media Communication* 22, no. 5 (December 2005): 351–371.

Brown, Candy Gunther. *The Word in the World; Evangelical Writing, Publishing, and Reading in America, 1789–1880*. Chapel Hill, N.C.: University of North Carolina Press, 2004.

Brown, Lorraine, ed. *Liberty Deferred and Other Living Newspapers of the 1930s Federal Theatre Project*. Fairfax, Va.: George Mason University Press, 1989.

Brown, Merrill. "Abandoning the News." *Carnegie Reporter*, Spring 2005 <http://www.carnegie.org/reporter/10/news/> (10 Mar. 2007).

Brundage, Anthony. "Private Charity and the 1834 Poor Law." Pp. 99–119 in *With Us Always; A History of Private Charity and Public Welfare*, edited by Donald T. Critchlow and Charles H. Parker. Lanham, Md.: Rowman & Littlefield, 1998.

Bryan, Carter R. "Negro Journalism in America Before Emancipation," *Journalism Monographs* no. 12 (Sept. 1969).

Buck-Morss, Susan. *The Dialectics of Seeing; Walter Benjamin and the Arcades Project*. Cambridge, Mass.: MIT Press, 1989.

Buhle, Paul, ed. *History and the New Left: Madison, Wisconsin, 1950–1970*. Philadelphia, Pa.: Temple University Press, 1990.

———. *Marxism in the United States; Remapping the History of the American Left*. New York: Verso, 1987.

Bullert, B.J. *Public Television: Politics and the Battle over Documentary Film*. New Brunswick, N.J.: Rutgers University Press, 1997.

Bullock, Henry Allen. *A History of Negro Education in the South; From 1619 to the Present*. Cambridge, Mass.: Harvard University Press, 1967.

Bullock, Penelope. *The Afro-American Periodical Press, 1838–1909*. Baton Rouge, La.: Louisiana State University Press, 1981.

Bulmer, Martin. *The Chicago School of Sociology*. Chicago, Ill.: University of Chicago Press, 1984.

Burbach, Roger. *Globalization and Postmodern Politics: From Zapatistas to High-Tech Robber Barons*. London: Pluto Press, 2001.

Burrows, Arthur Richard. *The Story of Broadcasting*. London: Cassell and Company, 1924.

Burton, Julianne, ed. *The Social Documentary in Latin America*. Pittsburgh: University of Pittsburgh Press, 1990.

Cain, John. *The BBC: 70 Years of Broadcasting*. London: BBC, 1992.

Calhoun, Craig, ed. *Habermas and the Public Sphere*. Cambridge, Mass.: MIT Press, 1992.

Callahan, Sean, ed. *The Photographs of Margaret Bourke-White*. New York: New York Graphic Society, 1972.

Campbell, W. Joseph. *The Year that Defined American Journalism: 1897 and the Clash of Paradigms*. New York: Routledge, 2006.

Carey, James W. *Communication as Culture; Essays on Media and Society*. Boston, Mass.: Unwin Hyman, 1989.

———. "The Chicago School and the History of Mass Communication Research." Pp. 14–33 in *James Carey; A Critical Reader*, edited by Even Stryker Munson and Catherine A. Warren. Minneapolis, Minn.: University of Minnesota Press, 1997.

Carpentier, Nico. "The BBC's Video Nation as a Participatory Media Practice: Signifying Everyday Life, Cultural Diversity and Participation in an Online Community." *International Journal of Cultural Studies* 6, no. 4 (December 2003): 425–47.

Carson, Gerald. *The Old Country Store*. New York: Oxford University Press, 1954.

Carter, Susan B., Scott Sigmund Gartner, Michael R. Haines, Alan L. Olmstead, Richard Sutch, and Gavin Wright, eds. *Historical Statistics of the United States, Earliest Times to the Present: Millennial Edition*. New York: Cambridge University Press, 2006.

Carter, Susan. "A Mic of Her Own: Stations, Collectives, and Women's Access to Radio." *Journal of Radio Studies* 11, no. 2 (December 2004): 169–83.

Cassedy, James H. *Demography in Early America; Beginnings of the Statistical Mind, 1600–1800*. Cambridge, Mass.: Harvard University Press, 1969.

Castells, Manuel. *End of Millennium*. Malden, Mass.: Blackwell, 1998.

———. *The Power of Identity*. Malden, Mass.: Blackwell, 1997.

———. *The Rise of the Network Society*. Cambridge, Mass.: Blackwell, 1996.

"Challenging Big Media News And Censorship; A.C.M.E. Activities Guide." *Project Censored* <http://www.projectcensored.org/resources/teacherguide.htm> (15 Mar. 2007).

Chan, Mary. "From Essayist to Author: Roger North and Evolving Narrative Forms." *Seventeenth Century* 15, no. 2 (Autumn 2000): 266–82.

Chandler, Annmarie and Norie Newmark, eds. *At a Distance; Precursors to Art and Activism on the Internet*. Cambridge, Mass.: MIT Press, 2005.

Chartier, Roger. *The Cultural Uses of Print in Early Modern France*. Translated by Lydia G. Cochrane. Princeton: Princeton University Press, 1987.

Cheon, Yong-Cheol. "Internet Newspapers as Alternative Media: OhmyNews in South Korea." *Media Development* 51, no. 1 (2004): 28–31.

Cho, Dae-Youp. "The Generation of Social Movement in Korea: 386." *Ideology* 14, no. 3 (September 2002): 125–47.

Cho, Dal-Ho. *In the Age of US $10,000 Income, the Meaning and Task*. Seoul: Samsung Economic Research Institute, 1996.

Cho, Sung-Nam and Sook-Mi Park. "The Concept of Generation and the Social Change in Korea." *The Research of Social Science* 9, no. 1 (December 2002): 39–68.

Cirino, Robert. "An Alternative American Communications System." Pp. 568–76 in *American Media and Mass Culture; Left Perspectives*, edited by Donald Lazere. Berkeley, Calif.: University of California Press, 1987.

Claeys, Gregory. *Citizens and Saints; Politics and Anti-Politics in Early British Socialism*. Cambridge: Cambridge University Press, 1989.

———. *Machinery, Money, and the Millennium: From Moral Economy to Socialism, 1815–60*. Princeton, N.J.: Princeton University Press, 1987.

Clark, Jessica and Tracy Van Slyke. "Welcome to the Media Revolution; How Today's Media Makers are Shaping Tomorrow's News." *In These Times*, 28 June 2006. <http://www.inthesetimes.com/site/main/article/2687/> (1 Oct. 2006).

Clark, Sandra. *The Elizabethan Pamphleteers; Popular Moralistic Pamphlets 1580–1640*. Rutherford, N.J.: Fairleigh Dickinson University Press, 1983.

Clifford, Mark and Moon Ihlwan. "The Web Site that Elected a President." *Business Week*, 24 February 2003, 30.

Clinton, Hillary. Speech to Kaiser Family Foundation, 5 March 2005 <http://clinton.senate.gov/news/statements/details.cfm?id=233740> (25 Apr. 2007).

Cmiel, Kenneth. *Democratic Eloquence: the Fight Over Popular Speech in Nineteenth-Century America*. New York : W. Morrow, 1990.

"CML Media Kit." *Center for Media Literacy* <http://www.medialit.org/mlk_c.html> (25 Apr. 2007).

Coakley, Thomas F. "Preaching the Gospel by Wireless." *Catholic World*, January 1922, 517–18.

Cochran, Bert. *Labor and Communism; The Conflict that Shaped American Unions*. Princeton, N.J.: Princeton University Press, 1977.

Cohen, Barbara and Craig Kaczorowski. "Newspapers' Business Model Transformation." *Newspaper Association of America*. <http://www.naa.org/horizon/Biz_model/Biz_model_2.html> (14 Mar. 2007).

Cohen, Jeff and Norman Solomon. *Through the Media Looking Glass: Decoding Bias and Blather in the News*. Monroe, Maine: Common Courage Press, 1995.

Cohen, Lizabeth. "Encountering Mass Culture at the Grassroots: the Experience of Chicago Workers in the 1920s." *American Quarterly* 41, no. 1 (March 1989): 6–33.

———. *A Consumer's Republic: The Politics of Mass Consumption in Postwar America*. New York: Knopf, 2003.

Cohen, Mark R. *Poverty and Charity in the Jewish Community of Medieval Egypt*. Princeton, N.J.: Princeton University Press, 2005.

Cohen-Cruz, Jan, ed. *Radical Street Performance; An International Anthology*. New York: Routledge, 1998.

Colclough, David. *Freedom of Speech in Early Stuart England*. Cambridge: Cambridge University Press, 2005.

Collins, E.J.T. "The 'Machinery Question' in English Agriculture in the Nineteenth Century." Pp. 203–217 in *Research in Economic History*, Suppl. 5, Pt. A: *Agrarian Organization in the Century of Industrialization: Europe, Russia and North America*, edited by George Grantham and Carol Leonard. London: JAI Press, 1989.

Commission on Freedom of the Press. *A Free and Responsible Press; a General Report on Mass Communication: Newspapers, Radio, Motion Pictures, Magazines, and Books*. Chicago, Ill.: University of Chicago Press, 1947.

Compaine, Benjamin M. and Douglas Gomery. *Who Owns the Media? Competition and Concentration in the Mass Communications Industry.* Third edition. Mahwah, N.J.: Lawrence Erlbaum, 2000.

Conboy, Martin. *Journalism: A Critical History.* London: Sage, 2004.

Cone, Edward. "Customer-Created Content; Ready for Prime Time?" *CIO Insight*, January 2007, 21.

Conkin, Paul. *Tomorrow a New World: The New Deal Community Program.* Ithaca, N.Y.: Cornell University Press, 1959.

Conlin, Joseph, ed. *The American Radical Press 1880–1960.* 2 vols. Westport, Conn.: Greenwood Press, 1974.

Connery, Thomas B., ed. *A Sourcebook of American Literary Journalism: Representative Writers in an Emerging Genre.* New York: Greenwood, 1992.

Constantelos, Demetrios J. *Byzantine Philanthropy and Social Welfare.* Second revised edition. New Rochelle, N.Y.: Aristide D. Caratzas, 1991.

———. *Poverty, Society and Philanthropy in the Late Mediaeval Greek World.* New Rochelle, N.Y.: Aristide D. Caratzas, 1992.

Cooper, John P.D. *Propaganda and the Tudor State: Political Culture in the Westcountry.* New York: Oxford University Press, 2003.

Coote, Lesley A. *Prophecy and Public Affairs in Later Medieval England.* Rochester, N.Y.: Boydell Press, 2000.

Corns, Thomas N. and David Lowenstein, eds. *The Emergence of Quaker Writing; Dissenting Literature in Seventeenth-Century England.* London: Frank Cass, 1995.

Cottle, Simon. "Media Organization and Production: Mapping the Field." Pp. 3–20 in *Media Organization and Production*, edited by Simon Cottle. London: Sage, 2003.

Couldry, Nick and James Curran, eds. *Contesting Media Power: Alternative Media in a Networked World.* Lanham, Md.: Rowman and Littlefield, 2005.

Couldry, Nick. "Mediation and Alternative Media, or Relocating the Centre of Media and Communication Studies." *Media International Australia*, no. 103 (May 2002): 24–31.

———. *The Place of Media Power: Pilgrims and Witnesses of the Media Age.* New York: Routledge, 2000.

Covert, Catherine L. "'We May Hear Too Much': American Sensibility and the Response to Radio, 1919–1924." Pp. 199–220 in *Mass Media between the Wars: Perceptions of Cultural Tension, 1918–1941*, edited by Catherine L. Covert and John D. Stevens. Syracuse, N.Y.: Syracuse University Press, 1984.

Cowling, Paul. "An Earthly Enigma: The Role of Localism in the Political, Cultural and Economic Dimensions of Media Ownership Regulation." *Hastings Communications and Entertainment Law Journal* 27, no. 2 (Winter 2005): 257–358.

Coyer, Kate. "If It Leads It Bleeds: The Participatory Newsmaking of the Independent Media Centre." Pp. 165–78 in *Global Activism, Global Media*, edited by Wilma De Jong, Martin Shaw, and Neil Stammers. London: Pluto Press, 2005.

Coyle, Jake. "TV Networks Embrace the YouTube Model." *Associated Press*, 6 March 2007 <LexisNexis at http://web.lexis-nexis.com/> (26 Mar. 2007).

Cressy, David. "Books as Totems in Seventeenth-Century England and New England." *Journal of Library History, Philosophy and Comparative Librarianship* 21, no. 1 (Winter 1986): 92–106.

———. *Literacy and the Social Order: Reading and Writing in Tudor and Stuart England.* Cambridge: Cambridge University Press, 1980.

Critchlow, Donald T. and Charles H. Parker, eds. *With Us Always; A History of Private Charity and Public Welfare.* Lanham, Md.: Rowman & Littlefield, 1998.

Crocker, Ruth. "From Gift to Foundation: The Philanthropic Lives of Mrs. Russell Sage." Pp. 199–215 in *Charity, Philanthropy, and Civility in American History*, edited by Lawrence J. Friedman and Mark D. McGarvie. New York: Cambridge University Press, 2003.

Crossick, Geoffrey and Serge Jaumain, eds. *Cathedrals of Consumption: the European Department Store, 1850–1939*. Aldershot: Ashgate, 1999.

Crowder, George. *Classical Anarchism; The Political Thought of Godwin, Proudhon, Bakunin, and Kropotkin*. Oxford: Clarendon Press, 1991.

"C-SPAN – ABOUT US." *C-SPAN*. <http://c-span.org/about/index.asp?code=About> (31 Jan. 2007).

Curran, James and Myung-Jin Park. *De-Westernizing Media Studies*. New York: Routledge, 2000.

Curran, James. "Media and the Making of British Society, c. 1700–2000." *Media History* 8, no. 2 (December 2002): 135–154.

"Current TV // About Current." *Current TV*. <http://current.tv/about/> (24 Sept. 2005).

"Current TV // Make Video // V-CAM = Viewer Created Ad Message." *Current TV*. <http://www.current.tv/make/vc2/vcam> (21 Mar. 2007).

"Current TV // The TV Network Created by the People Who Watch It." *Current TV*. <http://www.current.tv/> (1 Mar. 2007).

Curti, Merle. *The Social Ideas of American Educators*. New York: Scribner's, 1935.

Cutting, Hunter and Makani Themba-Nixon. *Talking the Walk; A Communications Guide for Racial Justice*. Oakland, Calif.: AK Press, 2006.

Czitrom, Daniel. *Media and the American Mind; From Morse to McLuhan*. Chapel Hill, N.C.: University of North Carolina Press, 1982.

D'Arcy, Margaretta. "Playing With the Airwaves." *TDR: The Drama Review* 34, no. 4 (Winter 1990): 179–82.

———. *Galway's Pirate Women; A Global Trawl*. Galway: Women's Pirate Press, 1996.

"Dallas Cowboys Cheerleaders—Making the Commercial." *Country Music Channel*. <http:/www.makingthecommercial.cmt.com/> (16 Oct. 2006).

Daniel, Cletus E. *Bitter Harvest; A History of California Farmworkers, 1870–1941*. Ithaca, N.Y.: Cornell University Press, 1981.

Danius, Sara, Stefan Jonsso, and Gayatri Chakravorty Spivak. "An Interview with Gayatri Chakravorty Spivak." *boundary 2* 20, no. 2 (Summer, 1993), 34–44.

Dann, Martin ed. *The Black Press 1827–1890; The Quest for National Identity*. New York: G.P. Putnam, 1971.

"Dark Alliance." *The Lycaeum*. <http://www.lycaeum.org/drugwar/DARKALLIANCE/> (10 Jan. 2006).

Davidson, Philip. *Propaganda and the American Revolution, 1763–1783*. Chapel Hill, N.C.: University of North Carolina Press, 1941.

Davis, Allen F. *Spearheads for Reform: The Social Settlements and the Progressive Movement, 1890–1914*. New Brunswick, N.J.: Rutgers University Press, 1984.

Davis, Lennard. *Factual Fictions: the Origins of the English Novel*. New York: Columbia University Press, 1983.

Davis, Trevor Hugh. "Coda?" *Crisis Pictures*. <http://crisispictures.org/AFG/354/> (8 Mar. 2006).

Davison, Lee, Tim Hitchcock, Tim Keirn, and Robert B. Shoemaker, eds. *Stilling the Grumbling Hive; The Response to Social and Economic Problems in England, 1689–1750*. New York: St. Martin's Press, 1992.

de Certeau, Michel. *The Practice of Everyday Life*. Translated by Steven Rendall. Berkeley, Calif.: University of California Press, 1984.

de Forest, Lee. *Father of Radio: The Autobiography of Lee de Forest.* Chicago: Wilcox and Follett, 1950.

De Jong, Wilma, Martin Shaw, and Neil Stammers, eds. *Global Activism, Global Media.* London: Pluto Press, 2005.

De Krey, Gary S. *A Fractured Society: The Politics of London in the First Age of Party 1688–1715.* New York: Oxford University Press, 1985.

de Tocqueville, Alexis. *Democracy in America.* New York: A. Knopf, 1994.

De Vellis, Phil. "I Made the 'Vote Different' Ad." *The Huffington Post.* <http://www.huffingtonpost.com/phil-de-vellis-aka-parkridge/i-made-the-vote-differen_b_43989.html> (22 Mar. 2007).

Debord, Guy. *Society of the Spectacle.* Detroit, Mich.: Black and Red, 1983.

Della Porta, Donatello and Mario Diani. *Social Movements; An Introduction.* Second edition. Malden, Mass.: Blackwell, 2006.

DeLuca, Kevin. *Image Politics: The New Rhetoric of Environmental Activism.* New York: Guilford Press, 1999.

Demos, John. *A Little Commonwealth; Family Life in Plymouth Colony.* New York: Oxford University Press, 2000.

DeMott, Robert. *Working Days: The Journals of The Grapes of Wrath 1938–1941.* New York: Viking, 1989.

Denning, Michael. "The End of Mass Culture." *International Labor and Working Class History,* no. 37 (Spring 1990): 4–18.

———. *The Cultural Front; The Laboring of American Culture in the Twentieth Century.* New York: Verso, 1996.

Denton, Virginia Lantz. *Booker T. Washington and the Adult Education Movement.* Gainesville, Fla.: University Press of Florida, 1993.

Dery, Mark. "Brown Power." *Print,* September/October 2005, 94–101.

———. *Culture Jamming: Hacking, Slashing and Sniping in the Empire of the Signs.* Westfield, N.J.: Open Pamphlet Series, 1993.

Dettmar, Kevin J.H. and Stephen Watt, eds. *Marketing Modernisms; Self-Promotion, Canonization, Rereading.* Ann Arbor, Mich.: University of Michigan Press, 1996.

Dewey, John. *My Pedagogical Creed.* Washington, D.C.: Progressive Education Association, 1929.

———. *The Public and Its Problems.* New York: Henry Holt, 1927.

DiCenzo, Maria. "Feminist Media and History: A Response to James Curran." *Media History* 10, no. 1 (April 2004): 43–49.

Dickinson, Roger, Ramaswani Harindranath, and Olga Linné, eds. *Approaches to Audiences: a Reader.* London: Arnold, 1998.

Digby-Junger, Richard. "*The Guardian, Crisis, Messenger,* and *Negro World*: The Early-20th-Century Black Radical Press." *Howard Journal of Communications* 9, no. 3 (July–Sept. 1998): 263–82.

Diggins, John. *The Rise and Fall of the American Left.* New York: Norton, 1992.

"Distribution of Foundation Grants by Subject Categories, circa 2004." *Foundation Center.* <http://www.foundationcenter.org/findfunders/statistics/> (4 Sept. 2007).

"Distribution of Foundation Grants by Subject Categories, circa 2005." *Foundation Center.* <http://www.foundationcenter.org/findfunders/statistics/> (4 Sept. 2007).

Dixon, Penelope. *Photographers of the Farm Security Administration: An Annotated Bibliography, 1930–1980.* New York: Garland, 1983.

Douglas, Susan. *Inventing American Broadcasting, 1899–1922.* Baltimore, Md.: Johns Hopkins University Press, 1987.

Dow, Bonnie J. and Celeste M. Condit. "The State of the Art in Feminist Scholarship in Communication." *Journal of Communication* 55, no. 3 (September 2005): 448–78.

Dowie, Mark. *American Foundations; An Investigative History*. Cambridge, Mass.: MIT Press, 2001.

Downing, John D.H. "Audiences and Readers of Alternative Media; The Absent Lure of the Virtually Unknown." *Media, Culture & Society* 25, no. 5 (September 2003): 625–45.

———. "The Independent Media Center Movement and the Anarchist Socialist Tradition." Pp. 243–57 in *Contesting Media Power: Alternative Media in a Networked World*, edited by Nick Couldry and James Curran. Lanham, Md.: Rowman and Littlefield, 2003.

———. *Internationalising Media Theory: Transition, Power, Culture: Reflections on Media in Russia, Poland and Hungary, 1980–95*. London: Sage, 1996.

Downing, John D.H., Tamara Villarreal Ford, Genève Gil, and Laura Stein. *Radical Media; Rebellious Communication and Social Movements*. Thousand Oaks, Calif.: Sage, 2001.

Dowsett, H.M. *Wireless Telephony and Broadcasting*. Vol. 1. London: Gresham Publishing, 1924.

Doyle, Laura and Laura Winkiel, eds. *Geomodernisms: Race, Modernism, Modernity*. Bloomington, Ind.: Indiana University Press, 2005.

Dubofsky, Melvin. "The Rise and Fall of Revolutionary Syndicalism in the United States." Pp. 203–220 in *Revolutionary Syndicalism; An International Perspective*, edited by Marcel van der Linden and Wayne Thorpe. Aldershot: Scolar Press, 1990.

———. *We Shall Be All; A History of the Industrial Workers of the World*. Chicago: Quadrangle Books, 1969.

Dunbar, John. "FCC chairman says cities are blocking cable competition." *Associated Press Financial Wire*. 17 December 2006. Retrieved from Lexis-Nexis Academic <http://web.lexis-nexis.com/universe/> (18 Dec. 2006).

Duncombe, Stephen. *Notes from Underground; Zines and the Politics of Alternative Culture*. London: Verso, 1997.

Durston, Christopher. "Puritan Rule and the Failure of Cultural Revolution, 1645–1660." Pp. 210–33 in *The Culture of English Puritanism, 1560–1700*, edited by Christopher Durston and Jacqueline Eales. London: Macmillan, 1996.

"DVD Offer > Sierra Club Chronicles > Sierra Club." *Sierra Club*. <http://www. sierraclub.org/tv/dvdoffer/> (19 Oct. 2006).

Dyer, Carolyn Stewart. "Political Patronage of the Wisconsin Press, 1849–1861: New Perspectives on the Economics of Patronage." *Journalism Monographs* no. 109 (February 1989).

Eales, Jacqueline. "A Road to Revolution: The Continuity of Puritanism, 1559–1642." Pp. 184–209 in *The Culture of English Puritanism, 1560-1700*, edited by Christopher Durston and Jacqueline Eales. London: Macmillan, 1996.

"Echo Chamber Project." *Echo Chamber*. <http://www.echochamberproject.com/> (10 Mar. 2007).

Eckler, A. Ross. *The Bureau of the Census*. New York: Praeger, 1972.

Eco, Umberto. "Towards a Semiological Guerrilla Warfare." Pp. 135–44 in *Travels in Hyper Reality: Essays*. Edited by Helen Wolff and Kurt H. Wolff. Translated by William Weaver. London: Harcourt Brace Jovanovich, 1986.

———. *The Open Work*. Translated by Anna Cancogni. Cambridge, Mass.: Harvard University Press, 1989.

Edmondson, Richard. *Rising Up; Class Warfare in America from the Streets to the Airwaves.* San Francisco, Calif.: Librad Press, 2000.
Edwards, David and David Cromwell. *Guardians of Power: The Myth of the Liberal Media.* Ann Arbor, Mich.: Pluto Press, 2006.
Eisenstein, Elizabeth L. *The Printing Revolution in Early Modern Europe.* Reprint ed. Cambridge: Cambridge University Press, 1984.
Ekirch, Arthur. *Ideologies and Utopias; The Impact of the New Deal on American Thought.* Chicago, Ill.: Quadrangle, 1969.
Eksterowicz, Anthony J. and Robert N. Roberts, eds. *Public Journalism and Political Knowledge.* Lanham, Md.: Rowman & Littlefield, 2000.
Elam, Jr., Harry J. *Taking It to the Streets: The Social Protest Theater of Luis Valdez and Amiri Baraka.* Ann Arbor, Mich.: University of Michigan Press, 1997.
Eliasoph, Nina. "Routines and the Making of Oppositional News." *Critical Studies in Mass Communication* 5, no. 4 (December 1988): 313–34.
Endres, Kathleen and Therese L. Lueck. *Women's Periodicals in the United States; Social and Political Issues.* Westport, Conn.: Greenwood, 1996.
Enriquez, Sam. "Who's in Charge in Oaxaca?" *Los Angeles Times*, 28 August 2006, A7.
Epstein, Barbara. *Political Protest and Cultural Revolution.* Berkeley, Calif.: University of California Press, 1991.
Eschle, Catherine. *Global Democracy, Social Movements, and Feminism.* Boulder, Colo.: Westview, 2001.
Evans, David and Sylvia Gohl. *Photomontage; A Political Weapon.* London: Gordon Fraser, 1986.
Evans, David. *John Heartfield AIZ; Arbeiter-Illustrierte Zeitung Volks Illustrierte 1930–38.* New York: Kent, 1992.
Ewen, Stuart. *All Consuming Images; The Politics of Style in Contemporary Culture.* New York: Basic Books, 1988.
———. *PR!: a Social History of Spin.* New York: Basic Books, 1996.
"Factiva Launches Most Sophisticated Reputation Intelligence Tool Available." *PR Newswire*, 4 August 2005> (from LexisNexis at http://web.lexis-nexis.com/, 22 February 2007.
Fairchild, Charles. *Community Radio and Public Culture: Being an Examination of Media Access and Equity in the Nations of North America.* Cresskill, N.J.: Hampton Press, 2001.
Fairclough, Adam. *A Class of Their Own; Black Teachers in the Segregated South.* Cambridge, Mass.: Balknap Press, 2007.
Fanuzzi, Robert. *Abolition's Public Sphere.* Minneapolis, Minn.: University of Minnesota Press, 2003.
Federal Communications Commission. "FCC Adopts Rules to Ensure Reasonable Franchising Process for New Video Market Entrants." News release, 20 December 2006.
———. *In the Matter of 2006 Quadrennial Regulatory Review—Review of the Commission's Broadcast Ownership Rules and Other Rules Adopted Pursuant to Section 202 of the Telecommunications Act of 1996.* Further Notice of Proposed Rulemaking. MB Docket Nos. 06-121, 02-277, MM Docket Nos. 01-235, 01-317, 00-244, 24 July 2006.
———. "Low Power FM (LPFM) Radio Service." Federal Communications Commission <http://www.fcc.gov/mb/audio/lpfm/> (1 Feb. 2007).
———. *Low-Power FM Radio: An Applicant's Guide.* Washington, D.C.: GPO, 2000.
———. *Memorandum Order and Reconsideration: In the Matter of Creation of Low Power Radio Service.* MM Docket No. 99-25, 20 September 2000.

———. *Notice of Proposed Rulemaking; In the Matter of 2002 Biennial Regulatory Review – Review of the Commission's Broadcast Ownership Rules and other Rules Adopted Pursuant to Section 202 of the Telecommunications Act of 1996.* MB Docket Nos. 02-277, 01-235, 01-317, 00-244, 23 September 2002.

———. *Report and Order and Notice of Proposed Rulemaking; In the Matter of 2002 Biennial Regulatory Review—Review of the Commission's Broadcast Ownership Rules and Other Rules Adopted Pursuant to Section 202 of the Telecommunications Act of 1996.* MB Docket Nos. 02-277, 01-235, 01-317, 00-244, 2 July 2002.

———. *Report and Order and Notice of Proposed Rulemaking, In the Matter of 2002 Biennial Regulatory Review.* MB Docket 02-277, 01-235, 01-317, 00-244, and 03-130, 2 June 2003.

———. *Report and Order: In the Matter of the Creation of Low-Power Radio Service.* MM Docket No. 99-25, 20 January 2000, 2.

Ferguson, Sarah. "The Inconvenient Death of Brad Will." *Village Voice.* <http://villagevoice.com/news/0646,ferguson,75024,2.html> (27 Nov. 2006).

Ferrell, Jeff. "'The Song the Capitalist Never Sings': The Brotherhood of Timber Workers and the Culture of Conflict." *Labor History* 32, no. 3 (Summer 1991): 422–31.

Fideler, Paul A. *Social Welfare in Pre-Industrial England: The Old Poor Law Tradition.* New York: Palgrave Macmillan, 2006.

"Fifty-Five People in Korea." *KyungHyang Newspaper*, 8 October 2001, 42–47.

"Film Aid International's FAQs." *Film Aid.* <http://filmaid.org/who/faq.shtml> (19 Oct. 2006).

Finkle, Lee. *Forum for Protest: The Black Press During World War II.* Rutherford, N.J.: Farleigh Dickinson University Press, 1975.

Finnegan, Margaret. *Selling Suffrage; Consumer Culture and Votes for Women.* New York: Columbia University Press, 1999.

Fisher, Donald. "The Role of Philanthropic Foundations in the Reproduction and Production of Hegemony." *Sociology* 17 (1983): 206–223.

Fisher, Marc. "Low-Power to the People," *American Journalism Review*, 25 October 2000, 43–48.

Fisher, Shelley Fishkin. *From Fact to Fiction: Journalism and Imaginative Writing in America.* Baltimore, Md.: Johns Hopkins University Press, 1985.

"Five Key Questions That Can Change the World." *Center for Media Literacy* <http://www.medialit.org/reading_room/article661.html> (25 Apr. 2007).

Flacks, Richard. "The Question of Relevance in Social Movement Studies." Pp. 3–19 in *Rhyming Hope and History; Activists; Academics, and Social Movement Scholarship.* Edited by David Croteau, Williams Hoynes, and Charlotte Ryan. Minneapolis, Minn.: University of Minnesota Press, 2005.

Flanagan, Maureen A. *America Reformed: Progressives and Progressivisms, 1890s–1920s.* New York: Oxford University Press, 2007.

Fleischhauer, Carl and Beverly W. Brannan, eds. *Documenting America, 1935–1943.* Berkeley, Calif.: University of California Press, 1988.

Fleming, Juliet. *Graffiti and the Writing Arts of Early Modern England.* Philadelphia, Pa.: University of Pennsylvania Press, 2001.

Fletcher, Anthony and Diarmaid MacCulloch. *Tudor Rebellions.* 4th ed. London: Longman, 1997.

Flores-Koulish, Stephanie A. *Teacher Education for Critical Consumption of Mass Media and Popular Culture.* New York: RoutledgeFalmer, 2005.

Fogarty, Robert S. *All Things New; American Communes and Utopian Movements, 1860–1914.* Chicago: University of Chicago Press, 1990.

Foley, Barbara. *Radical Representations; Politics and Form in US Proletarian Fiction, 1929–1941*. Durham, N.C.: Duke University Press, 1993.

Foner, Eric. *Tom Paine and Revolutionary America*. New York: Oxford University Press, 1976.

Foner, Philip. *Women and the American Labor Movement: From the First Trade Unions to the Present*. New York : Free Press, 1982.

———. *Organized Labor and the Black Worker, 1619–1981*, 2nd ed. New York: International Publishers, 1981.

Fones-Wolf, Elizabeth. "Defending Listeners' Rights: Labor and Media Reform in Postwar America." *Canadian Journal of Communication* 31, no. 3 (2006): 499–518.

———. *Waves of Opposition; Labor and the Struggle for Democratic Radio*. Urbana, Ill.: University of Illinois Press, 2006.

Foran, John, ed. *The Future of Revolutions: Rethinking Radical Change in the Age of Globalization*. London: Zed Books, 2003.

Formisano, Ronald P. *The Birth of Mass Political Parties, Michigan, 1827–1861*. Princeton, N.J.: Princeton University Press, 1971.

Foster, Hal. *The Anti-Aesthetic: Essays on Postmodern Culture*. Port Townsend, Wash.: Bay Press, 1983.

Fox, Adam. *Oral and Literate Culture in England 1500–1700*. Oxford: Oxford University Press, 2000.

Frank, Thomas. "Liberation Marketing and the Culture Trust." Pp. 173–90 in Patricia Aufderheide, Erik Barnouw, Richard M. Cohen, Thomas Frank, Todd Gitlin, David Lieberman, Mark Crispin Miller, Gene Roberts, and Thomas Schatz, *Conglomerates and the Media*. New York: New Press, 1997.

———. *The Conquest of Cool; Business Culture, Counterculture, and the Rise of Hip Consumerism*. Chicago, Ill.: University of Chicago Press, 1997.

"Free Press: Media Reform Through Outreach, Activism, Lobbying and Networking." *Free Press*. <http://www.freepress.net/> (12 Jan. 2006).

"Free Speech Radio News/Pacifica Reporters Against Censorship." *Free Speech Radio News*. <http://www.fsrn.org/> (19 Oct. 2006).

"Free Speech TV: About Us." *Free Speech TV*. <http://www.freespeech.org/html/ aboutus.shtml> (21 Jan. 2007).

Freire, Paulo. *Education for Critical Consciousness*, translated and edited by Myra Bergman Ramos. New York: Seabury, 1973.

———. *Pedagogy of the Oppressed*. Translated by Myra Bergman Ramos. New York: Herder and Herder, 1970.

———. *The Politics of Education; Culture, Power, and Liberation*. South Hadley, Mass.: Bergin and Garvey, 1985.

Freist, Dagmar. *Governed by Opinion; Politics, Religion and the Dynamics of Communication in Stuart London 1637–1645*. London: Tauris Academic Studies, 1997.

Fritzsche, Peter. *Reading Berlin 1900*. Cambridge: Harvard University Press, 1996.

Frizot, Michel. "The Ancestry of Photomontage." [Unpaginated] in *Photomontage; Experimental Photography Between the Wars*, translated by Shaun Whiteside. London: Thames and Hudson, 1991.

Frumkin, Peter. *On Being Nonprofit; A Conceptual and Policy Primer*. Cambridge, Mass.: Harvard University Press, 2002.

Fumerton, Patricia. *Unsettled; The Culture of Mobility and the Working Poor in Early Modern England*. Chicago, Ill.: University of Chicago Press, 2006.

Fuoss, Kirk W. *Striking Performances/Performing Strikes*. Jackson, Miss.: University Press of Mississippi, 1997.

Fussell, G.E. *The Farmer's Tools, 1500–1900; The History of British Farm Implements, Tools and Machinery Before the Tractor Came.* London: A. Melrose, 1952.

———. *The Old English Farming Books; From Fitzberbert to Tull, 1523 to 1730.* London: Crosby Lockwood & Son, 1947.

Gadotti, Moacir. *Reading Paulo Freire; His Life and Work.* Translated by John Milton. Albany, N.Y.: State University of New York Press, 1994.

Gaither, Gerald H. *Blacks and the Populist Movement: Ballots and Bigotry in the New South.* Revised edition. Tuscaloosa, Ala.: University of Alabama Press, 2005.

Galbraith, John Kenneth. *The Affluent Society.* Boston, Mass.: Houghton Mifflin, 1958.

Gans, Herbert. *Popular Culture and High Culture; An Analysis and Evaluation of Taste.* New York: Basic Books, 1974.

Garcelon, Marc. "The 'Indymedia' Experiment: The Internet as Movement Facilitator Against Institutional Control." *Convergence: The International Journal of Research into New Media Technologies* 12, no. 1 (February 2006): 55–82.

Geist, Johann Friedrich. *Arcades, the History of a Building Type.* Cambridge, Mass.: MIT Press, 1983.

Gettleman, Marvin E. "Philanthropy as Social Control in Late Nineteenth-Century America: Some Hypotheses and Data on the Rise of Social Work." *Societas* 5, no. 1 (Winter 1975): 49–59.

Gillett, James. "The Challenges of Institutionalisation for AIDS Media Activism." *Media, Culture and Society* 25, no. 5 (September 2003): 607–624.

Giroux, Henry. "Forward." Pp. ix–xii in Sut Jhally, *The Spectacle of Accumulation; Essays in Culture, Media & Politics.* New York: Peter Lang, 2006.

Gitelman, Lisa and Geoffrey B. Pingree, eds. *New Media, 1740–1915.* Cambridge, Mass.: MIT Press, 2003.

Gitlin, Todd. *The Sixties; Years of Hope, Days of Rage.* New York: Bantam, 1989.

———. *The Whole World Is Watching; Mass Media in the Making and Unmaking of the New Left.* Berkeley, Calif.: University of California Press, 1980.

Giugni, Marco, Doug McAdam, and Charles Tilly, eds. *How Social Movements Matter.* Minneapolis, Minn.: University of Minnesota Press, 1999.

"Give Peace a Chance." *Harper's Bazaar*, March 2005, 328–39.

Godfried, Nathan. *WCFL: Chicago's Voice of Labor, 1926–1978.* Urbana, Ill.: University of Illinois Press, 1997.

Goldberg, Bernard. *Bias: A CBS Insider Exposes How the Media Distort the News.* Washington, D.C.: Regnery Publishers, 2002.

Goodwyn, Lawrence. *The Democratic Promise.* New York: Oxford University Press, 1976.

Gramsci. Antonio. *Prison Notebooks.* New York: Columbia University Press, 1991.

Grant, Nancy L. "Adult Education for Blacks during the New Deal and World War II: The Federal Program." Pp. 211–31 in *Education of the African American Adult; An Historical Overview*, edited by Harvey G. Neufeldt and Leo McGee. Westport, Conn.: Greewood, 1990.

Greek, Cecil E. *The Religious Roots of American Sociology.* New York: Garland, 1992.

Green, James. *Grass Roots Socialism.* Baton Rouge, La.: Louisiana State University Press, 1983.

Green, Martin. *New York 1913; The Armory Show and the Paterson Strike Pageant.* New York: Scribner's, 1988.

Gregory, James. *American Exodus; The Dust Bowl Migration and Okie Culture in California.* New York: Oxford University Press, 1989.

Griffen-Foley, Bridget. "From Tit-Bits to Big Brother: A Century of Audience Participation in the Media." *Media, Culture and Society* 26, no. 4 (July 2004): 533–48.

Grossberg, Lawrence, Ellen Wartella, D. Charles Whitney, and J. Macgregor Wise. *Mediamaking; Mass Media in a Popular Culture*. 2nd. ed. Thousand Oaks, Calif.: Sage, 2006.

"Ground Truth Events | Create Event." *Ground Truth*. <http://events.thegroundtruth .net/page/event/create> (19 Oct. 2006).

Guarneri, Carl J. *The Utopian Alternative; Fourierism in Nineteenth-Century America*. Ithaca, N.Y.: Cornell University Press, 1991.

Guerin-Gonzales, Camille. *Mexican Workers and American Dreams; Immigration, Repatriation, and California Farm Labor, 1910–1939*. New Brunswick, N.J.: Rutgers University Press, 1994.

Gunster, Shane. *Capitalizing on Culture; Critical Theory for Cultural Studies*. Toronto: University of Toronto Press, 2004.

Gunther, Barry. *Measuring Bias on Television*. Luton: University of Luton Press, 1997.

Gupta, Akhil and James Ferguson. "Beyond 'Culture': Space, Identity, and the Politics of Difference." *Cultural Anthropology* 7, no. 1 (February 1992): 6–23.

Guthrie, Woody. *Pastures of Plenty; A Self-Portrait*. Edited by Dave Marsh and Harold Leventhal. New York: HarperPerennial, 1992.

Gutman, Herbert G. *Power and Culture; Essays on the American Working Class*. New York: Pantheon, 1987.

Haas, Tanni and Linda Steiner. "Public Journalism; A Reply to the Critics." *Journalism: Theory, Practice, Criticism* 7, no. 2 (May 2006): 238–54.

Haber, Samuel. *Efficiency and Uplift; Scientific Management in the Progressive Era, 1890–1920*. Chicago, Ill.: University of Chicago Press, 1964.

Habermas, Jürgen. "Rationalism Divided Into Two." Pp. 195–223 in *Positivism and Sociology*. Edited by Anthony Giddens. Revised edition. Aldershot: Gower, 1987.

———. *The Structural Transformation of the Public Sphere: an Inquiry into a Category of Bourgeois Society*. Translated by Thomas Burger and Frederick Lawrence. Cambridge, Mass.: MIT Press, 1989.

Haedicke, Susan C. and Tobin Nellhaus, eds. *Performing Democracy; International Perspectives on Urban Community-Based Performance*. Ann Arbor, Mich.: University of Michigan Press, 2001.

Hagen, Ingunn and Janet Wasko, eds. *Consuming Audiences? Production and Reception in Media Research*. Cresskill, N.J.: Hampton Press, 2000.

Hall, Stuart and Paul du Gay, eds. *Questions of Cultural Identity*. London: Sage, 1996.

Hall, Stuart. "'The Rediscovery of 'Ideology': Return of the Repressed in Media Studies." Pp. 56–90 in *Culture, Society and the Media*, edited by Michael Gurevitch, Tony Bennett, James Curran, and Janet Woollacott. New York: Routledge, 1988.

———. "The Problem of Ideology: Marxism Without Guarantees." in *Marx, A Hundred Years On*, edited by Betty Matthews. London: Lawrence and Wishart, 1983), 57–85.

Halleck, DeeDee. "Deep Dish TV: Community Video from Geostationary Orbit." *Leonardo* 26, no. 5 (October 1993): 415–20.

Hamilton, James F. "Rationalizing Dissent? Challenging Conditions of Low-Power FM Radio." *Critical Studies in Media Communication* 21, no. 1 (March 2004): 44–63.

———. "Remaking Media Participation in Early Modern England." *Journalism: Theory, Practice, Criticism* 4, no. 3 (August 2003): 293–313.

———. "Theory Through History: Exploring Scholarly Conceptions of U.S. Alternative Media." *The Communication Review* 4, no. 3 (2001): 305–326.

Hamilton, James F. and Chris Atton, "Theorizing Anglo-American Alternative Media: Toward a Contextual History and Analysis of US and UK Scholarship." *Media History* 7, no. 2 (December 2001): 119–35.

Hamlin, Jesse. "Gore's New Media Venture Seeks to Blend TV, Internet; Former VP Says his Current Network Gives Viewers a Voice." *San Francisco Chronicle*, 5 April 2005, A1.

Hammack, David C. and Stanton Wheeler. *Social Science in the Making: Essays on the Russell Sage Foundation, 1907–1972.* New York: Russell Sage Foundation, 1994.

———. *Making the Nonprofit Sector in the United States; A Reader.* Bloomington, Ind.: Indiana University Press, 1998.

Hammond, Brean S. *Professional Imaginative Writing in England, 1670–1740; "Hackney for Bread."* Oxford: Clarendon Press, 1997.

Handy, Robert T., ed. *The Social Gospel in America, 1870–1920.* New York: Oxford University Press, 1966.

Hankins, Barry. *The Second Great Awakening and the Transcendentalists.* Westport, Conn.: Greewood Press, 2004.

Harcourt, Wendy, ed. *Women@Internet: Creating New Cultures in Cyberspace.* London: Zed Books, 1999.

Hardt, Hanno and Bonnie Brennen. "Introduction: Communication and the Question of History." *Communication Theory* 3, no. 2 (May 1993): 130–36.

Hardt, Hanno. "Conflicts of Interest: Newsworkers, Media, and Patronage Journalism." Pp. 209–224 in *Media Power, Professionals and Policies*, edited by Howard Tumber. New York: Routledge, 2000.

———. *Critical Communication Studies; Communication, History & Theory in America.* New York: Routledge, 1992.

———. *Social Theories of the Press; Early German and American Perspectives.* Beverly Hills, Calif.: Sage, 1979.

Harold, Christine. *OurSpace; Resisting the Corporate Control of Culture.* Minneapolis, Minn.: University of Minnesota Press, 2007.

Harper, R.D. "The Church—Her Danger and Duty." *The Christian Recorder*, 7 June 1862

Harris, Michael. "The Structure, Ownership and Control of the Press, 1620–1780." Pp. 82–97 in *Newspaper History from the Seventeenth Century to the Present Day*, edited by George Boyce, James Curran, and Pauline Wingate. London: Constable, 1978.

Harris, Tim. "Introduction: Revising the Restoration." Pp. 1–28 in *The Politics of Religion in Restoration England*, edited by Tim Harris, Paul Seaward, and Mark Goldie. Oxford: Basil Blackwell, 1990.

———. "The Leveller Legacy: From the Restoration to the Exclusion Crisis." Pp. 219-240 in *The Putney Debates of 1647; The Army, the Levellers and the English State*, edited by Michael Mendle. Cambridge: Cambridge University Press, 2001.

———. *London Crowds in the Reign of Charles II; Propaganda and Politics from the Restoration Until the Exclusion Crisis.* Cambridge: Cambridge University Press, 1987.

Harrison, Mark. *Crowds and History: Mass Phenomena in English Towns, 1790–1835.* Cambridge: Cambridge University Press, 1988.

Hartsock, John C. *A History of American Literary Journalism: the Emergence of a Modern Narrative Form.* Amherst, Mass.: University of Massachusetts, 2000.

Harvey, David. *Paris, Capital of Modernity.* New York: Routledge, 2003.

———. *Social Justice and the City*. Baltimore, Md.: Johns Hopkins University Press, 1973.

———. *Spaces of Global Capitalism; Towards a Theory of Uneven Geographical Development.* New York: Verso, 2006.

———. *The Condition of Postmodernity; An Enquiry into the Origins of Cultural Change*. Cambridge, Mass.: Basil Blackwell, 1989.

Hawkes, David. *Idols of the Marketplace; Idolatry and Commodity Fetishism in English Literature, 1580–1680*. New York: Palgrave, 2001.

Hawkes, Ellison. *The Romance and Reality of Radio*. London: T. C. & E. C. Jack, 1923.

Hay, Douglas and Nicholas Rogers. *Eighteenth Century English Society; Shuttles and Swords*. Oxford: Oxford University Press, 1997.

Hayden, Dolores. *Seven American Utopias; The Architecture of Communitarian Socialism, 1790–1975*. Cambridge, Mass.: MIT Press, 1976.

Hayden, Tom and Dick Flacks. "The Port Huron Statement at 40." *The Nation*, 5 August 2002, 18–21.

Haywood, William. *Bill Haywood's Book; The Autobiography of William D. Haywood*. Reprint edition. New York: International Publishers, 1966.

Heartfield, John. *Photomontages of the Nazi Period*. New York: Universe Books, 1977.

Hebdige, Dick. *Hiding in the Light; On Images and Things*. New York: Routledge, 1988.

———. *Subculture, the Meaning of Style*. London: Methuen, 1979.

Hegeman, Susan. *Patterns for America: Modernism and the Concept of Culture*. Princeton, N.J.: Princeton University Press, 1999.

Henderson, Linda Dalrymple. *The Fourth Dimension and Non-Euclidean Geometry in Modern Art*. Princeton, N.J.: Princeton University Press, 1983.

Hendricks, Geoffrey, ed. *Critical Mass: Happenings, Fluxus, Performance, Intermedia, and Rutgers University, 1958–1972*. New Brunswick, N.J.: Rutgers University Press, 2003.

Henretta, James A. "Social History as Lived and Written," *American Historical Review* 84, no. 5 (December 1979): 1293–1323.

Herman, Edward S. and Noam Chomsky. *Manufacturing Consent; The Political Economy of the Mass Media*. New York: Pantheon, 1988.

Hiken, Louis. "Forward." Pp. ix–xii in Lawrence Soley, *Free Radio: Electronic Civil Disobedience*. Boulder, Colo.: Westview Press, 1999.

Hill, Christopher. *The World Turned Upside Down; Radical Ideas During the English Revolution*. New York: Viking, 1972.

Hilliard, Robert L. and Michael Keith. *The Broadcast Century and Beyond: A Biography of American Broadcasting*. 3rd ed. Boston, Mass.: Focal Press, 2001.

———. *The Quieted Voice: the Rise and Demise of Localism in American Radio*. Carbondale, Ill.: Southern Illinois University Press, 2005.

Hilmes, Michele. *Hollywood and Broadcasting; From Radio to Cable*. Urbana, Ill.: University of Illinois Press, 1990.

Hindle, Steve. *The State and Social Change in Early Modern England, c. 1550–1640*. New York: St. Martin's, 2000.

Hirsch, Jerrold. *Portrait of America: a Cultural History of the Federal Writers' Project*. Chapel Hill, N.C.: University of North Carolina Press, 2003.

Hirsch, Jr., E.D., Joseph Kett, and James Trefil. *Cultural Literacy: What Every American Needs to Know*. Boston, Mass.: Houghton Mifflin, 1987.

Hirsch, Susan Eleanor. *After the Strike: A Century of Labor Struggle at Pullman*. Urbana, Ill.: University of Illinois Press, 2003.

Hobsbawm, Eric and George Rudé. *Captain Swing*. London: Lawrence and Wishart, 1969.

Hobsbawm, Eric. *Social Bandits and Primitive Rebels; Studies in Archaic Forms of Social Movement in the 19th and 20th Centuries*. Glencoe, Ill.: Free Press, 1959.

———. *The Age of Extremes*. New York: Pantheon, 1994.

———. *The Age of Revolution, 1789–1848*. Cleveland, Ohio: World Publishing Co., 1962.

Hoerder, Dirk and Christiane Harzig. *The Press of Labor Migrants in Europe and North America, 1880s to 1930s*. Bremen: Labor Newspaper Preservation Project, 1985.

Hoffman, Katherine. "Collage in the Twentieth Century: An Overview." Pp. 1–37 in *Collage: Critical Views*, edited by Katherine Hoffman. Ann Arbor, Mich.: UMI Research Press, 1989.

Holloway, John and Eloína Peláez, eds. *Zapatista!: Reinventing Revolution in Mexico*. London: Pluto Press, 1998.

Holloway, John. *Change the World Without Taking Power*. London: Pluto Press, 2002.

Holloway, Mark. *Heavens on Earth; Utopian Communities in America, 1680–1880*. New York: Library Publishers, 1951.

"Home Page :: My Dream." *Lincoln (Ford Motor Company)*. <http://mydream.tv> (14 Mar. 2007).

Hood, Stuart. "John Grierson and the Documentary Film Movement." Pp. 99–112 in *British Cinema History*, edited by James Curran and Vincent Porter. Totowa, N.J.: Barnes and Noble Books, 1983.

Hopkins, Charles Howard. *The Rise of the Social Gospel in American Protestantism 1865–1915*. New Haven, Conn.: Yale University Press, 1940.

Horkheimer, Max and Theodor Adorno. *Dialectic of Enlightenment*. Translated by John Cumming. New York: Continuum, 1972.

Horowitz, Daniel. *Vance Packard and American Social Criticism*. Chapel Hill, N.C.: University of North Carolina Press, 1994.

Horwitz, Robert Britt. "U.S. Media Policy Then and Now." Pp. 25–50 in *Converging Media, Diverging Politics: A Political Economy of News Media in the United States and Canada*. Edited by David Skinner, James R. Compton and Michael Gasher. Lanham, Md.: Lexington Books, 2005.

———. *The Irony of Regulatory Reform; The Deregulation of American Telecommunications*. New York: Oxford University Press, 1989.

"How to Host a Screening." *Brave New Theaters*. <http://bravenewtheaters.com/guide/host> (19 Oct. 2006).

Hower, Ralph M. *The History of an Advertising Agency; N.W. Ayer & Son at Work 1869–1949*. Cambridge, Mass.: Harvard University Press, 1949.

Howley, Kevin. *Community Media; People, Places, and Communication Technologies*. Cambridge: Cambridge University Press, 2005.

Huba, Jackie and Ben McConnell. *Citizen Marketers; When People Are the Message*. Chicago, Ill.: Kaplan, 2007.

———. *Creating Customer Evangelists: How Loyal Customers Become a Volunteer Sales Force*. Chicago, Ill.: Dearborn, 2003.

Hunt, Lynn, ed. *The New Cultural History*. Berkeley, Calif.: University of California Press, 1989.

Hurt, R. Douglas. *The Dust Bowl: An Agricultural and Social History*. Chicago, Ill.: Nelson-Hall, 1981.

Hutchins, Brett and Libby Lester. "Environmental Protest and Tap-Dancing with the Media in the Information Age." *Media, Culture and Society* 28, no. 3 (May 2006): 433–51.

Hutton, Frankie. *The Early Black Press in America, 1827–1860*. New York: Greenwood Press, 1993.

Huyssen, Andreas. *After the Great Divide: Modernism, Mass Culture, Postmodernism*. Bloomington, Ind.: Indiana University Press, 1986.

Hyman, Colette A. *Staging Strikes: Workers' Theatre and the American Labor Movement*. Philadelphia, Pa.: Temple University Press, 1997.

Ihlwan, Moon and Kenji Hall. "Ohmynews: Voices From the Street." *BusinessWeek.com*, 4 May 2006 <http://www.businessweek.com/globalbiz/ content/may2006/gb20060504_995206.htm> (10 Mar. 2007).

Ihlwan, Moon. "OhmyNews' Oh My Biz Problem." *BusinessWeek.com*, 1 November 2006 <http://www.businessweek.com/globalbiz/content/nov2006/gb20061101_539412.htm> (10 Mar. 2007).

Ilchman, Warren F., Stanley N. Katz and Edward L. Queen II, eds. *Philanthropy in the World's Traditions*. Bloomington, Ind.: Indiana University Press, 1998.

"Illegal-art.org :: A Project of Stay Free! Magazine." *Illegal Art*. <http://www.illegal-art.org/> (16 Feb. 2007).

Im, Yung-Ho. "Critical Communication Studies in South Korea." *Journal of Communication Inquiry* 17, no. 1 (Winter 1993), 108–123.

"Intergovernmental Panel on Climate Change." *Intergovernmental Panel on Climate Change*. <http://www.ipcc.ch/> (7 Feb. 2007).

International Telecommunication Union. *ITU Internet Reports 2003—Birth of Broadband*. Geneva: International Telecommunication Union, 2003.

"Internet Newspaper, Development Possibility is the Biggest." *Media Today*, 7 January 2004 <http://www.mediatoday.co.kr/> (20 Jun. 2004).

Irwin, Will. *The American Newspaper*. Ames, Iowa: Iowa State University Press, 1969.

Italia, Iona. *The Rise of Literary Journalism in the Eighteenth Century: Anxious Employment*. New York: Routledge, 2005.

Jacobs, Lewis, ed. *The Documentary Tradition*. New York: W.W. Norton, 1979.

Jacobson, Matthew Frye. *Whiteness of a Different Color: European Immigrants and the Alchemy of Race*. Cambridge, Mass.: Harvard University Press, 1998.

James, Robert. "'A Very Profitable Enterprise': South Wales Miners' Institute Cinemas in the 1930s." *Historical Journal of Film, Radio and Television* 27, no. 1 (March 2007): 27–61.

Jameson, Fredric. *Archaeologies of the Future; The Desire Called Utopia and Other Science Fictions*. London: Verso, 2005.

———. "Reflections in Conclusion." Pp. 196–213 in Ernst Bloch, Georg Lukács, Bertolt Brecht, Walter Benjamin, and Theodor Adorno, *Aesthetics and Politics*. Translated and edited by Ronald Taylor. London: New Left Books, 1977.

Janis, Harriet and Rudi Blesh, *Collage: Personalities, Concepts, Techniques*. Philadelphia, Pa.: Chilton, 1967.

Jansen, Sharon L. *Political Protest and Prophecy Under Henry VIII*. Rochester, N.Y.: Boydell Press, 1991.

Jayadev, Raj. "Making Media Make Change with Young Temps in Silicon Valley." *Media Alliance*, 14 May 2004 <http://www.media-alliance.org/article.php?story=20040514190157379> (12 Jan. 2006).

Jenemann, David. *Adorno in America*. Minneapolis, Minn.: University of Minnesota Press, 2007.

Jenkins, Henry. *Textual Poachers: Television Fans and Participatory Culture*. New York: Routledge, 1992.

Jenkins, Richard. *Rethinking Ethnicity: Arguments and Explorations*. Thousand Oaks, Calif.: Sage, 1997.

Jennings, Jeremy. *Syndicalism in France; A Study of Ideas*. New York: St. Martin's, 1990.

Jeong, Yun-Su. "Brave News World." *Joins.com*, 30 March 2003, accessed from Factiva at http://global.factiva.com/> (10 Apr. 2004).

Jeynes, William H. *American Educational History: Schools, Society, and the Common Good*. Thousand Oaks, Calif: Sage, 2007.

Jhally, Sut, and Jeremy Earp. "Empowering Literacy: Media Education as a Democratic Imperative." Pp. 239–67 in Sut Jhally, *The Spectacle of Accumulation; Essays in Culture, Media, and Politics*. New York: Peter Lang, 2006.

Jhally, Sut, and Justin Lewis. "The Struggle for Media Literacy." *Journal of Communication* 48, no. 1 (Winter 1998): 109–20.

John, Richard. *Spreading the News: The American Postal System from Franklin to Morse*. Cambridge, Mass.: Harvard University Press, 1995.

John, Steve and Stuart Thompson, eds. *New Activism and the Corporate Response*. New York: Palgrave Macmillan, 2003.

Johnson, Bobbie and Michael Fitzpatrick. "Korean Website's 40,000 Citizen Journalists Get Shot at Byline in Herald Tribune." *Guardian (UK) Unlimited*, 31 May 2006 <http://business.guardian.co.uk/story/0,,1786356,00.html> (26 Mar. 2007).

Johnson, Bradley. "Adwatch." *Advertising Age*, 12 February 2007, 8.

Johnston, Garth. "YouNewsTV Turns Viewers Into Stringers." *Broadcasting & Cable*, 5 March 2007 <http://www.broadcastingcable.com/article/CA6421029.html> (12 Mar. 2007).

Jones, Gareth Steadman. "Rethinking Chartism." Pp. 90–178 in *Languages of Class; Studies in English Working Class History 1832–1982*. Cambridge: Cambridge University Press, 1983.

Jones, Paul. *Raymond Williams's Sociology of Culture; A Critical Reconstruction*. New York: Palgrave Macmillan, 2004.

Jones, Van. "Beyond Eco-Apartheid; Is the Green Movement Too White?." *Common Ground*, April 2007 <http://www.commongroundmag.com/2007/04/eco-apartheid0704.html> (27 Apr. 2007).

Jung, Hong-Sub. "Gap in Generations and the Role of Education." *Education Science* 8, no. 1 (February 2003): 131–48.

Justice, Steven. *Writing and Rebellion; England in 1381*. Berkeley, Calif.: University of California Press, 1994.

Kadlec, David. *Mosaic Modernism; Anarchism, Pragmatism, Culture*. Baltimore, Md.: Johns Hopkins University Press, 2000.

Kaempffert, Waldemar. "The Progress of Radio Broadcasting." *Review of Reviews*, September 1922, 305.

Kahn, Douglas and Gregory Whitehead, eds. *Wireless Imagination: Sound, Radio, and the Avant-Garde*. Cambridge, Mass.: MIT Press, 1992.

Kahn, Douglas. *John Heartfield: Art and Mass Media*. New York: Tanam Press, 1985.

Kang, Myungkoo. "There Is No South Korea in South Korean Cultural Studies: Beyond the Colonial Condition of Knowledge Production." *Journal of Communication Inquiry* 28, no. 3 (July 2004): 253–269.

Kaniss, Phyllis. *Making Local News*. Chicago, Ill.: University of Chicago Press, 1991.

Kaprow, Allan. *Assemblages, Environments and Happenings*. New York: Harry N. Abrams, 1966.

Karena, Cynthia. "Get Heard: Get Involved in Community Radio." *Australian Screen Education* no. 37 (Summer 2004): 44–51.

Karl, Barry D. and Stanley N. Katz. "Foundations and Ruling Class Elites." *Daedalus* 116, no. 1 (Winter 1987): 1–40.

Katz, Philip M. *From Appomattox to Montmartre; Americans and the Paris Commune.* Cambridge, Mass.: Harvard University Press, 1998.

Kaufman, Anthony. "For a Few Dollars More; As Technologies Evolve and Priorities Shift, Media Nonprofits Struggle to Stay Afloat," *Village Voice*, 11 April 2006. <http://www.villagevoice.com/generic/show_print.php?id=72822> (31 Aug. 2006).

Kazin, Michael. *The Populist Persuasion: an American History*. Revised edition. Ithaca, N.Y.: Cornell University Press 1998.

Keane, John. *The Media and Democracy*. Cambridge, Mass.: Basil Blackwell, 1989), 94–114.

Keeble, N.H. *The Literary Culture of Nonconformity in Later Seventeenth-Century England.* Athens, Ga.: University of Georgia Press, 1987.

Kellner, Douglas and Jeff Share. "Toward Critical Media Literacy: Core Concepts, Debates, Organizations, and Policy." *Discourse: Studies in the Cultural Politics of Education* 26, no. 3 (September 2005): 369–386.

Kellner, Douglas. *Critical Theory, Marxism, and Modernity*. Baltimore, Md.: Johns Hopkins University Press, 1989.

———. *Television and the Crisis of Democracy*. Boulder: Westview, 1990.

Kellner, Hans. *Language and Historical Representation; Getting the Story Crooked.* Madison, Wisc.: University of Wisconsin Press, 1989.

Kemper, Bob. "Something Wiki This Way Comes in Partisan Politics." *Atlanta Journal-Constitution*, 13 March 2006, 1A.

Kern, Stephen. *The Culture of Time and Space 1880–1918*. Cambridge, Mass.: Harvard University Press, 1983.

Kerr, Robert. *History of the Government Printing Office. at Washington, D.C.); With a Brief Record of the Public Printing for a Century, 1789–1881*. New York: Burt Franklin, 1970.

Kessler, Lauren. *The Dissident Press; Alternative Journalism in American History*. Newbury Park, Calif.: Sage, 1984.

Kett, Joseph F. *The Pursuit of Knowledge Under Difficulties; From Self-Improvement to Adult Education in America, 1750–1990*. Stanford, Calif.: Stanford University Press, 1994.

Kidd, Dorothy. "Indymedia.org: A New Communication Commons." Pp. 47–69 in *Cyberactivism: Online Activism in Theory and Practice*, edited by Martha McCaughey and Michael D. Ayers. New York: Routledge, 2003.

Kielbowicz, Richard B. *News in the Mail; The Press, Post Office, and Public Information, 1700–1860s*. Westport, Conn.: Greenwood Press, 1989.

Kiger, Joseph C. *Philanthropic Foundations in the Twentieth Century*. Westport, Conn.: Greenwood, 2000.

Kim, Hae-Sik. "A Sociological Study on the Transformation of the Korean Press Since 1960s." Unpublished Ph.D. diss., Seoul University, 1993.

Kim, Ho-Ki. "The Pattern of Civil Society and Dual Civil Society." *Citizens and the World* 1, no. 1 (February 2002): 38–53.

Kim, Jung-Hun. "A Theoretical Research for Progressive Civil Society in Korea." Pp. 144–63 in *Civil Society and Civic Movement*, edited by Pal-Moo Yu and Jung-Hun Kim. Seoul: Hanul, 2001.

Kim, Seung-Su. *Media Reform for People*. Seoul: Segyesa, 2002.

King, John N. *English Reformation Literature; The Tudor Origins of the Protestant Tradition*. Princeton, N.J.: Princeton University Press, 1982.

King, Samantha. *Pink Ribbons, Inc.; Breast Cancer and the Politics of Philanthropy*. Minneapolis, Minn.: University of Minnesota Press, 2006.

Kingsley, Margery A. *Transforming the Word; Prophecy, Poetry, and Politics in England, 1650–1742*. Newark, N.J.: University of Delaware Press, 2001.

Kinter, Charles V. "The Changing Pattern of the Newspaper Publishing Industry." *American Journal of Economics and Sociology* 5, no. 1 (October 1945): 43–63.

Kirschner, Don S. *The Paradox of Professionalism; Reform and Public Service in Urban America, 1900–1940*. Westport, Conn.: Greewood Press, 1986.

Kivisto, Peter, ed. *The Ethnic Enigma; The Salience of Ethnicity for European-Origin Groups*. Philadelphia, Pa.: Balch Institute Press, 1989.

Klehr, Harvey. *The Heyday of American Communism; The Depression Decade*. New York: Basic Books, 1984.

Klinenberg, Eric. *Fighting for Air: The Battle to Control America's Media*. New York: Metropolitan, 2007.

Kling, Jr., Robert E. *The Government Printing Office*. New York: Praeger, 1970.

Kluger, Richard. *The Paper: The Life and Death of the New York Herald Tribune*. New York: Knopf, 1986.

Ko, Sung-Guk. *Formation and Development of Civil Society in Korea*. Chuncheon: Asian Culture Research Institute, Hanlim University, 1994.

Kolmerten, Carol A. *Women in Utopia: The Ideology of Gender in the American Owenite Communities*. Bloomington, Ind.: Indiana University Press, 1990.

"Konscious: Video/Art/Action." *Konscious*. <http://konscious.com/> (19 Oct. 2006).

Korea Press Foundation. *A Survey of Audience*. Seoul: Korea Press Foundation, 1998.

"Korea's Influential Media." *Sisa Journal*, 18 October 2000.

"Korea's Influential Media." *Sisa Journal*, 25 October 2001.

"Korea's Influential Media." *Sisa Journal*, 22 October 2002.

"Korea's Influential Media." *Sisa Journal*, 30 October 2003.

"Korea's Influential Media." *Sisa Journal*, 28 October 2004.

Krieger, Susan. *Hip Capitalism*. Beverly Hills, Calif.: Sage, 1979.

Kubey, Robert, ed. *Media Literacy in the Information Age*. New Brunswick, N.J.: Transaction, 1997.

Kuhn, Thomas S. *The Structure of Scientific Revolutions*. Chicago, Ill.: University of Chicago Press, 1962.

Labaton, Stephen. "Court Orders F.C.C. to Rethink New Rules on Growth of Media." *New York Times*, 25 June 2004, A1.

Lacey, Marc. "Mexican Federal Police and Army Move to Retake Oaxaca." *New York Times*, 30 October 2006, A3.

Laclau, Ernesto and Chantal Mouffe. *Hegemony and Socialist Strategy: Towards a Radical Democratic Politics*. Translated by Winston Moore and Paul Cammack. London: Verso, 1985.

Laclau, Ernesto. *New Reflections on the Revolution of Our Time*. London: Verso, 1990.

Lagemann, Ellen Condliffe. *The Politics of Knowledge; the Carnegie Corporation, Philanthropy, and Public Policy*. Middletown, Conn.: Wesleyan University Press, 1989.

Lake, Peter. "Deeds Against Nature: Cheap Print, Protestantism and Murder in Early Seventeenth-Century England." Pp. 257–83 in *Culture and Politics in Early Stuart England*, edited by Kevin Sharpe and Peter Lake. Stanford, Calif.: Stanford University Press, 1993.
Laliberté, Norman and Alex Mogelon, *Collage, Montage, Assemblage; History and Contemporary Techniques*. New York: Van Nostrand Reinhold, 1971.
Lambert, Josiah Bartlett. *"If the Workers Took a Notion"; The Right to Strike and American Political Development*. Ithaca, N.Y.: Cornell University Press, 2005.
Lancaster, Bill. *The Department Store: A Social History*. New York: Leicester University Press, 1995.
Land, Jeff. *Active Radio; Pacifica's Brash Experiment*. Minneapolis, Minn.: University of Minnesota Press, 1999.
Lander, Dan and Micah Lexier. *Sound By Artists*. Toronto: Art Metropole, 1990.
Lane, Jill. "Reverend Billy; Preaching, Protest, and Postindustrial Flânerie." Pp. 299–310 in *The Performance Studies Reader*, edited by Henry Bial. New York: Routledge, 2004.
Lange, Dorothea and Paul S. Taylor. *An American Exodus; A Record of Human Erosion in the Thirties*, revised edition. New Haven, Conn.: Yale University Press, 1969.
Langlois, Andrea, and Frédéric Dubois, eds. *Autonomous Media; Activating Resistance & Dissent*. Montréal, Québec: Cumulus Press, 2005.
Lasar, Matthew. *Pacifica Radio: the Rise of an Alternative Network*. Philadelphia, Pa.: Temple University Press, 1999.
Lasch, Christopher. "Journalism, Publicity, and the Lost Art of Argument." *Media Studies Journal* 9, no. 1 (Winter 1995): 81–91.
———. *The True and Only Heaven; Progress and Its Critics*. New York: W.W. Norton, 1991.
Lasch-Quinn, Elisabeth. *Black Neighbors: Race and the Limits of Reform in the American Settlement House Movement, 1890–1945*. Chapel Hill, N.C.: University of North Carolina Press, 1993.
Lasn, Kalle. *Culture Jam: How to Reverse America's Suicidal Consumer Binge—and Why We Must*. New York: Quill/HarperCollins, 2000.
"Last Video of Brad Will." *Friends of Brad Will*. <http://www.friendsofbradwill.org/category/multimedia/> (12 Nov. 2006).
Laurie, Bruce. *Artisans into Workers; Labor in Nineteenth-Century America*. New York: Noonday Press, 1989.
Lavin, Maud. *Cut With the Kitchen Knife; The Weimar Photomontages of Hannah Höch*. New Haven and London: Yale University Press, 1993.
Law, Lisa. "Transnational Cyberpublics; New Political Spaces for Labour Migrants in Asia." *Ethnic and Racial Studies* 26, no. 2 (March 2003): 234–52.
Leach, William. *Land of Desire: Merchants, Power, and the Rise of a New American Culture*. New York: Pantheon, 1993.
Lears, Jackson. *Fables of Abundance; A Culture History of Advertising in America*. New York: Basic Books, 1994.
Ledbetter, James. *Made Possible By– : the Death of Public Broadcasting in the United States*. New York: Verso, 1997.
Lee, Alfred McClung. *The Daily Newspaper in America; the Evolution of a Social Instrument*. New York: Macmillan, 1937.
Lee, Dong-Min. "Internet Campaigns Race Ahead of Election Laws." *Korea Herald*, 15 February 2002, accessed from Factiva: http://global.factiva.com/> (10 Apr. 2004).
Lee, Martyn J., ed. *The Consumer Society Reader*. Oxford: Blackwell, 2000.

Lee, Richard. *Life and Times of Cultural Studies; The Politics and Transformation of the Structures of Knowledge*. Durham, N.C.: Duke University Press, 2003.

Lefebvre, Henri. *The Urban Revolution*. Translated by Robert Bononno. Minneapolis, Minn.: University of Minnesota Press, 2003.

Leja, Michael. *Looking Askance; Skepticism and American Art from Eakins to Duchamp*. Berkeley, Calif.: University of California Press, 2004.

Leonard, Thomas. *News for All; America's Coming-of-Age with the Press*. New York: Oxford University Press, 1995.

Levenstein, Harvey A. *Communism, Anti-Communism, and the C.I.O.* Westport: Greenwood, 1981.

Levin, Carole. *Propaganda in the English Reformation: Heroic and Villainous Images of King John*. Lewiston, Maine: Edward Mellen Press, 1988.

Levy, Jonathan, Marcelino Ford-Livene, and Anne Levine. "Broadcast Television: Survivor in a Sea of Competition." Federal Communications Commission OPP Working Paper Series No. 37. September 2002.

Lieberman, Joseph. "Prepared Statement of Senator Joe Lieberman, Media Research Forum." News release, 9 April 2003 <http://lieberman.senate.gov/newsroom/ release.cfm?id=207585> (25 Apr. 2007).

Lieberman, Sima. *Labor Movements and Labor Thought; Spain, France, Germany, and the United States*. New York: Praeger, 1986.

Lieberman, Trudy. "The Epidemic." *Columbia Journalism Review*, March/April 2007 <http://www.cjr.org/issues/2007/2/Leiberman.asp> (14 Mar. 2007).

Lippmann, Walter. *Public Opinion*. New York: Harcourt, Brace, 1922.

Lipset, Seymour M. *Agrarian Socialism*. New York: Doubleday, 1968.

"'Listening In', Our New National Pastime." *Review of Reviews*, January 1923, 52.

Livingston, James. *Pragmatism, Feminism, and Democracy: Rethinking the Politics of American History*. New York: Routledge, 2001.

Loades, David. *England's Maritime Empire: Seapower, Commerce, and Policy, 1490–1690*. New York: Longman, 2000.

Lomax, Alan ed. *Hard-Hitting Songs for Hard-Hit People*. New York: Oak Publications, 1967.

Lomax, Susan. "The View From the Shop: Window Display, the Shopper and the Formulation of Theory." Pp. 265–292 in *Cultures of Selling: Perspectives on Consumption and Society Since 1700*, edited by John Benson and Laura Ugolini. Aldershot: Ashgate, 2006.

Long, Huey B. *Early Innovators in Adult Education*. New York: Routledge, 1991.

Love, Harold. *Scribal Publication in Seventeenth-Century England*. Oxford: Clarendon Press, 1993.

Lubove, Roy. *The Professional Altruist; The Emergence of Social Work as a Career 1880–1930*. Cambridge, Mass.: Harvard University Press, 1965.

Lunn, Eugene. *Marxism and Modernism; An Historical Study of Lukács, Brecht, Benjamin, and Adorno*. Berkeley, Calif.: University of California Press, 1982.

Lury, Celia. *Brands: the Logos of the Global Economy*. New York: Routledge, 2004.

Lynd, Staughton. "Towards a History of the New Left." Pp. 1–13 in *The New Left; A Collection of Essays*, edited by Priscilla Long. Boston, Mass.: Porter Sargent, 1969.

Main Page – Wikipedia, the Free Encyclopedia." *Wikipedia*. <http://en.wikipedia.org/> (16 Feb. 2007).

Majka, Linda C. and Theo J. Majka. *Farm Workers, Agribusiness, and the State*. Philadelphia, Pa.: Temple University Press, 1982.

Mander, Mary. "The Public Debate about Broadcasting in the Twenties: An Interpretive History." *Journal of Broadcasting* 28, no. 2 (Spring 1984): 167–85.

Manes, Sheila Goldring. "Depression Pioneers: The Conclusion of an American Odyssey; Oklahoma to California, 1930-1950, A Reinterpretation." PhD dissertation, University of California, Los Angeles, 1982.

Manheim, Jarol B. *Biz-War and the Out-of-Power Elite: the Progressive-Left Attack on the Corporation.* Mahwah, N.J.: Lawrence Erlbaum, 2004.

———. *The Death of a Thousand Cuts: Corporate Campaigns and the Attack on the Corporation.* Mahwah, N.J.: Lawrence Erlbaum, 2001.

Mann, Horace. *Lectures and Annual Reports on Education.* Cambridge, Mass., 1867.

Marchand, Roland. *Advertising the American Dream; Making Way for Modernity 1920–1940.* Berkeley, Calif.: University of California Press, 1985.

Marcuse, Herbert. *One-Dimensional Man; Studies in the Ideology of Advanced Industrial Society.* Boston, Mass.: Beacon Press, 1964.

Martin, Bradford D. *The Theater Is in the Street: Politics and Public Performance in Sixties America.* Amherst, Mass.: University of Massachusetts Press, 2004.

Marvin, Carolyn. *When Old Technologies Were New: Thinking about Communications in the Late Nineteenth Century.* New York: Oxford University Press, 1987.

Marx, Karl. *A Contribution to the Critique of Political Economy.* Translated by S.W. Ryazanskaya. New York: International Publishers 1970.

Matani, Sheetal and Gerri Spilka. "Strengths, Challenges and Collaboration: Advocacy Groups Organizing Together on Media Ownership and Beyond." Media Diversity Convening, 12–13 September 2003, 29. <http://www.mediaactioncenter.org/?q= Publications> (10 Oct. 2006).

Mather, Anne. "A History of Feminist Periodicals, Pt. 1." *Journalism History* 1, no. 3 (Autumn 1974): 82–85.

———. "A History of Feminist Periodicals, Pt. 2." *Journalism History* 1, no. 4 (Winter 1974–75): 108–111.

———. "A History of Feminist Periodicals, Pt. 3." *Journalism History* 2, no. 1 (Spring 1975): 19–23, 31.

Mathews, Joe. "A Plan for Very Civil Disobedience; Police and Union Will Follow a Script, Which Even Specifies Who Will Be Arrested, in a March near LAX to Organize Hotel Workers." *Los Angeles Times*, 28 September 2006, A1.

———. "Labor Protest Targets Airport-Area Hotels; Hundreds Are Willingly Arrested After Sit-Ins Shut Century Boulevard; Few Travelers Miss Flights." *Los Angeles Times*, 29 September 2006, B3.

"Matters We Ought to Know." *The New York Times*, 4 Jan. 1891, 19.

Matthews, Fred. *Quest for an American Sociology: Robert E. Park and the Chicago School.* Montreál, Quebéc: McGill-Queens University Press, 1977.

Mattson, Eric and Nora Ganim Barnes. "The Hype is Real: Social Media Invades the Inc. 500." Center for Marketing Research, University of Massachusetts Dartmouth. <http://www.umassd.edu/cmr/studies/blogstudy2.cfm> (10 Mar. 2007).

Mayer, Robert N. *The Consumer Movement: Guardians of the Marketplace.* Boston, Mass.: Twayne Publishers, 1989.

Mayo, Peter. *Liberating Praxis; Paulo Freire's Legacy for Radical Education and Politics.* Westport, Conn.: Praeger, 2004.

McCamy, James L. *Government Publicity: Its Practice in Federal Administration.* Chicago: University of Chicago Press, 1939.

McCarthy, John D., Clark McPhail, and Jackie Smith. "Images of Protest: Dimensions of Selection Bias in Media Coverage of Washington Demonstrations, 1982 and 1991." *American Sociological Review* 61, no. 3 (June 1996): 478–99.

McCarthy, Kathleen D. *American Creed; Philanthropy and the Rise of Civil Society 1700–1865*. Chicago: University of Chicago Press, 2003.

McCaughey, Martha and Michael D. Ayers, eds. *Cyberactivism: Online Activism in Theory and Practice*. New York: Routledge, 2003.

McChesney, Robert W. and Ben Scott, eds. *Our Unfree Press: 100 Years of Radical Media Criticism*. New York: New Press, 2004.

McChesney, Robert W. and John Nichols, *Our Media, Not Theirs; The Democratic Struggle Against Corporate Media*. New York: Seven Stories Press, 2002.

McChesney, Robert W. *Rich Media, Poor Democracy; Communication Politics in Dubious Times*. Urbana, Ill.: University of Illinois Press, 1999.

———. *Telecommunications, Mass Media, and Democracy: The Battle for the Control of U.S. Broadcasting, 1928–1935*. New York: Oxford University Press, 1993.

———. *The Problem of the Media: U.S. Communication Politics in the Twenty-First Century*. New York: Monthly Review Press, 2004.

McConachie, Bruce A. and Daniel Friedman, eds. *Theatre for Working-Class Audiences in the United States, 1830–1980*. Westport, Conn. Greenwood, 1985.

McCord, Richard. *The Chain Gang: One Newspaper Versus the Gannett Empire*. Columbia, Mo.: University of Missouri Press, 1996.

McCormick, Richard L. *The Party Period and Public Policy: American Politics from the Age of Jackson to the Progressive Era*. New York: Oxford University Press, 1986.

McCormick, Richard P. *The Second American Party System; Party Formation in the Jacksonian Era*. New York: W. W. Norton, 1973.

McDermott, Terry. "Blogs Can Top the Presses; Talking Points Memo Drove the U.S. Attorneys Story, Proof That Web Writers With Input from Devoted Readers can Reshape Journalism." *Los Angeles Times*, 17 March 2007, A1.

McDonough, Tom, ed. *Guy Debord and the Situationist International*. Cambridge, Mass.: MIT Press, 2002.

McGerr, Michael. *A Fierce Discontent: the Rise and Fall of the Progressive Movement in America, 1870–1920*. New York: Oxford University Press, 2005.

McIlroy, John and Sallie Westwood, eds. *Border Country; Raymond Williams in Adult Education*. Leicester: National Institute of Adult Continuing Education, 1993.

McKay, George. *Senseless Acts of Beauty; Cultures of Resistance Since the Sixties*. New York: Verso, 1996.

McKendrick, Neil. *The Birth of Consumer Society; The Commercialization of Eighteenth-Century England*. Bloomington, Ind.: Indiana University Press, 1982.

McKercher, Catherine. *Newsworkers Unite; Labor, Convergence, and North American Newspapers*. Lanham, Md.: Rowman & Littlefield, 2002.

McKinley, Michelle A. and Lene O. Jensen. "In Our Own Voices: Reproductive Health Radio Programming in the Peruvian Amazon." *Critical Studies in Media Communication* 20, no. 2 (June 2003): 180–203.

McLaughlin, Lisa. "Beyond 'Separate Spheres': Feminism and the Cultural Studies/Political Economy Debate." *Journal of Communication Inquiry* 23, no. 4 (October 1999): 327–354.

McLeary, Paul. "How TalkingPointsMemo Beat the Big Boys on the U.S. Attorney Story." *CJR Daily*. <http://www.cjrdaily.org/behind_the_news/how_ talkingpoints-memo_beat_the.php> (23 Mar. 2007).

McLeod, Kembrew. *Owning Culture: Authorship, Ownership, and Intellectual Property Law.* New York: Peter Lang, 2001.

McLuhan, Marshall and Quentin Fiore. *The Medium is the Massage.* New York: Random House, 1967.

McLuhan, Marshall. *Understanding Media; the Extensions of Man.* New York: Signet Books, 1966.

McMath, Robert C. *American Populism: a Social History, 1877–1898.* New York: Noonday Press, 1993.

McMurtrie, Douglas. "The Photomontage—A New Illustrative Technique," *Printer's Ink,* 26 December 1929, 89–91.

McWhirter, Darien. *Freedom of Speech, Press, and Assembly.* Phoenix, Ariz.: Oryx Press, 1994.

McWilliams, Carey. *Factories in the Field; The Story of Migratory Farm Labor in California.* Boston, Mass.: Little, Brown, 1939.

"Media Collage Index." *DIYMedia.net.* <http://www.diymedia.net/collage/> (16 Feb. 2007).

"Media Empowerment Organizing Manual; A Guide to Understanding Media Power and Organizing for Media Justice in Your Community." *Office of Communication, United Church of Christ.* <http://www.ucc.org/ocinc/> (19 Oct. 2006).

Meehan, Eileen R. *Why TV is not Our Fault: Television Programming, Viewers, and Who's Really in Control.* Lanham, Md.: Rowman & Littlefield, 2005.

"Meet Brave New Films." *Brave New Films.* <http://www.bravenewfilms.org/ who.php> (31 Aug. 2006).

Meikle, Graham. *Future Active: Media Activism and the Internet.* New York: Routledge, 2002.

Meister, Dick and Anne Loftus. *A Long Time Coming: The Struggle to Unionize America's Farm Workers.* New York: Macmillan, 1977.

Melucci, Alberto. *Challenging Codes: Collective Action in the Information Age.* New York: Cambridge University Press, 1996.

"Membership Registration." *OhmyNews* International <http://english.ohmynews .com/member/member_step1.asp?> (7 Mar. 2007).

Mendle, Michael, ed. *The Putney Debates of 1647; The Army, the Levellers and the English State.* Cambridge: Cambridge University Press, 2001.

Mertes, Tom, ed. *A Movement of Movements; Is Another World Really Possible?.* London: Verso, 2004.

"Michael Albert." *Z Magazine.* <http://zena.secureforum.com/Znet/ZMag/ fralbert.htm> (22 Aug. 2006).

Mielants, Eric. *The Origins of Capitalism and the "Rise of the West."* Philadelphia, Penn.: Temple University Press, 2007.

Miller, Jason. "Liberating the 'Liberal Media'." *ZNet,* 4 May 2005. <http://www.zmag .org/content/print_article.cfm?itemID=7780&sectionID=21> (29 Jun. 2005).

Miller, John C. *Sam Adams; Pioneer in Propaganda.* New York: Little, Brown & Co., 1936.

Miller, Perry and Thomas H. Johnson, ed. *The Puritans.* Revised edition. New York: Harper and Row, 1963.

Mills, C. Wright. *The Marxists.* New York: Dell, 1962.

———. *The Power Elite.* New York: Oxford University Press, 1956.

———. *The Sociological Imagination.* New York: Oxford University Press, 1959.

Mingay, G.E. *A Social History of the English Countryside.* Reprint ed. London: Routledge, 2002.

———. *Land and Society in England 1750–1980.* London: Longman, 1994.

Ministry of Finance and Economy. *Economic Indicator, 2004–6.* Seoul: Ministry of Finance and Economy, 2004.

Ministry of Information and Communication. *The Strategy of Informationization in Korea.* Seoul: Ministry of Information and Communication, 2003.

Minow, Newton N. *Equal Time; the Private Broadcaster and the Public Interest.* New York: Atheneum, 1964.

Miracle, Joe. "Regent Journalism Alum Making Waves in Korea; Yeon-Ho Oh Makes Participatory Journalism Work With Internet Newpaper OhmyNews." *News@ Regent University*, 5 January 2005 <http://www.regent.edu/news/ohmynews.html> (26 Mar. 2007).

Mitchell, Caroline, ed. *Women and Radio: Airing Differences.* London: Routledge, 2000.

Mommsen, Wolfgang J. *The Age of Bureaucracy; Perspectives on the Political Sociology of Max Weber.* Oxford: Blackwell, 1974.

Montano, John Patrick. *Courting the Moderates: Ideology, Propaganda, and the Emergence of Party, 1660–1678.* Newark, N.J.: University of Delaware Press, 2002.

Montgomery, David. *The Fall of the House of Labor: The Workplace, the State, and American Labor Activism, 1865-1925.* New York: Cambridge University Press, 1987.

Moor, Liz. *The Rise of Brands.* Oxford: Berg, 2007.

Moore, Jason W. "Remaking Work, Remaking Space: Spaces of Production and Accumulation in the Reconstruction of American Capitalism, 1865–1920." *Antipode* 34, no. 2 (March 2002): 176–204.

Moreno, Paul D. *Black Americans and Organized Labor: A New History.* Baton Rouge, La.: Louisiana State University Press, 2006.

Morris, Brian. *Kropotkin; The Politics of Community.* Amherst, N.Y.: Humanity Books, 2004.

Morris, Douglas. "Globalization and Media Democracy: The Case of Indymedia." Pp. 325–352 in *Shaping the Network Society*, edited by Douglas Schuler and Peter Day. Cambridge, Mass.: MIT Press, 2004.

Morrow, Raymond and David Brown. "Postempiricist Critiques of Positivism and Empiricism." Pp. 62–81 in *Critical Theory and Methodology.* Thousand Oaks, Calif.: Sage, 1994.

Mulhern, Francis. "Towards 2000, or News From You-Know-Where." *New Left Review* no. 148 (November/December 1984): 5–30.

Munro, Ian. *The Figure of the Crowd in Early Modern London: the City and Its Double.* New York: Palgrave Macmillan, 2005.

Murdoch, Rupert. Speech to the American Society of Newspaper Editors on 13 April 2005. *News Corporation.* <http://www.newscorp.com/news/news_247.html> (23 June 2005).

Myatt, Alyce. "Report on the Public Media Caucus." *Center for Digital Democracy*, 3 May 2005, San Francisco, Calif. 1 June 2005, 9. <http://www.democraticmedia .org/ddc/PublicMediaCaucus.php#_Toc105405351> (19 Oct. 2006).

Natanson, Nicholas. *The Black Image in the New Deal: the Politics of FSA Photography.* Knoxville, Tenn.: University of Tennessee Press, 1992.

"National Progressive Media: Who's Left?." *Media File.* <http://www.media-alliance.org/article.php?story=20031109003346211> (31 Aug. 2006).

Nealon, Jeffrey T. and Caren Irr, eds. *Rethinking the Frankfurt School; Alternative Legacies of Cultural Critique.* Albany, N.Y.: State University of New York Press, 2002.

Nelson, Daniel. *Frederick W. Taylor and the Rise of Scientific Management*. Madison, Wisc.: University of Wisconsin Press, 1980.

Nerone, John. "Theory and History." *Communication Theory* 3, no. 2 (May 1993): 148–157.

New America Foundation. *Citizen's Guide to the Airwaves; A Graphic Depiction of the Uses—and Misuses—of the Radiofrequency Spectrum*. Washington, D.C.: Spectrum Policy Program, New America Foundation, 2003.

Newman, Kathy M. *Radio Active: Advertising and Consumer Activism, 1935–1947*. Berkeley, Calif.: University of California Press, 2004.

Newman, Michael. "Revising Modernism, Representing Postmodernism." Pp. 95–154 in *Postmodernism: ICA Documents*. Edited by Lisa Appignanesi. London: Free Association Books, 1989.

"News Hounds: News Hounds Manifesto." *Newshounds*. <http://www.newshounds.us/2004/07/02/news_hounds_manifesto.php> (19 Oct. 2006).

Newspaper Association of America. *The Source; Newspapers by the Numbers 2006*. <http://www.naa.org/thesource/> (9 Mar. 2007).

"Newspaper Next: Blueprint for Transformation." *American Press Institute* <http://www.newspapernext.org/> (10 Mar. 2007).

Nichols, John and Robert W. McChesney. *Tragedy and Farce: How the American Media Sell Wars, Spin Elections, and Destroy Democracy*. New York: New Press, 2005.

Nielsen, Waldemar A. *Inside American Philanthropy; The Dramas of Donorship*. Norman, Okla.: University of Oklahoma Press, 1996.

Norbrook, David. *Poetry and Politics in the English Renaissance*, revised edition. Oxford: Oxford University Press, 2002.

Nord, David Paul. "The Evangelical Origins of Mass Media in America, 1815–1835." *Journalism Monographs* no. 88, May 1984.

———. *Faith in Reading; Religious Publishing and the Birth of Mass Media in America*. New York: Oxford University Press, 2004.

Norwood, Janet. *Organizing to Count; Change in the Federal Statistical System*. Washington, D.C.: The Urban Institute, 1995.

Novek, Eleanor M. "'Heaven, Hell, and Here': Understanding the Impact of Incarceration through a Prison Newspaper." *Critical Studies in Media Communication* 22, no. 4 (October 2005): 281–301.

———. "'The Devil's Bargain'; Censorship, Identity and the Promise of Empowerment in a Prison Newspaper." *Journalism: Theory, Practice, Criticism* 6, no. 1 (February 2005): 5–23.

Nye, Russel. *The Unembarrassed Muse: the Popular Arts in America*. New York: Dial Press, 1970.

Oh, Yeon-Ho. "10 Preconditions for the Value of User-Generated Content." *OhmyNews* International, 26 February 2007 <http://www.ohmynews.com/ articleview/article_view.asp?menu=c10400&no=347268&rel_no=1> (7 Mar. 2007).

———. "Internet Media and Korean Society." Speech given at the Korea Institute of Future Strategy, 18 May 2002.

———. *The Speciality of Korea: OhmyNews*. Seoul: Humanist, 2004.

"OhmyNews Goes into the Black." *OhmyNews*, 31 December 2003 <http://www.ohmynews/board/bbs_view.asp> (20 Jun. 2004).

"OhmyNews Japan Debuts." *OhmyNews International*, 28 August 2006 <http://english.ohmynews.com/articleview/article_view.asp?article_class=8&no=313808&rel_no=1> (10 Mar. 2007).

"OhmyNews Runs to the Third Generation Internet Newspaper." *OhmyNews*, 23 February 2004 <http://www.ohmynews.com/articleview/article_view.asp?menu=0446&no=153582&rel_no=1> (10 Jun. 2004).

Olmstead, Alan L. and Paul Rhode. "An Overview of California Agricultural Mechanization, 1870-1930." *Agricultural History* 62, no. 3 (Summer 1988): 86–112.

"On These Sites, We're All Reporters," *Joins.com*, 5 March 2001, retrieved from Factiva at http://www.factiva.com/> (18 Mar. 2004).

"Online Ad Spending to Total $19.5 Billion in 2007." *eMarketer.com*, 28 February 2007 <http://www.emarketer.com/Article.aspx?1004635> (10 Mar. 2007).

Opel, Andy and Donnalyn Pompper, eds. *Representing Resistance: Media, Civil Disobedience, and the Global Justice Movement*. Westport, Conn.: Praeger, 2003.

"OpinionWare Establishes New Benchmark for Online Opinion-Expression." *PR Newswire*, 6 September 2000> (from LexisNexis at http://web.lexis-nexis.com/, 22 February 2007.

Orenstein, Claudia. *Festive Revolutions: the Politics of Popular Theater and the San Francisco Mime Troupe*. Jackson, Miss.: University Press of Mississippi, 1998.

Organisation for Economic Co-Operation and Development. "OECD Broadband Statistics to December 2006." *OECD* <http://www.oecd.org/sti/ict/broadband/> (30 Apr. 2007).

"Organs of Sound That Do and Do Not Broadcast." *Current Opinion*, July 1922, 68–69.

Outing, Steve. "Citizen-Media Start-ups Ply Innovative Business Models." *Poynter Online*, 6 April 2006 <http://www.poynter.org/content/content_view.asp? id=97601> (8 Mar. 2007).

———. "CitJ's National Networks: Will They Bloom?" *Poynter Online*, 6 April 2006 <http://www.poynter.org/content/content_view.asp?id=96566> (8 Mar. 2007).

———. "Independent CitJ: Web sites and Networks." *Poynter Online*, 6 April 2006 <http://www.poynter.org/content/content_view.asp?id=96563> (8 Mar. 2007).

———. "Innovative Thoughts from CitJ's Gurus." *Poynter Online*, 6 April 2006 <http://www.poynter.org/content/content_view.asp?id=97298> (8 Mar. 2007).

———. "The 11 Layers of Citizen Journalism." *Poynter Online*, 13 June 2005 <http://www.poynter.org/content/content_view.asp?id=83126> (10 Mar. 2007).

Oxford English Dictionary, 2nd edition. 1989.

Packard, Vance. *The Hidden Persuaders*. New York: D. McKay, 1957.

Palmer, Rex. "Future of Broadcasting as a Social Force." *English Review*, August 1923, 210–214.

Papke, David Ray. *The Pullman Case: The Clash of Labor and Capital in Industrial America*. Lawrence, Kan.: University Press of Kansas, 1999.

Park, Myung-Jin, Chang-Nam Kim and Byung-Woo Sohn. "Modernization, Globalization, and the Powerful State: The Korean Media." Pp. 111–123 in *De-Westernizing Media Studies*, edited by James Curran and Myung-Jin Park. New York: Routledge, 2000.

Peacey, Jason. *Politicians and Pamphleteers; Propaganda During the English Civil Wars and Interregnum*. Burlington, Vt.: Ashgate, 2004.

Peck, Abe. *Uncovering the Sixties; The Life and Times of the Underground Press*. New York: Pantheon, 1985.

"People's Choice Awards." *People's Choice Awards*. <http://www.pcavote.com/> (30 Dec. 2006).

"People's Choice Awards—Belong." *People's Choice Awards*. <http://www.pcavote.com/belong/> (31 Dec. 2006).

Perez, Juan Carlos. "Online Ad Spending had Healthy Growth in 2006." *InfoWorld*, 7 March 2007 <http://www.infoworld.com/article/07/03/07/Hnonlineadspending_1.html> (10 Mar. 2007).

Peters, John Durham. "'The Only Proper Scale of Representation': The Politics of Statistics and Stories." *Political Communication* 18, no. 4 (October–December 2001): 433–449.

———. "Witnessing." *Media, Culture and Society* 23, no. 6 (November 2001): 707–723.

———. *Speaking into the Air; A History of the Idea of Communication*. Chicago, Ill.: University of Chicago Press, 1999.

Peters, Kate. *Print Culture and the Early Quakers*. Cambridge: Cambridge University Press, 2005.

Peterson, Anna Lisa. *Seeds of the Kingdom; Utopian Communities in the Americas*. New York: Oxford University Press, 2005.

Pew Research Center for People and the Press. "Public More Critical of Press, but Goodwill Persists." *Project for Excellence in Journalism*. 26 June 2005 <http://people-press.org/reports/display.php3?ReportID=248> (29 Mar. 2007).

———. "Media Consumption and Believability Study." 8 June 2004 <http://people-press.org/reports/pdf/215.pdf> (12 Mar. 2007).

Phelan, Craig. *Grand Master Workman; Terence Powderly and the Knights of Labor*. Westport, Conn.: Greenwood Press, 2000.

Phillips, Paul T. *A Kingdom on Earth: Anglo-American Social Christianity, 1880–1940*. University Park, Pa.: Pennsylvania State University Press, 1996.

Pickard, Victor W. "Assessing the Radical Democracy of Indymedia: Discursive, Technical, and Institutional Constructions." *Critical Studies in Media Communication* 23, no. 1 (March 2006): 19–38.

———. "United Yet Autonomous: Indymedia and the Struggle to Sustain a Radical Democratic Network." *Media, Culture and Society* 28, no. 3 (May 2006): 315–336.

Pinkus, Philip. *Grub St. Stripped Bare*. London: Constable, 1968.

Pitzer, Donald, ed. *America's Communal Utopias*. Chapel Hill, N.C.: University of North Carolina Press, 1997.

Platon, Sara and Mark Deuze. "Indymedia Journalism: A Radical Way of Making, Selecting and Sharing News?" *Journalism: Theory, Practice, Criticism* 4, no. 3 (August 2003): 336–355.

Poggi, Christine. *In Defiance of Painting: Cubism, Futurism, and the Invention of Collage*. New Haven, Conn.: Yale University Press, 1992.

Pollack, Norman. *The Just Polity: Populism, Law, and Human Welfare*. Urbana, Ill.: University of Illinois Press, 1987.

Pope, Daniel. *The Making of Modern Advertising*. New York: Basic Books, 1983.

Popkin, Jeremy, ed. *Media and Revolution; Comparative Perspectives*. Lexington, Ky.: University of Kentucky Press, 1995.

Popper, Karl. *The Logic of Scientific Discovery*. New York: Basic Books, 1959.

———. *The Open Society and Its Enemies*. Revised edition. Princeton: N.J.: Princeton University Press, 1950.

Population Estimates Program, Population Division. *U.S. Census Bureau*. <http://www.census.gov/popest/> (9 Mar. 2007).

Potter, David M. *People of Plenty: Economic Abundance and the American Character*. Chicago, Ill.: University of Chicago Press, 1954.

"Preaching by Wireless." *Current Opinion*, May 1922, 649–50.

Pred, Allan. *Urban Growth and City Systems in the United States, 1840–1860*. Cambridge, Mass.: Harvard University Press, 1980.

Proffitt, Jennifer M. and Michael Brown. "Regulating the Radio Monopoly: Ewin Davis and His Legislative Debates, 1923–1928." *Journal of Radio Studies* 11, no. 1 (June 2004): 100–115.

Project for Excellence in Journalism. *The State of the News Media 2007; An Annual Report on American Journalism* <http://www.stateofthemedia.org/2007/> (13 Mar. 2007).

Quandt, Jean. "Religion and Social Thought; The Secularization of Postmillennialism." *American Quarterly* 25, no. 4 (October 1973): 390–409.

———. *From the Small Town to the Great Community; The Social Thought of Progressive Intellectuals.* New Brunswick, N.J.: Rutgers University Press, 1970.

"Questioning Media; Ten Basic Principles of Media Literacy Education." *Action Coalition for Media Education.* <http://www.acmecoalition.org/page.cfm?ID =102> (25 Apr. 2007).

Rabinowitz, Paula. *They Must Be Represented; The Politics of Documentary.* London: Verso, 1994.

Raboy, Marc, ed. *Public Broadcasting for the 21st Century.* Luton: University of Luton Press, 1996.

Radford, Benjamin. *Media Mythmakers: How Journalists, Activists, and Advertisers Mislead Us.* Amherst, N.Y.: Prometheus Books, 2003.

Radio Corporation of America. *The Place of Radio in World Communications.* New York: Radio Corporation of America, 1928.

"Radio Sermons for Pastorless Churches." *The Literary Digest*, 17 June 1922, 27.

Randall, Adrian and Andrew Charlesworth, eds. *Moral Economy and Popular Protest: Crowds, Conflict and Authority.* New York: St. Martin's, 2000.

Raphael, Chad. *Investigated Reporting; Muckrakers, Regulators, and the Struggle Over Television Documentary.* Urbana, Ill.: University of Illinois Press, 2005.

Raymond, Joad. *The Invention of the Newspaper; English Newsbooks 1641–1649.* Oxford: Clarendon Press, 1996.

Reay, Barry. *The Last Rising of the Agricultural Laborers; Rural Life and Protest in Nineteenth-Century England.* Oxford: Clarendon, 1990.

Reed, T.V. *The Art of Protest; Culture and Activism from the Civil Rights Movement to the Streets of Seattle.* Minneapolis, Minn.: University of Minnesota Press, 2005.

Reese, William J. *America's Public Schools; From the Common School to "No Child Left Behind".* Baltimore, Md.: Johns Hopkins University Press, 2005.

Reisler, Mark. "Mexican Unionization in California Agriculture, 1927–1936." *Labor History* 14, no. 4 (Fall 1973): 562–79.

"Remembering Independence." *Morning Edition, National Public Radio*, 14 Aug. 1997. <http://www.npr.org/> (accessed 2 Feb. 2001).

Reverend Billy and the Church of Stop Shopping." *Reverend Billy and the Church of Stop Shopping.* <http://www.revbilly.com/> (16 Feb. 2007).

Rhodes, Jane. *Mary Ann Shadd Cary: The Black Press and Protest in the Nineteenth Century.* Bloomington, Ind.: Indiana University Press, 1998.

Riesman, David. *The Lonely Crowd: A Study of the Changing American Character.* New Haven, Conn.: Yale University Press, 1950.

Riis, Jacob. *How the Other Half Lives; Studies among the Tenements of New-York.* Reprint edition. New York: Hill and Wang, 1957.

Risen, James and Eric Lichtblau. "Bush Lets U.S. Spy on Callers Without Courts." *New York Times*, 15 December 2005, A1.

Ritter, Alan. *Anarchism: A Theoretical Analysis.* Cambridge: Cambridge University Press, 1980.

Ritter, Gretchen. *Goldbugs and Greenbacks: the Antimonopoly Tradition and the Politics of Finance in America*. New York: Cambridge University Press, 1997.
Roberts, John Michael. *The Aesthetics of Free Speech; Rethinking the Public Sphere*. New York: Palgrave Macmillan, 2003.
Rocker, Rudolf. *Anarcho-Syndicalism*. 6th edition. Edinburgh: AK Press, 2004.
Rodriguez, Clemencia. *Fissures in the Mediascape; An International Study of Citizens' Media*. Cresskill, N.J.: Hampton, 2001.
Rogers, Nicholas. *Crowds, Culture, and Politics in Georgian Britain*. Oxford: Clarendon Press, 1998.
Rogers, Richard. *Information Politics on the Web*. Cambridge, Mass.: MIT Press, 2004.
Roggenkamp, Karen. *Narrating the News: New Journalism and Literary Genre in Late Nineteenth-Century American Newspapers and Fiction*. Kent, Ohio: Kent State University Press, 2005.
Roscigno, Vincent J. and William F. Danaher. *The Voice of Southern Labor; Radio, Music, and Textile Strikes, 1929–1934*. Minneapolis, Minn.: University of Minnesota Press, 2004.
Rose, Carol M. "The Ancient Constitution vs. the Federalist Empire: Anti-Federalism from the Attack on 'Monarchism' to Modern Localism." *Northwestern University Law Review* 84, no. 1. (1989–1990): 74–105.
Rosen, Jay. *Getting the Connections Right: Public Journalism and the Troubles in the Press*. New York: Twentieth Century Fund, 1996.
Rosenberg, Bernard and David Manning White, eds. *Mass Culture; The Popular Arts in America*. Glencoe, Ill.: Free Press, 1957.
Rosenzweig, Roy. *Eight Hours for What We Will: Workers and Leisure in an Industrial City, 1870–1920*. Cambridge: Cambridge University Press, 1983.
Rosewater, Victor. *History of Cooperative News-Gathering in the United States*. New York: D. Appleton and Company, 1930.
Ross, Elizabeth. "To Win, Don't Hide Behind False Brand Pretenses." *Advertising Age*, 26 February 2007, 18.
Roy, Tonya Couch. "Coke is It? (Dis)Locating Identity in Corporate Identity Discourse." Unpublished Ph.D. diss., University of Georgia, 2004.
Royle, Edward. *Robert Owen and the Commencement of the Millennium: a Study of the Harmony Community*. New York: Manchester University Press, 1998.
Rubin, Jerry. *Do It!; Scenarios of the Revolution*. New York: Simon and Schuster, 1970.
Rubin, Nan and Sharon Maeda. "Funding Media for Social Change." *MediaWorks Initiative*. November 2003. <http://www.centerforsocialmedia.org/documents/ FinalMWIreport.pdf> (11 Nov. 2006).
Rudé, George. *The Crowd in History; A Study of Popular Disturbances in France and England, 1730–1848*. New York: Wiley, 1964.
Ruff, Allen. *"We Called Each Other Comrade": Charles H. Kerr & Company, Radical Publishers*. Urbana, Ill.: University of Illinois Press, 1997.
Ruiz, Vicki L. *Cannery Women, Cannery Lives; Mexican Women, Unionization, and the California Food Processing Industry, 1930–1950*. Albuquerque, N.M.: University of New Mexico Press, 1987.
Rury, John L. *Education and Social Change; Themes in the History of American Schooling*. Mahwah, N.J.: Lawrence Erlbaum, 2005.
Rusk, Ralph L., ed. *The Letters of Ralph Waldo Emerson*. 2 vols. New York: Columbia University Press, 1939.
Rydell, Robert W. and Rob Kroes. *Buffalo Bill in Bologna: The Americanization of the World, 1869–1922*. Chicago, Ill.: University of Chicago Press, 2005.

Safley, Thomas Max. *The Reformation of Charity: The Secular and Religious in Early Modern Poor Relief.* Boston, Mass.: Brill Academic Publishers, 2003.

Said, Edward. *Covering Islam; How the Media and the Experts Determine How We See the Rest of the World.* Revised edition. New York: Vintage, 1997.

Salerno, Salvatore. *Red November, Black November; Culture and Community in the Industrial Workers of the World.* Albany, N.Y.: State University of New York Press, 1989.

Salter, Lee. "Structure and Forms of Use; A Contribution to Understanding the 'Effects' of the Internet on Deliberative Democracy." *Information, Communication and Society* 7, no. 2 (June 2004): 185–206.

Samantrai, Ranu. *AlterNatives; Black Feminism in the Postimperial Nation.* Stanford, Calif.: Stanford University Press, 2002.

Samson, Gloria Garrett. *The American Fund for Public Service; Charles Garland and Radical Philanthropy, 1922–1941.* Westport, Conn.: Greenwood, 1996.

Samuel, Raphael, Ewan MacColl, and Stuart Cosgrove. *Theatres of the Left, 1880–1935: Workers' Theatre Movements in Britain and America.* Boston, Mass.: Routledge & Kegan Paul, 1985.

Sanders, Elizabeth. *Roots of Reform; Farmers, Workers, and the American State, 1877–1917.* Chicago: University of Chicago Press, 1999.

Sanders, Lisa. "Creative License; Consumer-Generated Content Is the Future, but as Chevy Discovered, It's Also a Scary Prospect." *Advertising Age*, 17 April 2006, 12.

Saveth, Edward. "Patrician Philanthropy in America: The Late Nineteenth and Early Twentieth Centuries." *Social Service Review* 54, no. 1 (March 1980): 76–91.

Scannell, Paddy. "Introduction: The Relevance of Talk." Pp. 1–13 in *Broadcast Talk*, edited by Paddy Scannell. Newbury Park, Calif.: Sage, 1991.

Schechner, Richard. *The Future of Ritual; Writings on Culture and Performance.* New York: Routledge, 1993.

Schehr, Robert C. *Dynamic Utopia; Establishing Intentional Communities as a New Social Movement.* Westport, Conn.: Bergin & Garvey, 1997.

Schiller, Daniel. *Theorizing Communication: A History.* New York: Oxford University Press, 1996.

Schiller, Herbert. *Culture, Inc.* New York: Oxford University Press, 1989.

Schlossman, David A. *Actors and Activists; Politics, Performance, and Exchange Among Social Worlds.* New York: Routledge, 2002.

Schmeckebier, Laurence. *The Government Printing Office, Its History, Activities, and Organization.* Baltimore, Md.: Johns Hopkins University Press, 1925.

Schneewind, J.B. ed. *Giving; Western Ideas of Philanthropy.* Bloomington, Ind.: Indiana University Press, 1996.

Schneider, Gary. *The Culture of Epistolarity; Vernacular Letters and Letter Writing in Early Modern England, 1500–1700.* Newark, Del.: University of Delaware Press, 2005.

Schofield, Phillipp R. *Peasant and Community in Medieval England, 1200–1500.* New York: Palgrave Macmillan, 2003.

Schudson, Michael. *Advertising, the Uneasy Persuasion: Its Dubious Impact on American Society.* New York: Basic Books, 1984.

———. *Discovering the News; A Social History of American Newspapers.* New York: Basic Books, 1978.

———. "The Objectivity Norm in American Journalism." *Journalism: Theory, Practice, Criticism* 2, no. 2 (August 2001): 149–70.

Schwarzlose, Richard A. *The Nation's Newsbrokers, Volume 1: The Formative Years, from Pretelegraph to 1865*. Evanston, Ill.: Northwestern University Press, 1989.

———. *The Nation's Newsbrokers, Volume 2: The Rush to Institution, from 1865 to 1920*. Evanston, Ill.: Northwestern University Press, 1990.

Schweingruber, David. "Mob Sociology and Escalated Force: Sociology's Contribution to Repressive Police Tactics." *Sociological Quarterly* 41, no. 3 (Summer 2000): 371–389.

Sealander, Judith. "Curing Evils at Their Source: The Arrival of Scientific Giving." Pp. 217–239 in *Charity, Philanthropy, and Civility in American History*, edited by Lawrence J. Friedman and Mark D. McGarvie. New York: Cambridge University Press, 2003.

———. *Private Wealth and Public Life; Foundation Philanthropy and the Reshaping of American Social Policy from the Progressive Era to the New Deal*. Baltimore, Md.: Johns Hopkins University Press, 1997.

Sedgwick, C.M. "Scenes from Life in Town," *The Lady's Book*, April 1843, 159.

"Seeing Is Believing; Using New Technologies to Change the World." *Seeing Is Believing*. <http://www.seeingisbelieving.ca/> (12 Sept. 2006).

Sell, Mike. *Avant-Garde Performance & the Limits of Criticism: Approaching the Living Theatre, Happenings/Fluxus, and the Black Arts Movement*. Ann Arbor, Mich.: University of Michigan Press, 2005.

Shaaber, Matthias A. *Some Forerunners of the Newspaper in England. 1476–1622*. Philadelphia. Pa.: University of Pennsylvania Press, 1929.

Shafer, Richard. "Comparing Development Journalism and Public Journalism as Interventionist Press Models." *Asian Journal of Communication* 8, no. 1 (March 1998): 31–52.

Shagan, Ethan A. "Rumors and Popular Politics in the Reign of Henry VIII." Pp. 30–66 in *The Politics of the Excluded, c. 1500-1850*, edited by Tim Harris. New York: Palgrave, 2001.

Shah, Hemant. "Modernization, Marginalization, and Emancipation: Toward a Normative Model of Journalism and National Development." *Communication Theory* 6, no. 2 (May 1996): 143–166.

Shapin, Steven. *A Social History of Truth: Civility and Science in Seventeenth-Century England*. Chicago: University of Chicago Press, 1994.

Shapiro, Barbara J. *A Culture of Fact; England, 1550–1720*. Ithaca, N.Y.: Cornell University Press, 2000.

Share, Jeff, Tessa Jolls, and Elizabeth Thoman. "Five Key Questions That Can Change the World; Classroom Activities for Media Literacy." *Center for Media Literacy* <http://www.medialit.org/pdf/mlk/02_5KQ_ClassroomGuide.pdf> (25 Apr. 2007).

"Sharkle.com—Free Online Video Sharing Community." *Sharkle*. <http://www.sharkle.com/> (16 Feb. 2007).

Sheehan, Cindy. *Not One More Mother's Child*. Kihei, Hawaii: Koa Books, 2005.

Sheehan, Seán M. *Anarchism*. London: Reaktion Books, 2003.

Shefrin, Elana. "Lord of the Rings, Star Wars, and Participatory Fandom: Mapping New Congruencies Between the Internet and Media Entertainment Culture." *Critical Studies in Media Communication* 21, no. 3 (September 2004): 261–281.

Sheppard, Richard. *Modernism–Dada–Postmodernism*. Evanston, Ill.: Northwestern University Press, 2000.

"Sheri Herndon." *Olympic Movement for Justice and Peace* <http://www.omjp.org/ herndon.html> (25 Apr. 2007).

Shillington, V. George. *Jesus and His Parables: Interpreting the Parables of Jesus Today*. Edinburgh: T&T Clark, 1997.

Shindo, Charles J. "The Dust Bowl Myth." *Wilson Quarterly* 2000 24, no. 4 (Autumn 2000): 25–30.

———. *Dust Bowl Migrants in the American Imagination*. Lawrence, Kan.: University Press of Kansas, 1997.

Sholle, David, and Stan Denski. *Media Education and the (Re)Production of Culture*. Westport, Conn.: Bergin and Garvey, 1994.

Shore, Elliot. "Selling Socialism: The *Appeal to Reason* and the Radical Press in Turn-of-the-Century America." *Media, Culture and Society* 7, no. 2 (April 1985): 147–68.

———. *Talkin' Socialism: J.A. Wayland and the Role of the Press in American Radicalism, 1890–1912*. Lawrence, Ks.: University of Kansas Press, 1988.

Shuger, Debora. *Censorship and Cultural Sensibility: The Regulation of Language in Tudor-Stuart England*. Philadelphia, Pa.: University of Pennsylvania Press, 2006.

Siebert, Fredrick Seaton. *Freedom of the Press in England 1476–1776; the Rise and Decline of Government Controls*. Urbana, Ill.: University of Illinois Press, 1952.

Siers-Poisson, Judith. "Funding the Left." *Center for Media and Democracy*, 28 July 2006. <http://www.prwatch.org/node/5034> (13 Oct. 2006).

Silverman, David and David N. Tobenkin. "The FCC's Main Studio Rule: Achieving Little for Localism at a Great Cost to Broadcasters." *Federal Communications Law Journal* 53, no. 3 (May 2001): 469–507.

Silverstone, Roger. "Television and Everyday Life: Towards an Anthropology of the Television Audience." Pp. 173–89 in *Public Communication; the New Imperatives*, edited by Marjorie Ferguson. Thousand Oaks, Calif.: Sage, 1990.

Sinyai, Clayton. *Schools of Democracy; A Political History of the American Labor Movement*. Ithaca, N.Y.: ILR Press, 2006.

Sinclair, Upton. *The Jungle*. New York: Jungle Publishing Company, 1906.

Sipress, Alan. "FCC Vote A Victory For Phone Companies; TV Service Rule Hurts Cable Firms." *Washington Post*, 21 December 2006, D1.

Skowronek, Stephen. *Building a New American State; The Expansion of National Administrative Capacities, 1877–1920*. New York: Cambridge University Press, 1982.

Smith, Adam. *An Inquiry into the Nature and Causes of the Wealth of Nations*. 6th ed. Edited by Edwin Cannan. London: Methuen, 1961.

Smith, Culver H. *The Press, Politics, and Patronage; The American Government's Use of Newspapers 1789–1875*. Athens, Ga.: University of Georgia Press, 1977.

Smith, Jeffery A. *Printers and Press Freedom: The Ideology of Early American Journalism*. New York: Oxford University Press, 1988.

Smith, Marilyn Crafton. "Women's Movement Media and Cultural Politics." Pp. 278–98 in *Women and Mass Communication; Challenging Gender Values*, edited by Pamela J. Creedon. Newbury Park, Calif.: Sage, 1989.

Smith, Nigel. *Literature and Revolution in England 1640–1660*. New Haven, Conn.: Yale University Press, 1994.

———. *Perfection Proclaimed; Language and Literature in English Radical Religion 1640–1660*. Oxford: Clarendon Press, 1989.

Smith, Owen F. "Fluxus Praxis: An Exploration of Connections, Creativity, and Community." Pp. 116–38 in *At a Distance; Precursors to Art and Activism on the Internet*, edited by Annmarie Chandler and Norie Newmark. Cambridge, Mass.: MIT Press, 2005.

Smulyan, Susan. *Selling Radio: The Commercialization of American Broadcasting, 1920–1934*. Washington, D.C.: Smithsonian Institution Press, 1994.

Smuts, R. Malcolm. *Culture and Power in England 1585–1685*. New York: St. Martin's Press, 1999.

Snider, J.H. *An Explanation of the Citizen's Guide to the Airwaves*. Washington, D.C.: Spectrum Policy Program, New America Foundation, 2003.

"Sniggle.net: The Culture Jammer's Encyclopedia." *Sniggle.net*. <http://www.sniggle.net/> (16 Feb. 2007).

Society of Professional Journalists. "SPJ Code of Ethics." <http://www.spj.org/ ethicscode.asp> (14 Jul. 2006).

Solomon, Julie Robin. *Objectivity in the Making; Francis Bacon and the Politics of Inquiry*. Baltimore, Md.: Johns Hopkins University Press, 1998.

Solomon, William. "The Site of Newsroom Labor; The Division of Editorial Practices." Pp. 110–134 in *Newsworkers; Toward a History of the Rank and File*, edited by Hanno Hardt and Bonnie Brennen. Minneapolis: University of Minnesota Press, 1995.

Sommerville, C. John. *The News Revolution in England; Cultural Dynamics of Daily Information*. New York: Oxford University Press, 1996.

"SourceCode | Reboot the System | Submission." *SourceCode*. <http://sourcecode.freespeech.org/submission/> (19 Oct. 2006).

"SourceCode: About." *SourceCode*. <http://sourcecode.freespeech.org/about> (19 Oct. 2006).

Spann, Edward K. *Brotherly Tomorrows; Movements for a Cooperative Society in America, 1820–1920*. New York: Columbia University Press, 1989.

Spinelli, Martin. "Democratic Rhetoric and Emergent Media: The Marketing of Participatory Community on Radio and the Internet." *International Journal of Cultural Studies* 3, no. 2 (August 2000): 268–78.

Spurr, John. "'Virtue, Religion and Government': The Anglican Uses of Providence." Pp. 29–47 in *The Politics of Religion in Restoration England*, edited by Tim Harris, Paul Seaward, and Mark Goldie. Oxford: Basil Blackwell, 1990.

———. "From Puritanism to Dissent, 1660–1700." Pp. 234–65 in *The Culture of English Puritanism, 1560–1700*, edited by Christopher Durston and Jacqueline Eales. Basingstoke, UK: Macmillan, 1996.

"SSRC: Necessary Knowledge for a Democratic Public Sphere: Program Background Papers." *Social Science Research Council*. <http://www.ssrc.org/programs/ media/background/> (19 Sept. 2006).

Stark, Ryan J. "From Mysticism to Skepticism: Stylistic Reform in Seventeenth-century British Philosophy and Rhetoric." *Philosophy and Rhetoric* 34, no. 4 (2001): 322–334.

"State of the Union—George Bush Remix." *EBaum's World*. <http://www.ebaumsworld.com/presaddress.html> (31 Jan. 2007).

Stavitsky, Alan G. "The Changing Conception of Localism in U.S. Public Radio." *Journal of Broadcasting & Electronic Media* 38, no. 1. (Winter 1994): 19–33.

Steckel, Richard and Robin Simons. *Doing Best by Doing Good; How to Use Public Purpose Partnerships to Boost Corporate Profits and Benefit Your Community*. New York: Dutton, 1992.

Steel, David. "Surrealism, Literature of Advertising, and the Advertising of Literature in France, 1910–1930." *French Studies* 41, no. 3 (July 1987): 283–97.

Stein, Walter J. "A New Deal Experiment with Guided Democracy: The F.S.A. Migrant Camps in California." *Canadian Historical Papers 1970*, 132–46.

———. *California and the Dust Bowl Migration*. Westport, Conn.: Greenwood Press, 1973.

Steiner, Linda. "The Feminist Cable Collective as Public Sphere Activity." *Journalism: Theory, Practice, Criticism* 6, no. 3 (August 2005): 313–34.

"Step It Up 2007: Index." *Step It Up.* <http://www.stepitup2007.org//index.php> (7 Feb. 2007).

"Step It Up 2007: Organizer FAQs." *Step It Up.* <http://stepitup2007.org/article.php?id=76> (7 Feb. 2007).

"Step It Up 2007: Sign Up to Organize a Day of Action." *Step It Up.* <http://www.stepitup2007.org//article.php?id=38> (7 Feb. 2007).

Stephens, Julie. *Antidisciplinary Protest; Sixties Radicalism and Postmodernism.* Cambridge: Cambridge University Press, 1998.

Sterling, Christopher H. "Television and Radio Broadcasting." Pp. 299–371 in *Who Owns the Media?: Concentration of Ownership in the Mass Communications Industry*, edited by Benjamin Compaine, Christopher H. Sterling, Thomas Guback, and J. Kendrick Noble, Jr. Second edition. White Plains, N.Y.: Knowledge Industry Publications, 1982.

Stevens, Lisa Patel and Thomas W. Bean. *Critical Literacy; Context, Research, and Practice in the K–12 Classroom.* Thousand Oaks, Calif.: Sage, 2007.

Stevenson, Laura Caroline. *Praise and Paradox; Merchants and Craftsmen in Elizabethan Popular Literature.* Cambridge: Cambridge University Press, 1984.

Stewart, Lianne. "Promo Partner Personals: House Party Raises the Roof on In-Home Marketing." *Marketing*, 1 January 2006, 33.

Stimson, Black and Gregory Sholette, eds. *Collectivism After Modernism; The Art of Social Imagination after 1945.* Minneapolis, Minn.: University of Minnesota, 2007.

Stoianovich, Traian. *French Historical Method: The Annales Paradigm.* Ithaca, N.Y.: Cornell University Press, 1976.

Stole, Inger L. *Advertising on Trial: Consumer Activism and Corporate Public Relations in the 1930s.* Urbana, Ill.: University of Illinois Press, 2006.

Stone, Geoffrey R. *Perilous Times: Free Speech in Wartime from the Sedition Act of 1798 to the War on Terrorism.* New York: W.W. Norton, 2004.

Stoney, George C., Barbara Abrash, Lynne Jackson, and Cara Mertes. *A Festschrift in Honor of George Stoney.* Baltimore, Md.: Johns Hopkins University Press, 2001.

Storey, John. *Cultural Consumption and Everyday Life.* New York: Oxford University Press, 1999.

Story, Louise. "A Boston Marketing Stunt That Bombed, or Did It?." *New York Times*, 2 February 2007, C1.

"StoryCorps — Interview Your Loved Ones." *StoryCorps.* <http://www .storycorps.net/> (19 Jan. 2007).

"StoryCorps: About." *StoryCorps.* <http://www.storycorps.net/about/> (19 Oct. 2006.

Strangelove, Michael. *The Empire of Mind; Digital Piracy and the Anti-Capitalist Movement.* Toronto, Ontario: University of Toronto Press, 2005.

Stranman, Kerry. "Consumer Created Content." *Word of Mouth Marketing Association.* <http://www.womma.org/content/consumercreatedcontent.pdf> (19 Sept. 2005).

Strasser, Susan. *Satisfaction Guaranteed; The Making of the American Mass Market.* New York: Pantheon, 1989.

Street, Richard Steven. *Beasts of the Field; A Narrative History of California Farmworkers, 1769–1913.* Stanford, Calif. Stanford University Press, 2004.

———. *Photographing Farmworkers in California.* Stanford, Calif. Stanford University Press, 2004.

———. "Poverty in the Valley of Plenty: The National Farm Labor Union, DiGiorgio Farms, and Suppression of Documentary Photography in California, 1947–66." *Labor History* 48, no. 1 (February 2007): 25–48.

Streeter, Thomas. *Selling the Air; A Critique of the Policy of Commercial Broadcasting in the United States*. Chicago, Ill.: University of Chicago Press, 1996.

Streitmatter, Rodger. *Voices of Revolution: The Dissident Press in America*. New York: Columbia University Press, 2001.

Suggs, Henry Lewis. *The Black Press in the Middle West, 1865–1985*. Westport, Conn.: Greenwood Press, 1996.

———. *The Black Press in the South, 1865–1979*. Westport, Conn.: Greenwood Press, 1983.

Sullivan, Paul W. "News Piracy: Unfair Competition and the Misappropriation Doctrine." *Journalism Monographs*, no. 56. (May 1978).

Sumner, Gregory D. *Dwight MacDonald and the Politics Circle: The Challenge of Cosmopolitan Democracy*. Ithaca, N.Y.: Cornell University Press, 1996.

Sung, Dong-Gyu. *Cyber Communication*. Seoul: Segeysa, 2002.

"Surveillance Camera Players." *Factory School*. <http://www.factoryschool.org/pubs/scp/> (12 Feb. 2007).

Susman, Warren. *Culture as History; The Transformation of American Society in the Twentieth Century*. Reprint ed. Washington, D.C.: Smithsonian Institution Press, 2003.

Sutherland, James. *The Restoration Newspaper and Its Development*. Cambridge: Cambridge University Press, 1986.

"Sweatshopper.org." *Sweatshopper*. <http://www.sweatshopper.org/> (24 Oct. 2006).

Tagg, John. *The Burden of Representation: Essays on Photographies and Histories*. Amherst, Mass.: University of Massachusetts Press, 1988.

Tarleton, John. "Brad Will's Remarkable Life." *The Indypendent*, November 2006, 9.

Tarman, Glen. "Digital Activism, the WTO and International Trade Rules." *Digital Opportunity Channel*, 5 September 2003. <http://www.digitalopportunity.org/ article/view/67349/1/1134> (9 Feb. 2007).

Taylor, Barbara. *Eve and the New Jerusalem: Socialism and Feminism in the Nineteenth Century*. Cambridge, Mass.: Harvard University Press, 1993.

Taylor, Gary. *Buying Whiteness: Race, Culture, and Identity from Columbus to Hip Hop*. New York: Palgrave Macmillan, 2005.

Taylor, Paul S. and Tom Vasey. "Contemporary Background of California Farm Labor." *Rural Sociology* 1, no. 4 (December 1936): 410–19.

———. "Historical Background of California Farm Labor." *Rural Sociology* 1, no. 2 (June 1936): 281–295.

Taylor, Paul V. *The Texts of Paulo Freire*. Philadelphia, Pa.: Open University Press, 1993.

Teasdale, Parry D. *Videofreex; America's First Pirate TV Station & the Catskills Collective that Turned It On*. Hensonville, N.Y.: Black Dome Press, 1999.

Teitelbaum, Matthew, ed. *Montage and Modern Life, 1919–1942*. Cambridge, Mass.: MIT Press, 1992.

"The Coming of the Wireless Telephone." *The Spectator*, 13 May 1922, 582–83.

"The First Amendment Project." *The Sundance Channel*. <http://www. sundancechannel.com/firstamendment/> (19 Oct. 2006).

"The Gospel by Wireless." *The Literary Digest*, 4 February 1922, 32.

"The Ground Truth: About: About the Production." *The Ground Truth*. <http://thegroundtruth.org/> (19 Oct. 2006).

"The H-Bomb Secret: How We Got It—Why We're Telling It." *The Progressive*, November 1979.

The Holy Bible, Containing The Old And New Testaments, In The Common Version. With Amendments Of The Language, by Noah Webster. New Haven: Durrie & Peck, 1833.

"The Telephone Newspaper." *Scientific American*, October 1895, 267.

"The Yes Men." *The Yes Men*. <http://www.theyesmen.org/> (16 Feb. 2007).

"thepartyparty.com." *Thepartyparty.com*. <http://www.thepartyparty.com/> (16 Feb. 2007).

Thoman, Elizabeth and Jessica Jolls. "Literacy for the 21st Century; An Overview and Orientation Guide to Media Literacy Education." *Center for Media Literacy* <http://www.medialit.org/pdf/mlk/01_MLKorientation.pdf> (25 Apr. 2007).

Thomas, W.I. and Florian Znaniecki. *The Polish Peasant in Europe and America: Monograph of an Immigrant Group*. 5 vols. Boston: Gorham Press, 1918–1920.

Thompson, Dorothy. *The Chartists; Popular Politics in the Industrial Revolution*. New York: Pantheon, 1984.

Thompson, E.P. *The Making of the English Working Class*. New York: Vintage, 1966.

———. "The Moral Economy of the English Crowd in the Eighteenth Century." *Past and Present* no. 50 (February 1971): 76–136.

Thompson, Julius E. *The Black Press in Mississippi, 1865–1985*. Gainesville, Fla.: University Press of Florida, 1993.

Thompson, Margaret E., Katerina Anfossi Gómez, and María Suárez Toro. "Internet Audience Perceptions of Feminist International Radio Endeavour (FIRE)." *Feminist Media Studies* 5, no. 2 (July 2005): 215–236.

Thompson, Nato and Gregory Sholette, eds. *The Interventionists; Users' Manual for the Creative Disruption of Everyday Life*. North Adams, Mass.: MASS MoCA, 2004.

Thorburn, David and Henry Jenkins, eds. *Rethinking Media Change; The Aesthetics of Transition*. Cambridge, Mass.: MIT Press, 2003.

Thoreau, Henry David. *Walden, or, Life in the Woods*. Boston: Ticknor and Fields, 1854.

Thussu, Daya Kishan, ed. *Electronic Empires: Global Media and Local Resistance*. London: Arnold, 1998.

Todd, Charles L. "Trampling Out the Vintage." *Common Sense*, July 1939, 7–8, 30.

Tomlinson, John. *Cultural Imperialism: a Critical Introduction*. Baltimore, Md.: Johns Hopkins University Press, 1991.

———. *Globalization and Culture*. Chicago, Ill.: University of Chicago Press, 1999.

"TomPaine.com—About." *TomPaine.com*. <http://tompaine.com/about/> (19 Oct. 2006).

Tönnies, Ferdinand. *Community and Civil Society*. Translated by Jose Harris and Margaret Hollis. New York: Cambridge University Press, 2001.

"Top 35 Countries With the Highest Internet Penetration Rate." *Internet World Stats*. <http://www.internetworldstats.com/top25.htm> (26 Mar. 2007).

"Top 50 Recipients of Foundation Grants for Media and Communications, circa 2005." *Foundation Center*. <http://foundationcenter.org/findfunders /statistics/) > (4 Sept. 2007).

Tracey, Michael. *The Decline and Fall of Public Service Broadcasting*. New York: Oxford University Press, 1998.

Trachtenberg, Alan. *The Incorporation of America; Culture and Society in the Gilded Age*. New York: Hill and Wang, 1982.

"Tremor—Step 1." *Tremor*. <http://www.tremor.com/teenscreener/Page1.aspx> (14 Mar. 2007).

Trey, George. *Solidarity and Difference; The Politics of Enlightenment in the Aftermath of Modernity*. Albany, N.Y.: State University of New York Press, 1998.
Trickett, Eleanor. "Inside the Mix—Spoof Ads." *PR Week*, 14 November 2005, 10.
Trolander, Judith Ann. *Settlement Houses and the Great Depression*. Detroit, Mich.: Wayne State University Press, 1975.
"TruthOut—Donations." *TruthOut*. <https://secure.entango.com/donate/> (24 Sept. 2005).
Tucker, Jr., Kenneth H. *French Revolutionary Syndicalism and the Public Sphere*. Cambridge: Cambridge University Press, 1996.
Tunstall, Jeremy and David Machin. *The Anglo-American Media Connection*. Oxford: University of Oxford Press, 1999.
Tunstall, Jeremy. *Communications Deregulation: the Unleashing of America's Communication*. New York: Basil Blackwell, 1986.
Turner, Catherine. *Marketing Modernism Between the Two World Wars*. Amherst, Mass.: University of Massachusetts Press, 2003.
U.S. Department of Agriculture, Farm Security Administration. *Report of the Farm Security Administration, 1941*. Washington D.C.: U.S. Department of Agriculture, Farm Security Administration, 1941.
U.S., Congress, House, Committee on the Judiciary, Antitrust Task Force. Hearing on Competition and the Future of Digital Music. Testimony of Mel Karmazin. 110th Cong., 1st. sess., 28 February 2007.
U.S., Congress, House, Minority Staff. "The Constitution in Crisis; The Downing Street Minutes and Deception, Manipulation, Torture, Retribution, Coverups in the Iraq War, and Illegal Domestic Surveillance." Minority Report Prepared at the Direction of Rep. John Conyers, Jr. August 2006.
U.S., Congress, Senate, Committee on Commerce, Science, and Transportation. *Preservation of Localism, Program Diversity, and Competition in Television Broadcast Services Act of 2003*. 108th Cong., 1st sess. Washington, D.C.: Government Printing Office, 2003.
U.S., Congress, Senate, Committee on Education and Labor. *Violations of Free Speech and Rights of Labor*. Reprint edition. New York: Arno Press, 1975.
U.S., Congress, Senate, Committee on Health, Education, Labor, and Pensions. Hearing on "Elementary and Secondary Education Act Reauthorization: Improving NCLB To Close The Achievement Gap." Testimony of Reg Weaver. 13 March 2007, 109th Cong., 1st sess. <http://help.senate.gov/Hearings /2007_03_13/Weaver.pdf> (25 Apr. 2007).
U.S., Congress, Senate, Committee on the Judiciary, Subcommittee on the Constitution, Civil Rights and Property Rights. Hearing on "What's in a Game? Regulation of Violent Video Games and the First Amendment." 109th Cong., 2nd sess., 29 March 2006. Washington, D.C.: GPO, 2006.
———. Hearing on "Why the Government Should Care About Pornography: The State Interest in Protecting Children and Families." 109th Cong., 1st sess., 10 November 2005. Washington, D.C.: GPO, 2006.
U.S., Department of Labor, Bureau of Labor Statistics, *Labor Unionism in American Agriculture*, by Stuart M. Jamieson, Bulletin No. 836. Washington: GPO, 1945; reprint New York: Arno Press, 1975.
Underwood, Doug. *From Yahweh to Yahoo!: The Religious Roots of the Secular Press*. Urbana, Ill.: University of Illinois Press, 2002.
———. *When MBAs Rule the Newsroom*. New York: Columbia University Press, 1993.

Urstad, Tone Sundt. *Sir Robert Walpole's Poets: The Use of Literature as Pro-Government Propaganda, 1721–1742.* Newark, N.J.: University of Delaware Press, 1999.

"Usage and Population Statistics," *Internet World Stats.* <http://www.internetworldstats.com/stats.htm> (26 Mar. 2007).

Usborne, David. "What Went Wrong in Oaxaca?." *The Independent*, 31 October 2006, 26.

Vaidhyanathan, Siva. *Copyrights and Copywrongs; The Rise of Intellectual Property and How It Threatens Creativity.* New York: New York University Press, 2001.

van de Donk, Wim, Brian D. Loader, Paul G. Nixon, and Dieter Rucht, eds. *Cyberprotest; New Media, Citizens and Social Movements.* New York: Routledge, 2004.

van der Linden, Marcel and Wayne Thorpe. *Revolutionary Syndicalism; An International Perspective.* Aldershot: Scolar Press, 1990.

Veblen, Thorstein. T*he Theory of the Leisure Class; An Economic Study in the Evolution of Institutions.* New York: Macmillan, 1899.

"Video Activist Network." *Video Activist Network.* <http://www.videoactivism.org/> (16 Feb. 2007).

Viguerie, Richard A. and David Franke. *America's Right Turn: How Conservatives Used New and Alternative Media to Take Power.* Chicago, Ill.: Bonus Books, 2004.

Voss, Kim. *The Making of American Exceptionalism: The Knights of Labor and Class Formation in the Nineteenth Century.* Ithaca, N.Y.: Cornell University Press, 1993.

Vranica, Suzanne. "Best (and Worst) Ads of '06." *Associated Press.* 22 December 2006. Retrieved from LexisNexis Academic <http://web.lexis-nexis.com/> (12 Feb. 2007).

Waldman, Diane. *Collage, Assemblage, and the Found Object.* New York: Harry N. Abrams, 1992.

Walljasper, Jay. "Age of the Mega-Alternatives." *Utne Reader*, July–August 1997, 91–94.

Waring, Stephen P. *Taylorism Transformed: Scientific Management Theory Since 1945.* Chapel Hill, N.C.: University of North Carolina Press, 1991.

Wark, Jayne. *Radical Gestures; Feminism and Performance Art in North America.* Montreál, Quėbec: McGill-Queens University Press, 2006.

Warner, Michael. *Publics and Counterpublics.* New York: Zone Books, 2002

Waters, Malcolm ed. *Modernity: Critical Concepts.* 4 vols. New York: Routledge, 1999.

Watkins, William H. "A Marxian and Radical Reconstructionist Critique of American Education." Pp. 107–135 in *Black Protest Thought and Education*, edited by William H. Watkins. New York: Peter Lang, 2005.

Waugh, Thomas, ed. *"Show Us Life"; Toward a History and Aesthetics of the Committee Documentary.* Metuchen, N.J.: Scarecrow Press, 1984.

Wayne, Cynthia. *Dreams, Lies, and Exaggerations; Photomontage in America.* College Park: Art Gallery at the University of Maryland, 1991.

"Web Analytics Pioneer Brett Hurt Starts New Company To Help Online Brands Build Their Business Through Word of Mouth." *Business Wire*, 11 January 2006> (from LexisNexis at http://web.lexis-nexis.com/, 22 February 2007.

Webb, Gary. *Dark Alliance: The CIA, the Contras, and the Crack Cocaine Explosion.* New York: Seven Stories Press, 1998.

Weber, Devra. *Dark Sweat, White Gold; California Farm Workers, Cotton, and the New Deal.* Berkeley, Calif.: University of California Press, 1994.

Weir, Robert E. *Beyond Labor's Veil; The Culture of the Knights of Labor.* University Park, Pa.: Pennsylvania State University Press, 1996.

Weisman, John. *Guerrilla Theater; Scenarios for Revolution.* Garden City, N.Y.: Anchor Press, 1973.

"Welcome to Counterpunch." *Counterpunch.* <http://www.counterpunch.org/ aboutus.html> (19 Oct. 2006).

"Welcome to Tremor." *Tremor.* <http://www.tremor.com/login/about_us.aspx> (14 Mar. 2007).

Wenocur, Stanley and Michael Reisch. *From Charity to Enterprise; The Development of American Social Work in a Market Economy.* Urbana, Ill.: University of Illinois Press, 1989.

Westerkamp, Hildegard. "The Soundscape on Radio." Pp. 87–94 in *Radio Rethink; Art, Sound, Transmission,* edited by Daina Augaitis and Don Lander. Banff: Walter Phillips Gallery, 1994.

"What is DAMN?." *Direct Action Media Network.* <http://www.tao.ca/earth/damn /index2.html> (13 May 2002).

"What is Indymedia?" *The Indypendent,* Nov. 2006, 2.

"What is Media Literacy?." *Center for Media Literacy* <http://www.medialit.org/gifs /MML_404.jpg> (25 Apr. 2007).

White, Jr., Ronald C. and C. Howard Hopkins. *The Social Gospel; Religion and Reform in Changing America.* Philadelphia, Pa.: Temple University Press, 1976.

White, Shirley A., ed. *Participatory Video: Images that Transform and Empower.* Thousand Oaks, Calif.: Sage, 2003.

Whiteaker, L.H. "Adult Education within the Slave Community." Pp. 3–10 in *Education of the African American Adult; An Historical Overview,* edited by Harvey G. Neufeldt and Leo McGee. Westport, Conn.: Greenwood, 1990.

Whitfield, Stephen J. *A Critical American: The Politics of Dwight MacDonald.* Hamden, Conn.: Archon Books, 1984.

"Why Media Matters." *Office of Communication, United Church of Christ.* <http://www.ucc.org/ocinc/> (19 Oct. 2006).

Whyte, William H. *The Organizational Man.* New York: Doubleday, 1957.

Wicke, Jennifer. *Advertising Fictions: Literature, Advertisement & Social Reading.* New York: Columbia University Press, 1988.

Wicklein, John. *Electronic Nightmare: The New Communications and Freedom.* New York: Viking, 1981.

Wiebe, Robert H. *Self-Rule: a Cultural History of American Democracy.* Chicago, Ill.: University of Chicago Press, 1995.

Wiener, Jon. "Radical Historians and the Crisis in American History, 1959–1980." Pp. 175–216 in *Professors, Politics and Pop.* New York: Verso, 1991.

Wilentz, Sean. *Chants Democratic: New York and the Rise of the American Working Class, 1788–1850.* Oxford: Oxford University Press, 1984.

Will, Brad. "'One More Martyr in a Dirty War—One More Time to Cry and Hurt." *The Indypendent,* November 2006, 8.

———. "Fragments of a Shattered Hope; Brazil Makes War on Sohno Real Squatters." *Brad Will in Boog City,* 25 April 2005. <http://bradwillboogcity.blogspot.com/> (28 Oct. 2006).

Williams, Heather Andrea. *Self-Taught; African-American Education in Slavery and Freedom.* Chapel Hill, N.C.: University of North Carolina Press, 2005.

Williams, Lisa, Dan Gillmor, and Jane MacKay. "Frontiers of Innovation in Community Engagement: News Organizations Forge New Relationships with Communities." *Center for Citizen Media.* <http://citmedia.org/frontiers/> (8 Mar. 2007).

Williams, Raymond. "Base and Superstructure in Marxist Cultural Theory." *New Left Review* 82 (November/December 1973): 5–16.

———. *Culture and Society, 1780–1950.* New York: Columbia University Press, 1958.

———. *Keywords; A Vocabulary of Culture and Society*. New York: Oxford University Press, 1976.

———. *Marxism and Literature*. Oxford: Oxford University Press, 1977.

———. *Politics and Letters*. London: New Left Books, 1979.

———. *Problems in Materialism and Culture*. London: Verso, 1980.

———. *Resources of Hope*. Edited by Robin Gable. London: Verso, 1989.

———. *Television; Technology and Cultural Form*. New York: Schocken, 1975.

———. *The Country and the City*. London: Chatto & Windus, 1973.

———. *The Long Revolution*. New York: Columbia University Press, 1961.

———. *The Politics of Modernism; Against the New Conformists*. Edited by Tony Pinkney. London: Verso, 1996.

———. "The Press and Popular Culture: An Historical Perspective." Pp. 41–50 in *Newspaper History from the Seventeenth Century to the Present Day*, edited by George Boyce, James Curran and Pauline Wingate. London: Constable, 1978.

———. *The Year 2000*. New York: Pantheon, 1983.

———. *What I Came To Say*. London: Hutchinson Radius, 1989.

Willis, Fran. "Citizen Journalism, the Next Generation of Local News and Advertising." Presentation to Growing Audience Seminar, Newspaper Association of America, 5 June 2006 <http://www.naa.org/conferences/ gas/program.html> (9 Mar. 2007).

Wilson, James Bright. "Social Attitudes of Certain Migratory Agricultural Workers in Kern County, California." MA thesis, University of Southern California, 1942.

Wilson, Raymond. *In Quest of Community: Social Philosophy in the United States. 1860–1920*. New York: John Wiley, 1968.

"Windows Vista – Show Us Your 'Wow'." *Microsoft Corp*. <http://showusyourwow .msn.com/> (22 Jan. 2007).

Windschuttle, Keith. "Steinbeck's Myth of the Okies." *The New Criterion* 20, no. 10 (June 2002): 24–32.

Winston, Brian. *Claiming the Real: The Griersonian Documentary and its Legitimations*. London: British Film Institute, 1995.

Winters, Donald E. *The Soul of the Wobblies; The I.W.W. Religion, and American Culture in the Progressive Era, 1905–1917*. Westport, Conn.: Greenwood, 1985.

Withington, Phil. *The Politics of Commonwealth; Citizens and Freemen in Early Modern England*. Cambridge: Cambridge University Press, 2005.

"Witness Performance Dashboard Mid-Year Report (Jul.–Dec., FY06)." *Witness*. <http://witness.org/images/stories/pdf/dashboardfinal4.pdf> (20 Mar. 2006).

"Witness: See it. Film it. Change it." *Witness*. <http://www.witness.org/> (20 Mar. 2006).

Wolfram, Eddie. *History of Collage; An Anthology of Collage, Assemblage, and Event Structures*. New York: Macmillan, 1975.

Wollenberg, Charles. "Introduction." Pp. v–xvii in *The Harvest Gypsies*, by John Steinbeck. Berkeley, Calif.: Heyday Books, 1988.

Wolseley, Roland E. *The Black Press, U.S.A.* Ames, Iowa: Iowa State University Press, 1971.

Wood, Andy. *The Politics of Social Conflict: The Peak Country, 1520–1770*. New York: Cambridge University Press, 1999.

Wood, Stephen. *The Degradation of Work?: Skill, Deskilling, and the Labour Process*. London: Hutchinson, 1982.

Woodson, C.G. *The Education of the Negro Prior to 1861; a History of the Education of the Colored People of the United States from the Beginning of Slavery to the Civil War.* New York: G.P. Putnam's, 1915.

"Word of Mouth Marketing Association." *Word of Mouth Marketing Association* <http://www.womma.org/> (10 Mar. 2007).

Wreszin, Michael. *A Rebel in Defense of Tradition: The Life and Politics of Dwight MacDonald.* New York: Basic Books, 1994.

Wrightson, Keith. *English Society 1580-1680.* London: Routledge, 2003.

"WTO | World Trade Organization: WTO / GATT." *WTO / GATT.* <http://gatt.org/> (9 Feb. 2007).

"YOUR Video—A Practical Guide to Planning and Creating Effective Videos." *Video Toolkit* <http://www.ourvideo.org/toolkit/> (25 Apr. 2007).

"YouTube – Broadcast Yourself." *YouTube.* <http://www.youtube.com/> (19 Feb. 2007).

Ytreberg, Espen. "Formatting Participation Within Broadcast Media Production." *Media, Culture and Society* 26, no. 5 (September 2004): 677–92.

Zimmerman, Patricia. *States of Emergency: Documentaries, Wars, Democracies.* Minneapolis: University of Minnesota Press, 2000.

Zinn, Howard. *A People's History of the United States: 1492–Present.* New York: HarperCollins, 1999.

Zunz, Oliver. "The Synthesis of Social Change; Reflections on American Social History." Pp. 81–92 in *Reliving the Past; The Worlds of Social History*, edited by Oliver Zunz. Chapel Hill, N.C.: University of North Carolina, 1985.

Manuscript Collections

Records of the Farm Security Administration, Record Group 96. National Archives–Pacific Sierra Region, San Bruno, Calif.

Simon J. Lubin Society Papers. Bancroft Library, University of California, Berkeley, Calif.

National Agricultural Library, Bethesda, Md.

Archived Newspapers and Magazines

Frederick Douglass Paper
Godey's Lady's Book
The Christian Recorder
The Colored American
The National Era
The North Star
Tow-Sack Tattler
Weedpatch Cultivator

Low-Power FM Radio Station Applications

All are filed responses to "FCC 318—Application for Construction Permit for a Low Power FM Broadcast Station," and available from the Federal Communications Commission.

Anchor City Radio, Anchor City, Alaska. File no. BNPL-20000605AAA, 1 June 2000.

Brunswick High Pirate Pride Booster Club, Brunswick, GA. File no. BNPL-20000531AAF, 24 June 2003.

City Heights Community Development Corporation, San Diego, Calif. File no. BNPL-20000605ADB, 1 June 2000.

City of Logan, Logan, Utah. File no. BNPL-20000605AAS, 1 June 2000.

First United Pentecostal Church of Redlands, Redlands, Calif. File no. BNPL-20000605AAK, 1 June 2000.

Glenn Austin, Minneapolis, Minn. File no. BNPL-20000901ABU, 1 September 2000.

Governing Board Grossmont College, El Cajon, Calif. File no. BNPL-20000531AAI, 31 May 2000.

Harry Bridges Institute, San Pedro, Calif. File no. BNPL-20000605ABQ, 5 June 2000.

House Of Gold Foundation, East Providence, R.I. File no. BNPL-20000608ABN, 8 June 2000.

Howell Mountain Christian Radio, Angwin, Calif. File no. BNPL-20000608ACN, 8 June 2000.

Joseph M. Futrelle, Urbana, Ill. File no. BNPL-20000828ACF, 28 August 2000.

Kyle Cetrulo, San Francisco, Calif. File no. BNPL-20000531AAW, 31 May 2000.

Liberty Creek Homeowners Assoc., Indianapolis, Ind. File no. BNPL-20000605AAG, 8 June 2000.

LSNET Educational Foundation, Galax, Va. File no. BNPL-20000828ABS, 28 August 2000.

Maranatha Fellowship Of Saddleback Valley, Mission Viejo, Calif. File no. BNPL-20000605ADZ, 5 June 2000.

Midwestern Theater Troupe, Inc., Tulsa, Okla. File no. BNPL-20000607AAB, 7 June 2000.

Monroe County Public Library, Bloomington, Ind. File no. BNPL-20000602ACS, 2 June 2000.

National Jazz Hall of Fame Foundation, Palm Springs, Calif. File no. BNPL-20000531ABG, 31 May 2000.

National Multi-Ethnic Families Association, Oxnard, Calif. File no. BNPL-20000530AAI, 24 May 2000.

Prophecy in the News, Inc., Oklahoma City, Okla. File no. BNPL-20000602ADC, 2 June 2000.

Rondaradio, Inc., Round Lake, Ill. File no. BNPL-20000828AAW, 21 October 2003.

Sacramento Community Cable Foundation DBA Access Sacramento, Sacramento, Calif. File no. BNPL-20000607AAY, 7 June 2000.

Severe Storms Intercept Team, Stillwater, Okla. File no. BNPL-20000605ADE, 5 June 2000.

Shishmaref School, Shishmaref, Alaska. File no. BNPL-20000606AAL, 6 June 2000.

Startree Foundation, Inc., Costa Mesa, Calif. File no. BNPL-20000606AAF, 28 October 2002.

Stockton Good Samaritan Service Center, Inc., Stockton, Calif. File no. BNPL-20000605AAC, 2 June 2000.

Tulsa Free Radio Project, Tulsa, Okla. File no. BNPL-20000530AAY, 30 May 2000.

Unity World Production, Incorporated, Fort Wayne, Ind. File no. BNPL-20000608ACU, 5 June 2000.

Washtenaw Community College, Ann Arbor, Mich. File no. BNPL-20000901ABZ, 31 August 2000.

Index

activism, 16, 145. *See also* broadcasting; communication, evangelical mass media; social movements
Adams, Samuel, 8
AdBusters, 183
Addams, Jane, 242. *See also* settlement-house movement
Ader, Clément, 77
adult education, 241–2
Advertising Council, 108. *See also* philanthropic media
advertising, 173, 210–11, 212–14, 268. *See also* modernism
aestheticization, 169–71, 178–9, 268. *See also* critical modernism; dissent; modernism; incorporation
agrarian populism. *See* populism, agrarian
Albert, Michael, 61
Alliance for Community Media, 121, 122
alternative culture, 8, 9
alternative media, and mainstream, 1, 2–4, 9, 13, 79, 94–6, 134; study of, 4–12, 29, 31, 62. *See also* broadcasting; ; community media; evangelical mass media; mainstream media; philanthropic media; vanguardism
Alternative Press Syndicate, 8
American Civil Liberties Union (A.C.L.U.), 181
American Federation of Labor (A.F.L.), 132. *See also* unions, labor
American Society of Newspaper Editors, 2. *See also* Murdoch, Rupert; News Corporation
anarchism, 13, 131, 133–4, 145–6, 167. *See also* Paris Commune
anarcho-syndicalism. *See* anarchism
Anderson, Benedict, 123. *See also* community
Anderson, Perry, 5. *See also* materialism
Anglicanism. *See* Christianity; Church of England
Annales history, 7, 32n8
Anti-mercuries, 49. *See also* reportage
Appeal to Reason, 134–5, 166. *See also* community media; unions, labor
argumentation. *See* culture of truth; polemic; rationalist empiricism;
Aristotelian–Thomist scholasticism, 43. *See also* rationalist empiricism
Armory Show, 175
Arvin (Calif.) Migratory Labor Camp, 240
Associated Press, 2, 206. *See also* TruthOut.org; alternative media; mainstream media
Association of Alternative Newsweeklies, 9
Atton, Chris, 13, 203. *See also* anarchism
attorneys, United States, 2

Bacon, Francis, 43–50, 105. *See also* capitalism; culture of truth; rationalist empiricism; reportage
Bale, John, 36, 47. *See also* providentialism; rationalist empiricism
Barnum and Bailey, 173
Barthes, Roland, 202. *See also* text; structuralism
Beard, Charles, 7

Bede, prophecies of, 37. *See also* prophecy
Beecher, Lyman, 68. *See also* broadcasting; evangelical mass media
Bell, Alexander Graham, 76
Berger, John, 173–4
bias, 33, 54n2, 110, 210. *See also* rationalist empiricism
Bible-tract societies, 62, 67–9, 75, 98. *See also* broadcasting; evangelical mass media
biography, 10, 12, 20n26
Birkhimer, Sam, 247. *See also* migrant farmworkers; unions, labor
Bourdieu, Pierre, 12, 265. *See also* alternative media; structuralism; vanguardism
Bourke-White, Margaret, 178
Braille Institute of America, 108. *See also* philanthropic media
Brave New Films, 221
Brave New Foundation, 111. *See also* documentary; philanthropic media; professionalization
Brecht, Bertolt, 255, 266
Breton, Andre, 174, 178
Broadcast Interactive, 200
broadcasting, 16, 61–85, 99, 110, 135, 141, 145, 179. *See also* culture of truth; evangelical mass media; philanthropic media; publics; technological determinism
broadsides, 39, 47. *See also* manuscripting; reportage
Bureau of Labor Statistics, Department of Labor, 105. *See also* documentary; reportage
bureaucratization, 65, 109–110, 179. *See also* rationalization
Bush, George W., 161–2. *See also* critical modernism

Cable News Network (C.N.N.), 210
capitalism, 16; and community media, 123–5, 130–139; and early modern England, 30–31, 40–41, 45–50, 53, 63–6; and market radicalism, 199–201, 204–11, 216; and media literacy, 252; and modernism, 163–7, 172–3, 176, 179, 183; and philanthropic media, 70, 75–9, 89, 95–103. *See also* empire; institutionalization; mass-culture critique; modernism; participation; professionalization; rationalization
Carnegie Institution of Washington, 103–4
Cartwright, Thomas, 39. *See also* prophecy; manuscripting
Castells, Manuel, 12. *See also* alternative media; structuralism; technological determinism; vanguardism
censorship, 38. *See also* manuscripting
Center for Media Literacy, 254
Central Intelligence Agency (C.I.A.), 2. *See also* "Dark Alliance"
charity. *See* philanthropic media
Charles I, and use of propaganda, 65. *See also* propaganda
Charles II, 42, 48
chartering, 41, 51
Chartier, Robert, 11, 38. *See also* culture
Chartism, 102, 131. *See also* Social Gospel
Chicago School of sociology, 107, 135–6, 141–2, 168. *See also* documentary; philanthropic media; reportage
Chideya, Farai, 94
cholo, 10. *See also* gang member
Christian Broadcasting Network, 216
Christian socialism. *See* Social Gospel
Christianity, 34, 42, 48–50, 67, 102, 125–6. *See also* evangelical mass media; hybridization; providentialism; Protestantism, radical; secularization
Church of England, 65. *See also* Christianity; dissent
Church of Stop Shopping, 183
citizen journalism, 17, 200, 214–20, 253. *See also* community media; market radicalism; participation
citizen marketers, 212. *See also* advertising; participation; rationalization
City Heights Community Development Corporation, 142. *See also* community media; low-power FM radio

"city upon a hill," 125–6. *See also* community; Protestantism, radical
civil wars, English, 30
Clark, Jessica, and Tracy Van Slyke, 94–5, 113. *See also* philanthropic media.
class, and early modern England, 37–9, 41, 44, 64, 68, 70–71; and modernism, 162; and nineteenth-century United States, 16, 74, 80, 100–103, 125–7. *See also* broadcasting; mass-culture critique; Marxist historiography, British; participation; publics
Clinton, Sen. Hillary, 214
Cole, Juan, 94
Collins, Thomas, 241
colonies, utopian. *See* utopian colonies
commentary. *See* polemic; rationalist empiricism
commercialization, 62, 64, 75–9, 98; of media, 130, 136–45. *See also* broadcasting; capitalism; community media; low-power FM radio; newspaper exchange; participation; publics; rationalization
Commission on Freedom of the Press, 209
commodification, 46–8, 50–53. *See also* capitalism; commercialization; rationalization
Common Dreams News Center, 111. *See also* philanthropic media
communication, mass. *See* mass communication
Communications Act of 1934, 137
Communist Party of the United States of America (C.P.U.S.A.) , 243
communitarianism. *See* Chicago School of sociology
communities, imagined. *See* community
communities, scribal. *See* scribal communities
community media, 121–59, 218–20. *See also* community; newspaper exchange
community, 17, 123–8, 130–4, 144. *See also* community media; cooperative commonwealth; radical republicanism; publics
community, utopian. *See* community
composition, direct autonomous. *See* direct autonomous composition
Comte, Auguste, 45. *See also* Bacon, Francis
conditions, historical, 12, 13, 16, 184, 203. *See also* cultures of truth; democratic communication; history; materialism
Congress of Industrial Organization (C.I.O.), 243. *See also* unions, labor
Constitution, U.S., 129. *See also* community media; newspaper exchange
consumer-generated content, 17, 212–14. *See also* citizen journalism; advertising; market radicalism; media industries; participation
consumerism. *See* commercialization; institutionalization; rationalization
Consumers Union, 210
consumption, of media. *See* materialism
content, consumer-generated. *See* consumer-generated content
contra rebels, 2. *See also* "Dark Alliance."
cooperative commonwealth, 125, 132. *See also* community; community media; Patrons of Husbandry (the Grange); radical republicanism; union, labor
Copyright Act of 1790, 129. *See also* newspaper exchange
corantos, 51. *See also* reportage
corporations, 68, 100–101, 130, 173, 268. *See also* broadcasting; evangelical mass media; incorporation; philanthropic media
counter-prophecy. *See* prophecy
Creative Revolution, 212. *See also* advertising
Crisis Pictures, 93–4, 96
critical media literacy, 252. *See also* media literacy
critical modernism, 17, 91, 105, 161–94, 237, 268. *See also* culture of truth; modernism; rationalist empiricism

critical pedagogy, 18, 255. *See also* democratic communications; direct autonomous composition; materialism; Williams, Raymond
Cromwell, Oliver, 40, 42, 51
C-SPAN, 162, 210. *See also* critical modernism
cubism, 173–4
cultural form. *See* form
Cultural Front, 243
culture of truth, 15–17, 30–34, 40, 48–51, 64–7, 79, 96, 161–94, 209–11. *See also* providentialism; critical modernism; rationalist empiricism; publics; rationalization; participation
culture, 12–13. *See also* alternative culture; culture of truth; democratic communications; history; materialism
culture-jamming, 180. *See also* technological determinism
Current TV, 1, 2, 182, 213

dada, 17, 175–6, 177. *See also* culture of truth; critical modernism; form; modernism; performance
"Dark Alliance," and series in *San Jose Mercury News*, 2
Davis, Trevor, 93–4. *See also* Crisis Pictures; philanthropic media
Day, Benjamin, 72
de Forest, Lee, 75, 77. *See also* broadcasting; telephone
De-Bug, 233, 254
Declaration of Independence, 104
democratic communications, 3, 14–18, 34, 96, 112, 166, 199, 255, 265. *See also* history; materialism
Denning, Michael, 243
Department of Education, 121
Dewey, John, 242. *See also* Chicago School of sociology
Diggers, 177. *See also* modernism; performance
Direct Action Media Network, 2
direct action. *See* anarchism
direct autonomous composition, 18, 233–64. *See also* critical pedagogy; democratic communications; participation; Williams, Raymond
dissent, 10, 17, 65, 67, 171, 179, 182. *See also* evangelical mass media; incorporation; modernism; protests; Thompson, E.P.
Dobson, John, 39. *See also* manuscripting
documentary, 12, 105–7, 110–111, 268. *See also* empiricism; rationalist empiricism; reportage
domestic spying, 2. *See also* alternative media
Downing, John D.H., 13, 203. *See also* alternative media; anarchism; history
DuBois, W.E.B. , 243
Dust-Bowl migration. *See* migrant farmworkers

Earl of Leicester, 39. *See also* manuscripting; prophecy
early modern England. *See* England, early modern
Echo Chamber, 221
Eco, Umberto, 202
Economic Research Service, Department of Agriculture, 105. *See also* documentary; reportage
Edison, Thomas, 77. *See also* broadcasting; telephone
education. *See* critical pedagogy; democratic communications; direct autonomous composition
Educational Broadcasting Corporation, 108. *See also* philanthropic media
Einstein, Albert, 168
Elizabeth I, 34–5, 40, 41, 63, 97. *See also* Henry VIII
empire, 40–42, 45, 130–131, 164, 177, 250. *See also* Bacon, Francis; capitalism; Freire, Paulo; modernism
empiricism, 30, 43–50, 106–7. *See also* Bacon, Francis; culture of truth; documentary; rationalist empiricism; reportage
England, early modern, 16, 30–59, 63–6. *See also* culture of truth; providentialism; rationalist empiricism

Enlightenment, 163, 166. *See also* critical modernism; modernism
epistolary culture, 47, 50, 205. *See also* reportage
essay, 45–7. *See also* polemic; rationalist empiricism
essentialism, 5–6, 8, 10–11, 17, 90, 123–5, 184, 201–4; and direct autonomous composition, 237–8, 241, 244, 251, 255. *See also* critical modernism; direct autonomous composition; dissent; modernism; incorporation
ether (æther [*sic*]), 76. *See also* broadcasting; mystification; naturalization
ethics, journalistic codes of, 33. *See also* rationalist empiricism
European Commission, 254
evangelical mass media, 61–85, 98–9, 141. *See also* broadcasting; Christianity; participation; political communication
experimentation. *See* Bacon, Francis; empiricism
expert. *See* professionalization; participation

fact. *See* Bacon, Francis; culture of truth; empiricism
Fallujah, Iraq, 93
Farm Security Administration (F.S.A.), 178, 240–4. *See also* migrant farmworkers
Federal Communications Commission (F.C.C.), 139, 208–9
Federal Writers Project, 168
Federalist Papers, 104
Federation of Churches and Christian Workers, 106. *See also* documentary; reportage
feminist media, 146. *See also* alternative media
Fenn, Humphrey, 39. *See also* manuscripting; prophecy
Fessenden, Reginald, 77. *See also* Ader, Clément; broadcasting; de Forest, Lee; Edison, Thomas
Fifth Monarchists, 40. *See also* prophecy
Film Aid International, 112. *See also* philanthropic media
Firelight Media, 108. *See also* philanthropic media
First Amendment, 68. *See also* broadcasting; evangelical mass media
First United Pentecostal Church of Redlands, 143. *See also* community media; low-power FM radio
Fluxus, 177. *See also* modernism; performance
Ford Foundation, 110. *See also* philanthropic media
form, 16, 36–8, 46–52, 62–6, 105–7, 123. *See also* community media; culture of truth; documentary; prophecy; reportage
formations, 16–18, 29, 80, 89–90, 95, 166–7. *See also* critical modernism; market formation; market radicalism; providentialism; rationalist empiricism
Fourierism, 127. *See also* community
Fox News, 112
franchise, 16, 125
Franken, Al, 94
Frankfurt School, 137, 169. *See also* bureaucratization; mass-culture critique; rationalization
Free Speech TV, 166
Friere, Paulo, 210, 250

gang member, 10. *See also cholo*
Garofalo, Janeane, 94
Garst, Jonathon, 241. *See also* Farm Security Administration (F.S.A.)
George I, 63
Gibson, Thomas, 38. *See also* prophecy
Gillett, James, 203
global village, 166. *See also* critical modernism; technological determinism
Glorious Revolution, 51, 63. *See also* chartering; licensing; monopolies
Gompers, Samuel, 132. *See also* unions, labor
Gore, Al, 1, 94
gossip, 248–9. *See also* direct autonomous composition; migrant farmworkers

Grange. *See* Patrons of Husbandry (the Grange).
grants. *See* philanthropic media; professionalization
Great Awakening, 98. *See also* evangelical mass media
Greene, Robert, 47. *See also* essay; professionalization; rationalist empiricism
Greenpeace, 180
guerrilla marketing, 182, 213
Guerrilla NewsVideo, 2
Guthrie, Woody, 240, 247

Habermas, Jürgen, 11, 64. *See also* culture
Harris, Tim, 64
Harry Bridges Institute, Inc., 142. *See also* community media; low-power FM radio
Harvey, David, 5, 170
Hausmann, Raoul, 175–6
Haward, Sir William, 40. *See also* manuscripting; prophecy
Haywood, William ("Big Bill"), 134, 135
Hearst, William Randolph, 130
Heartfield, John, 176. *See also* modernism; montage
Heisenberg, Werner, 168
Henry VIII, 34–8. *See also* Elizabeth I.
Herndon, Sheri, 255
Herrold, Charles, 78. *See also* broadcasting
hip consumerism, 212. *See also* advertising; incorporation
history, 5, 12–17, 20n17, 89, 171, 200–201, 203–4, 245, 249. *See also* cultures of truth; formations; Marxist historiography, British; materialism; Williams, Raymond
history, labor, new. *See* Marxist historiography, British; new social history
history, social, new. *See* new social history
Hobsbawm, Eric, 7, 64. *See also* Marxist historiography, British
Hoffman, Abbie, 177. *See also* modernism; performance
Holy Cross, 38. *See also* prophecy
homily, 46. *See also* form; providentialism; rationalist empiricism
House of Gold Foundation, 143. *See also* community media; low-power FM radio
house party, 112, 221. *See also* Brave New Foundation; participation
How the Other Half Lives, 107. *See also* documentary; reportage
Howell Mountain Christian Radio, 141. *See also* community media; low-power FM radio
humanism, 42
Hutchins Commission, 209
hybridization, 6, 20n14, 38, 47–50, 97–100, 104–5, 126, 130–34, 141, 163, 171, 214, 254. *See also* empiricism; form; philanthropic media; providentialism; rationalist manuscripting; reportage

imagined communities. *See* community
imperialism. *See* empire
In These Times, 94. *See also* Clark Jessica and Tracy Van Slyke; philanthropic media
incorporation, 10, 90, 100, 166, 183, 180. *See also* capitalism; *cholo*; corporations
Independent Media Centers, 122, 145–6, 211, 221, 254, 255. *See also* anarchism; community; community media
India, 177
Industrial Workers of the World (I.W.W.), 133–5. *See also* unions, labor
inspiration, 268–9
institutionalization, 16–18, 107–9. *See also* capitalism; commercialization; culture of truth; professionalization; rationalization
Intergovernmental Panel on Climate Change, 161
International Center for Journalists, 108. *See also* philanthropic media
International News Serv. v. Associated Press, 129, 150n60. *See also* newspaper exchange

Iowa Public Television, 108. *See also* philanthropic media

James I, 41, 43, 44, 48, 65
Jameson, Fredric, vii, 169
Jayadev, Raj, 233–4
Joachim of Fiore, 36. *See also* prophecy
John F. Slater Fund for the Education of Freedmen in the South, 103
jokes, 248–9. *See also* direct autonomous composition; migrant farmworkers
journalism, citizen. *See* citizen journalism

KCET, 108. *See also* philanthropic media
Keane, John, 235
Kellner, Douglas, 251–3, 254
Knewstub, John, 39. *See also* manuscripting; prophecy
Knights of Labor, 131, 133. *See also* cooperative commonwealth; radical republicanism; unions, labor
Knowledge Networks, 121
Kogawa, Tetsuo, vii
Korea, 214–16
Kropotkin, Peter, 134. *See also* anarchism
Kuhn, Thomas, 168

labor history, new. *See* new social history
labor union. *See* unions, labor
labor, 131, 213, 246. *See also* capitalism; class; participation; unions, labor
Laclau, Ernesto, 12. *See also* structuralism; vanguardism
lampoon, scribal. *See* scribal lampoon
Leicester, Earl of. *See* Earl of Leicester
letter-writing, 50. *See also* epistolary culture; form; reportage
Lewis, John L., 243. *See also* unions, labor
liberalism. *See* pluralism
Liberty Creek Homeowners Association, 145. *See also* community media; low-power FM radio
licensing, 41, 51
linguistics, 17, 168. *See also* critical modernism; dissent; materialism; modernism; reflection theory
Lippmann, Walter, 236. *See also* vanguardism
"listening in," 76, 78. *See also* broadcasting
literacy, media. *See* media literacy
literary journalism. *See* new journalism
localism, 17, 136–9, 206–7, 209, 219–20. *See also* community; community media; mass-culture critique
longue durée, 30, 32n8. *See also* Annales history
low-power FM radio (LPFM), 17, 122, 139–45. *See also* community media
Luddism, 71. *See also* broadcasting; rationalization

MacDonald, Dwight, 137. *See also* community media; localism; mass-culture critique
mainstream media, 1–4, 9, 13, 29, 62, 79, 94–6, 134, 204–11. *See also* alternative media; broadcasting; evangelical mass media; philanthropic media
Manifest Destiny, 75. *See also* broadcasting
Mann, Horace, 72, 241. *See also* adult education; broadcasting; evangelical mass media
manuscripting, 30, 38–40, 51, 65, 205. *See also* providentialism
market democracy. *See* capitalism
market formations, 16, 29, 79–80, 179, 254. *See also* capitalism; culture of truth; formations
market radicalism, 17, 199–231. *See also* capitalism; formations; culture of truth
market. *See* capitalism
marketplace of ideas, 235
Marxist historiography, British. *See* Hobsbawm, Eric; history; materialism; Rudé, George; Thompson, E.P.; Williams, Raymond

mass communication, 16, 17, 61, 63–7, 75, 173. *See also* broadcasting; culture of truth; evangelical mass media; vanguardism; participation; technological determinism
mass culture. *See* mass-culture critique
Massachusetts Bay Colony, 126
mass-culture critique, 6–8, 17, 74–6, 137–8, 101–103, 136–9, 143–4, 173, 205; and media literacy, 251–2, 254, 268. *See also* capitalism; Chicago School of sociology; class; community media; media literacy; participation
mass-society thesis. *See* mass-culture critique
materialism, 5, 12–16, 18, 123, 170, 204, 238, 245. *See also* Anderson, Perry; cultures of truth; democratic communications; Harvey, David; Williams, Raymond
McCarthyism, 209
McLuhan, Marshall, 177, 180. *See also* modernism; technological determinism
media industries, 204–11
media literacy, 18, 250–4. *See also* critical pedagogy; direct autonomous composition; institutionalization; rationalization
media, alternative. *See* alternative media
media, community. *See* community media
media, mainstream. *See* mainstream media
media, philanthropic. *See* philanthropic media
media, social-reform. *See* philanthropic media
media. *See* mass communication; technological determinism
Melucci, Alberto, 12. *See also* structuralism; vanguardism
Merlin, 37, 40. *See also* prophecy
Merton, Robert, 137. *See also* community media; localism; mass-culture critique;
Methodism, 62, 68, 79. *See also* broadcasting; evangelical mass media
Methodist Book Concern, 99. *See also* broadcasting; evangelical mass media; philanthropic media
Metropolitan Opera House (New York City), 77. *See also* broadcasting; de Forest, Lee; telephone
microradio. *See* LPFM radio
Midwestern Theater Troupe, 143. *See also* community media; low-power FM radio
migrant farmworkers, 18, 239–49. *See also* democratic communications; direct autonomous composition; popular culture; unions, labor
migrant labor camp newspapers, 240
millennial republicanism, 126. *See also* Christianity
Mills, C. Wright, 7, 138. *See also* localism; mass-culture critique
mimesis. *See* documentary; reflection theory; reportage
Minnesota Public Radio, 108. *See also* philanthropic media
modernism, 17–18, 163–171, 172–3, 209–11. *See also* culture of truth; critical modernism
monopolies, printing, 41, 51
montage, 172, 175–6, 178, 183. *See also* critical modernism
Montaigne, Michel de, 46. *See also* essay
Morse, Samuel, 127. *See also* community; technological determinism
Mother Shipton, 40. *See also* prophecy
Mouffe, Chantal, 12. *See also* alternative media; structuralism; vanguardism
Mulhern, Francis, 122. *See also* community media; materialism
Murdoch, Rupert, 1–2. *See also* American Society of Newspaper Editors; News Corporation
mystification, 72–6, 170. *See also* evangelical mass media; naturalization; technological determinism

Nashe, Thomas, 47. *See also* essay; professionalization; rationalist empiricism

National Aeronautical and Space Administration (N.A.S.A.), 121, 122. *See also* community media
National Agricultural Statistical Service, Department of Agriculture, 105. *See also* documentary; reportage
National Broadcasting Company (N.B.C.), 77. *See also* broadcasting
National Center for Educational Statistics, Department of Education, 105. *See also* documentary; reportage
National Day of Climate Action, 161, 170
National Intelligencer, 129–30. *See also* newspaper exchange
National Public Radio, 108. *See also* philanthropic media
National Security Administration (N.S.A.), 2
national tract societies. *See* Bible-tract society
natural rights, 124. *See also* community; community media
naturalism, 16, 18, 78, 105–7, 110–111, 172, 173. *See also* broadcasting; critical modernism; culture of truth; form; mystification; naturalization; rationalist empiricism
naturalization, 72–6, 170. *See also* broadcasting; evangelical mass media; mystification; technological determinism
Nedham, Marchamont, 47. *See also* essay; professionalization; rationalist empiricism
neoplatonism, 43. *See also* nominalism; empiricism; rationalist empiricism
New Journalism, 210
new labor history. *See* new social history
new social history, 7. *See also* Marxist historiography, British; popular culture
New Testament, 71. *See also* broadcasting; evangelical mass media
New World Information and Communications Order, 210. *See also* critical modernism; empire
New York City Tract Society, 71. *See also* broadcasting; evangelical mass media
New York Daily Graphic, 76
New York Sun, 72. *See also* broadcasting; evangelical mass media; penny press
News Corporation, 1. *See also* American Society of Newspaper Editors; Fox News; Murdoch, Rupert
news, 47–8, 52, 78. *See also* broadcasting; newsletters; reportage
newsbooks, 51. *See also* reportage
newsletters, 47. *See also* reportage
newspaper exchange, 128–30. *See also* community media
newspapers. *See* media industries
Newspoetry (organization), 144. *See also* community media; low-power FM radio
nominalism, 43. *See also* empiricism; neoplatonism; rationalist empiricism
Non-Aligned Movement, 210. *See also* critical modernism; empire
Nostradamus, 40. *See also* manuscripting; prophecy
novels, 12. *See also* new journalism

Oaxaca, Mexico, 122
objectivity. *See* documentary; empiricism; rationalist empiricism; reportage
Oh, Yeon-Ho, 199, 216. *See also OhmyNews*
OhmyNews, 199, 214–18. *See also* citizen journalism; market radicalism
Old St. Patrick's Church (Pittsburgh, Penn.), 76. *See also* broadcasting; evangelical mass media
Old Testament, 71. *See also* broadcasting; evangelical mass media
oral culture. *See* manuscripting
Outfoxed: Rupert Murdoch's War on Journalism, 112. *See also* Brave New Foundation; documentary; News Corporation
Owenites, 127. *See also* community

Packard, Vance, 138. *See also* localism; mass-culture critique
Paine, Tom, 8. *See also* radical republicanism
Paracelsus, 40. *See also* prophecy
Paris Commune, 102, 131. *See also* anarchism; Social Gospel
Parliament, English, 34
parody. *See* satire
participation, 17–18, 37–39, 46, 53, 63–6, 70–80, 97, 101–12, 133–7; and direct autonomous composition, 238; and market radicalism, 199, 202–11, 204–21. *See also* citizen journalism; consumer-generated content; direct autonomous composition; manuscripting; prophecy
participatory journalism. *See* citizen journalism
patronage, 41
Patrons of Husbandry (the Grange), 132. *See also* unions, labor
Patterson Strike, 175
PBS Foundation, 108. *See also* philanthropic media
Peabody Education Fund, 103
pedagogies, critical. *See* critical pedagogies
Pedagogy of the Oppressed, 250
penny press, 71. *See also* evangelical mass media
People's Choice Award, 121
People's Choice Community, 121, 122
performance art. *See* performance
performance, 171, 177
Perkins, William, 46–7.
Personal Democracy Forum, 94. *See also* philanthropic media
philanthropic media, 93–120, 163, 166. *See also* capitalism; documentary; naturalism; professionalization; rationalization; reportage; Social Gospel
philanthropy, 17, 97–104. *See also* democratic communications; formations; philanthropic media
photomontage. *See* montage
pluralism, 7, 17
pluralization, 41–2, 63. *See also* capitalism; participation; publics
podcasts, 182
polemic, 16, 30, 45–7, 65–6, 104, 111. *See also* broadcasting; culture of truth; form; propaganda; rationalist empiricism
political communications, 65–6, 72–6. *See also* broadcasting; mass-culture critique; rationalization
Pollock, Jackson, 177. *See also* modernism; performance
Poor Law of 1601, 98. *See also* philanthropic media
Popper, Karl, 168
popular culture, 7–8, 18, 78, 101, 236. *See also* broadcasting; capitalism; critical pedagogies; mass-culture critique; professionalization
Popular Front, 245
populism, agrarian, 132. *See also* capitalism; Patrons of Husbandry (the Grange); unions, labor
postmodernism, 169
postpositivism, 168
Potter, David, 138. *See also* localism; mass-culture critique
power, 35–7, 41–2, 45, 97. *See also* Elizabeth I; Henry VIII; pluralization; rationalization
practice. *See* culture; democratic communications; direct autonomous composition; materialism
print capitalism. *See* institutionalization
print culture, 51–3, 64. *See also* commercialization; commodification; manuscripting; publics; rationalist empiricism; rationalization
Procter and Gamble, 213
professionalization, 30–1, 50–3, 65–7, 78, 97, 101–112, 134–46; and critical modernism, 166–7, 169, 175–6, 182; and direct autonomous composition, 235–7, 253–4; and market radicalism, 199, 201–2, 209–21. *See also* bureaucratization; documentary; empiricism; participation; political communications; reportage
progressive radio. *See* low-power FM radio

Project Censored, 253
propaganda, 6–8, 17, 65. *See also* critical modernism; incorporation; modernism
prophecy, 36–8, 40, 49, 104. *See also* culture of truth; manuscripting; providentialism
Protectorate, 42
protest demonstration. *See* protests
Protestant Reformation. *See* Christianity; Reformation, Protestant; secularization
Protestantism, 124–6. *See also* Christianity; community
protests, 10, 171–2, 175, 181. *See also* modernism.
providentialism, 16, 30–31, 34–40, 62–5, 79, 104–5, 268. *See also* culture of truth; formations; manuscripting; rationalist empiricism
Public Radio International, 108. *See also* philanthropic media
public, educational, and governmental (P.E.G.) media centers, 121
publicity, 179–82, 211. *See also* rationalization
publics, 16, 63–66, 79, 72–6, 171. *See also* broadcasting; Habermas, Jürgen; participation; rationalization
Pulitzer, Joseph, 130
Puritan radicalism, 126. *See also* Christianity
Puritanism. *See* Protestantism; city upon a hill; community; dissent

radical republicanism, 17, 124–8, 181. *See also* community; community media; newspaper exchange
radicalism, agrarian. *See* populism, agrarian
radicalism, market. *See* market radicalism
Radio Act of 1927, 137
Radio Corporation of America (R.C.A.), 77, 79. *See also* broadcasting
rationalist empiricism, 16–18, 30, 43–48, 53, 62–6, 79, 104–11, 141, 268; and critical modernism, 163, 166–7, 174; and market radicalism, 209, 214. *See also* capitalism; critical modernism; providentialism
rationalization, 16–18, 30–31, 44–5, 53, 62, 75–9, 101–9, 130, 130–131; and critical modernism, 165–7, 169–70, 179–83; and market radicalism, 204–11, 217. *See also* Bacon, Francis; capitalism; participation; rationalist empiricism
Recording for the Blind and Dyslexic, 108. *See also* philanthropic media
reflection theory, 10–11, 17–18, 96, 105–7, 174, 209–10; and media literacy, 245–6. *See also* critical modernism; documentary; empiricism; reportage
Reformation, Protestant, 34, 67. *See also* broadcasting; Christianity; secularization
reformism, 101–104, 131, 138–9, 141–4
Regent University, 216. *See also* Oh, Yeon-Ho
reification, 76–7. *See also* broadcasting; mystification; naturalization; technological determinism
reportage, 31, 47–9, 50–53, 105–7, 109–111. *See also* documentary; empiricism; rationalist empiricism
Reporters Committee for Freedom of the Press, 108. *See also* philanthropic media
Resettlement Administration (R.A.), 239
Restoration, 42
Reverend Billy, 183. *See also* critical modernism
Rhymer, prophecies of, 37. *See also* prophecy
Riesman, David, 138. *See also* localism; mass-culture critique
Riis, Jacob, 107
Robertson, Pat, 216
romanticism, 170, 269
Rondaradio, Inc., 142. *See also* community media; low-power FM radio
Rubin, Jerry, 177. *See also* modernism; performance
Rudé, George, 7, 64. *See also* Marxist historiography, British

Russell Sage Foundation, 104

San Jose Mercury News, 2. *See also* "Dark Alliance."
satire, form of. *See* polemic
scientific method, 44. *See also* Bacon, Francis; rationalist empiricism
Scott, Thomas, 48. *See also* hybridization
scribal communities, 39. *See also* manuscripting
scribal lampoon, 48. *See also* satire
scribal production. *See* manuscripting
Scripps, E.W., 136. *See also* community media; commercialization; localism
Seattle, and World Trade Organization meeting, 122, 179, 266
secularization, 34, 40–43
seed-drilling, 70. *See also* broadcasting
Seeing Is Believing (organization), 96. *See also* empiricism; philanthropic media; rationalist empiricism
self-education, 233–64
sermon, 46, 49. *See also* providentialism; rationalist empiricism; reportage
settlement-house movement, 107, 131. *See also* philanthropic media
Severe Storms Intercept Team, 142. *See also* community media; low-power FM radio
Shakers, 127. *See also* community
Share, Jeff, 251–3, 254
Sheehan, Cindy, 111. *See also* philanthropic media
Shils, Edward, 137. *See also* mass-culture critique; community media; localism
Sierra Club Chronicles, 112. *See also* philanthropic media
Sierra Club. *See Sierra Club Chronicles*
Silfry, Micah, 94. *See also* philanthropic media
Sinclair, Upton, 134
Situationist International, 177. *See also* modernism; performance
slavery, U.S., 72–6. *See also* broadcasting; evangelical mass media
social composition. *See* direct autonomous composition
social documentary. *See* documentary
social engineering, 116n47. *See also* rationalization; reformism
Social Gospel, 101–2, 131. *See also* evangelical mass media; philanthropic media
social history, new. *See* new social history
social media, 200, 211. *See also* citizen journalism; consumer-generated content; participation; professionalism
social movements, 6–8, 10–12, 21n26, 54, 72–6, 98, 105–7; and critical modernism, 179, 181, 187n47; and market radicalism, 214–18. *See also* broadcasting; evangelical mass media; Marxist historiography, British
Social Reconstructionist Movement, 243
social reform, 17, 241–3, 246. *See also* mass communication; reformism
Social Science Research Council, 110. *See also* philanthropic media
socialism, Christian. *See* Social Gospel
Socialist Party, 244
social-reform communications. *See* philanthropic media
social-reform media. *See* philanthropic media
Society for Propagating the Gospel among the Indians and Others in North America, 98–9. *See also* Bible-tract societies; broadcasting; evangelical mass media; philanthropic media
Society for the Propagation of the Gospel in Foreign Parts, 68, 98. *See also* Bible-tract societies; broadcasting; evangelical mass media; philanthropic media
Sohno Real, 122
sola scriptura, 67. *See also* broadcasting; evangelical mass media
soundscapes. *See also* community media

Source Code (organization), 111. *See also* philanthropic media
South Wales Miners Institute, 200
sower, parable of the, 71. *See also* broadcasting; evangelical mass media
Star Tree Foundation, 140. *See also* community media; low-power FM radio
State of the Union speech, 161–2, 170. *See also* critical modernism
statistical profiling. *See* documentary; reportage
Statistics of Income Division of the Department of the Treasury, 105. *See also* documentary; reportage
statistics. *See* documentary; reportage
Steinbeck, John, 241
Step It Up (organization), 161. *See also* protest demonstration
StoryCorps, 111, 168. *See also* philanthropic media
Strangelove, Michael, 211
street theater. *See* theater
strikes, 131, 171, 175, 181, 242, 244. *See also* capitalism; class; protests; unions, labor
structuralism, 12, 237, 250–2
subcultures, 8, 10–11, 17. *See also* essentialism
Surrealism, 174
Surveillance Camera Players, 182–3
Sweatshopper (organization), 112, 182. *See also* philanthropic media
syndicalism. *See* anarchism

Tapp, George, 248. *See also* migrant farmworkers
taxation, 51. *See also* chartering; licensing; monopolies
Taylorism. *See* social engineering
technological determinism, 11–14, 18, 72–6, 127–8, 136. *See also* mass-culture critique; mystification; naturalization; Williams, Raymond
technology, 72, 75, 179. *See also* evangelical mass media; technological determinism
telegraphy, 77, 127–8, 173. *See also* broadcasting; technological determinism
Telephon Hírmondó, 78
telephone, 76–8. *See also* mass-culture critique; technological determinism
television. *See* media industries
testimony. *See* documentary; reportage
text, 12, 202, 250–2. *See also* culture; structuralism
The Grapes of Wrath, 240, 241
The Ground Truth, 112. *See also* philanthropic media
The Jungle, 134
The New York Times, 2
The North Star, 76. *See also* broadcasting; evangelical mass media; social movement
The Panther, 40. *See also* Fifth Monarchists; prophecy
The Progressive, 2. *See also* alternative media; mainstream media
The Women's Journal, 99. *See also* feminist media; philanthropic media; social movements
The Yes Men, 183
theater, 12, 172, 175, 182. *See also* alternative media.
Thompson, E.P., 7, 64, 68. *See also* history; Marxist historiography, British; materialism
Toleration Act of 1689, 67. *See also* broadcasting; evangelical mass media
TomPaine.com, 111. *See also* philanthropic media
Trachtenberg, Alan, 100. *See also* incorporation
trade unionism. *See* unions, labor
trade, international, 42. *See also* empire
Treason Acts of 1534, 37. *See also* prophecy
truth. *See* culture of truth
TruthOut.org, 2. *See also* alternative media; Associated Press; mainstream media
Tudor, reign of. *See* Elizabeth I; Henry VIII

Tulsa Free Radio Project, 144. *See also* community media; low-power FM radio
Tuskegee Institution, 242

U.S. Army, 121, 122
underground culture. *See* alternative culture
Underground Press Syndicate. *See* Alternative Press Syndicate
unions, labor, 17, 125, 131–5, 142, 181; and migrant farmworkers, 243–4, 247. *See also* American Federation of Labor (A.F.L.); class; Congress of Industrial Organization (C.I.O.); Industrial Workers of the World (I.W.W.); Knights of Labor; radical republicanism; United Cannery, Agricultural, Packing and Allied Workers of America (U.C.A.P.A.W.A.); Worker's Alliance of America
United Cannery, Agricultural, Packing and Allied Workers of America (U.C.A.P.A.W.A.), 243. *See also* unions, labor
United Church of Christ, 110. *See also* philanthropic media
United Nations Educational, Scientific and Cultural Organization (U.N.E.S.C.O.), 210. *See also* empire
Unity World Production, 143. *See also* community media; low-power FM radio
user-generated content. *See* consumer-generated content

utopia, 266–70
utopian colonies, 17, 126–7. *See also* community

Van Slyke, Tracy. *See* Clark, Jessica, and Tracy Van Slyke
vanguardism, 7, 10–18, 71–9, 110, 135, 267–8; and direct autonomous composition, 235–6, 249, 250, 252. *See also* broadcasting; evangelical mass media; social movements
Veblen, Thorstein, 167–8. *See also* capitalism; modernism
Video Nation, 203
video news releases, 211
videotext, 208
Vietnam, 209

wall-writing, 39. *See also* manuscripting
Wal-Mart, 112, 182. *See also* philanthropic media; Sweatshopper
Washington Post, 2
Washington, Booker T., 242
Wayland, J.A., 134–5, 166. *See also* *Appeal to Reason*; community media; unions, labor
Weber, Max, 169. See also bureaucratization; rationalization
weblogs, 211
Wesley, John. *See* Methodism
Western Federation of Miners, 134, 135. *See also* unions, labor
WETA, 108. *See also* philanthropic media
WGBH, 108. *See also* philanthropic media
Whitefield, George, 98
Whyte, William, 138. *See also* localism; mass-culture critique
Wikipedia, 182
Will, Brad, 122. *See also* anarchism; community media
Williams, Raymond, ix, 3, 13–14, 123, 170, 203–4; and direct autonomous composition, 234, 236–9, 252. *See also* cultures of truth; history; materialism; technological determinism
Winthrop, John, 126
wireless, 173. *See also* telegraph
Witness.org, 96, 109. *See also* documentary; philanthropic media; rationalist empiricism; reportage
witnessing. *See* documentary; reflection theory; reportage
Wittgenstein, Ludwig, 168
Wobblies. *See* Industrial Workers of the World (I.W.W.)
Word of Mouth Marketing Association, 1
word-of-mouth-marketing, 213

World Trade Organization, Seattle meeting, 122, 179, 266
Worker's Alliance of America, 244. *See also* migrant farmworkers; unions, labor
workers' theater. *See* theater
working class. *See* class; unions, labor
Works Progress Administration (W.P.A.), 168

Youth Radio, 108. *See also* philanthropic media

'zines, 203, 233

About the Author

James F. Hamilton is currently an Associate Professor in the Department of Advertising and Public Relations in the Grady College of Journalism and Mass Communication at the University of Georgia, where he has been since 1999. Prior to this appointment, he taught for five years (1995–1999) in the Department of Communication at the State University of New York College at Geneseo. Dr. Hamilton's research for the past fifteen years has focused on the history, theory and practice of alternative media, and on critical media pedagogy as assisted by digital technology. He is widely published, with work appearing in journals such as *Critical Studies in Mass Communication*, *Media, Culture & Society*, *Media History*, and *The Communication Review*. Dr. Hamilton also serves as co-editor of book reviews for *Social Movement Studies*. He has presented his work at a variety of academic conferences of such organizations as the National Communication Association, the Union for Democratic Communications, and the Association for Education in Journalism and Mass Communication. He holds a Ph.D. in Mass Communications from the University of Iowa, and an M.A. in Communications and a B.A. in English and Communications from the University of Washington.